ADVANCES IN AGROPHYSICAL RESEARCH

Edited by **Stanisław Grundas** and **Andrzej Stępniewski**

Advances in Agrophysical Research

http://dx.doi.org/10.5772/3341
Edited by Stanisław Grundas and Andrzej Stępniewski

Contributors

Seth Idowu Manuwa, Leszek Moscicki, Bing Si, Charles Igwe, Sunday Obalum, Elsa Paula Morgado Sampaio, Júlio César Lima, Alexander Demyanchuk, Stanisław Teodor Grundas, Leonid Velikanov, Agnieszka Nawrocka, Joanna Lamorska, Krzysztof Górnicki, Agnieszka Kaleta, Wojciech Skierucha, Agnieszka Szypłowska, Andrzej Wilczek, Asim Biswas, Natalya Buchkina, Elena Rizhiya, Eugene Balashov, Sergey Pavlik, Majda Sfiligoj-Smole, Tatjana Kreze, Silvo Hribernik, Karin Stana-Kleinschek, Andrzej Stępniewski

Published by InTech

Janeza Trdine 9, 51000 Rijeka, Croatia

Publishing Process Manager Petra Nenadic
Typesetting InTech DTP Team
Cover InTech Design Team

Additional hard copies can be obtained from orders@intechopen.com

Advances in Agrophysical Research, Edited by Stanisław Grundas and Andrzej Stępniewski
p. cm.
ISBN 978-953-51-1184-9

For EU product safety concerns:
IN TECH d.o.o., Prolaz Marije Krucifikse Kozulić 3, 51000 Rijeka, Croatia, info@intechopen.com or visit our website at intechopen.com.

Meet the editors

Prof. Grundas obtained the degree of MSc in 1971 at the University of Life Sciences, Lublin. Since this year he began working at the Institute of Agrophysics of the Polish Academy of Sciences in Lublin, and next obtained the degrees of PhD (1977), and DSc (1988) from this University. In 1994, he was granted the title professor of Agric. Sc. His scientific area is physical properties of plant materials. The list of his publications amounts to a total of 260, including articles in several encyclopedias on the cereal grains. Editor of the book "Advances in Induction and Microwave Heating of Minerals and Organic Materials" by InTech (2011). Life Member of the AACCInt, member of the Polish Society of Agrophysics, and the Scientific Committee of Agrophysics of the PAS.

Dr Eng. Andrzej Stępniewski graduated from Lublin University of Technology in 1985 and was awarded Engineer degree in mechanics. Since 1986 he has been working at the Institute of Agrophysics of the Polish Academy of Sciences in Lublin, Poland. In 1997 he was awarded PhD in agronomy, agrophysics. His main scientific interest concentrated on mechanical properties of plant material in the aspect of technological processes. Recently his scientific interest includes the properties of rapeseed and rapeseed oil used as biofuel. He participated in a number of scholarships: at Hohenheim University in 1991 (DAAD); at North Carolina State University in 2000 and at Instituto Agronomico do Parana 2002. He is a member of the International Society of Food Physicists, Polish Society of Agrophysics, Scientific Committee of Agrophysics of the PAS, and Polish Society of Agriculture Engineers. Since 1999 Dr. A. Stępniewski has been working as regional expert of the European Framework Programmes and holds the position of the Director of Regional Contact Point of Research Programmes of European Union.

Contents

Preface

The first edition of the book entitled "Advances in Agrophysical Research" is issued to honour the memory of the prominent researchers, founders of the institutes of agrophysics:

- prof. Abram Fedorovich Ioffe (AFI), the member of the Soviet Union Academy of Sciences, who in 1932 established the Scientific-Research Institute of Agrophysics in Leningrad (SU), widely known in the world as „AFI", and now officially named the Scientific-Research Institute of Agrophysics of the Russian Academy of Agricultural Science (S-RIA RAAS) in St. Petersburg (Russia), and

- prof. Bohdan Dobrzański, the member of the Polish Academy of Science (PAS), who in 1968 established the Agrophysical Departament belonged primary to the Institute of Plant Physiology PAS, next renamed as the Institute of Agrophysics PAS in Lublin (Poland).

In the 2012 the S-RIA RAAS in St. Petersburg (Russia) was celebrated 80th Jubilee, and the IA PAN in Lublin (Poland) is celebrated 45th Jubilee in 2013.

More information about Institutes can be find on pages: www.agrophys.ru and www.ipan.lublin.pl

The book offers comprehensive analysis of the wide range of scientific knowledge in agrophysics i.e. application of physics in agriculture. More precisely the book contains results of investigations focused on agronomy with special emphases on interdisciplinary field of agrophysics. The book is divided into three sections: first section A is an introduction, than section B and C. Introduction contains the general information about agrophysics as a branch of science, its history and some information about researchers and institutions which were/are involved in agrophysical research. The section B is devoted to physical properties of soil and environment, and section C contains data of physical properties of plant and its products. The Reader can find same interesting aspects on the role of agrophysics in agriculture, i.e. the physics of soil and agricultural environment as well as physical properties of reproductive materials and row materials up to physical properties of food.

The book is divided into 15 chapters, the first one is in the section A, 8 - in the section B, and 6 - in the section C. The authors are specialists in wide range of discipline what results in interdisciplinary approach to the investigation and interpretation of the results.

At the beginning the Reader finds the section A (Chapter 1) which is introduction to agrophysics. It presents the existing definitions of agrophysics, the history of agrophysical institutes in Russia and Poland as well as some basic activities which are in close relation to agrophysics. The fundamental mission of physics in agriculture is shortly described including societal activities related to agrophysics, which were undertaken mostly in Poland.

The next chapter opens section B. The 2nd chapter presents modern aquametry techniques applicable to soil, plant materials and food products. A special emphasis is placed on the promising dielectric spectroscopy methods. The presentation of the open research problems and possible further developments conclude this chapter.

In the 3rd chapter temporal stability of soil water content (SWC) with factors controlling this parameter is discussed. Soil, topography, vegetation, climate have been found to influence temporal stability of SWC. The object of this chapter is to provide comprehensive review on temporal stability of SWC, mainly focused on the associated concept, methodology, application, and controlling factors. Associated application of temporal stability concepts are also included in this chapter. In the 4th chapter knowledge of soil spatial variability in ecological modelling, environmental prediction, precision agriculture, soil quality assessment and natural resources management is presented. The model averaging for semivariogram models parameters is discussed. The performance of the models are analysed using Akaike Information Criterion (AIC). The proposed method has potential in improving experimental design, sampling, mapping and interpolation of the spatial processes such as krigging. The object of 5th chapter is devoted to examine the scale of time stability of no-stationary soil water spatial pattern at different seasons using global wavelet coherency. Soil water was measured up to 1.4 m depth using neutron probe and time domain reflectometry (TDR) along a 576 m long transect established within a hummocky landscape at central Saskatchewan. In the 6th chapter two fertile soils were managed under two alternatives: traditional mechanized system (TS) and direct-seeding system (DS). Both production systems during a ten-year period and their physical and chemical characteristics were compared in terms of agrophysical optimisation to produce dry matter (DM). In the Mediterranean areas the seasonal distribution of precipitation is harmful to plant (crop) DM production which can be increased by adequately combining a good management of all input factors with a special care to natural resources protection, namely soil and water conservation. These results are well in the scope of the international concerns on adopting improved common polices for food safety and problems associated to water scarcity. In chapter 7th soil strength is regarded as important characteristics that affects many aspects of agriculture, such as the performance of cultivation implements, root growth, least-limiting water range and traffic ability. The main object of the chapter is to evaluate the effects of applied pressure and moisture content on strength indices such as bulk density, penetration resistance and shear strength of the soil. The development of the relationships between the strength indices for prediction purposes required in soil management is also described. Chapter 8th is devoted to the review highlights of factors favouring micro aggregation as well as those causing dispersion in tropical soils. This review showed that micro aggregate stability might be a useful tool for assessing erosion hazards in tropical soils. It has exposed the rather low level of information on micro aggregation and erosion hazards in these soils. Therefore more research on this subject is encouraged in this chapter, same of which are suggested. In the last, 9th chapter of this section the nitrous oxide flux from agricultural soils on its physical properties is discussed. It is now widely accepted that agriculture is the main source of anthropogenic nitrous oxide. The agriculture contributes to 60% of the global emission of the nitrous oxide. In this chapter direct of the nitrous oxide emission from arable loamy sand Spodosols, which are typical for North-western Russia, under potato, cabbage and carrot are investigated both on ridges and in furrows during four growing seasons 2004-2007.

At the beginning of the section C in the chapter 10th main interest is the plant morphology as one of the oldest branch of knowledge which describes the external plant characteristics. Morphological systematic of wheat which we can find in academic books classified morphotypes of wheat on the base of vegetation period length and mass of 1000 grain. Authors are interested in the problem how the properties of wheat grain, correlate with its reproductive features. They tried to evaluate plant productivity from visualization of grain, especially of germ using X-ray method. A new approach to practical estimation of morphotype of grain wheat based on it's really existed properties are discussed in order to evaluate its indexes, which can be used for the identification of varieties as well as in the improvements of breading processes. Exemplary algorithms which allows practical utilisation of morphological indexes of kernels for particular purposes are also presented. Some physical properties of seeds and grain are described in chapter 11th. . They can be useful for proper harvesting and storage as well as for processing such as drying, freezing etc. They can also be utilised for the designing of machinery. Accurate design of machines and performance of processes in the food chain production from field to table requires an understanding of physical properties of raw agricultural material. The following features should be considered: shape, size, volume, density, specific gravity, surface area and other mechanical characteristics. The measurement techniques allow computation of these characteristics, which can inform about the effects of processing. On the other hand some of characteristics, such as colour, mechanical and rheological properties, thermal and electrical resistance, water content and other physical quantities give excellent description of product quality. Chapter 12th is devoted to evaluation of determination criteria for safe grain storage time. Nowadays grain is harvested with a combine harvester what result in quick gather. Therefore it is possible to delay the harvesting to gather ripe and dry grain. It eliminates some bigger losses caused by sprouting of grain, yet in certain batch polluted with green parts of plant, straws and seeds of weeds etc., moisture content can exceed 30% w.b., and temperature is often above 30ºC. The basic criteria of determination the length of the storage period are: CO2 production connected with loss of the dry matter of grain, appearance of visible moulds and others. The dependencies for determining the time of safe grain storage are discussed in this chapter. The chapter 13th contents review on the extrusion technology, well known in the plastic industry, but now become widely used in agri-food sector, where it is classified as extrusion-cooking operation. The rheological status of the starch is changed during conveying of raw material under high pressure and temperature through a die or a series of dies. During this process the product expands to its final shape. The chapter describes some physical aspects of extrusion-cooking of starch. In the 14th chapter authors give comprehensive review of spectroscopic methods used for determination of food quality. Spectroscopic methods have been historically very successful at evaluating the quality of agricultural products, especially foods. These methods are highly desirable for analysis of food components because they often require minimum or no sample preparation, provide rapid and on-line analysis, and have the potential to run multiple tests on a single sample. The aim of this chapter is to demonstrate applicability of four spectroscopic techniques, e.g. UV-VIS spectroscopy, fluorescence, infrared and Raman spectroscopy, as rapid analysis method to determine the quality of cereals, cereals products and oils. In the last 15th chapter of this section the comprehensive description of plant fibres for textile and technical application is presented. Recently great attention has been focused to the utilization of agricultural by-products, e.g. wheat straw, quinoa stems and sugarcane bagasse, hot stems, sponge, Muscaceae plants, and others. Authors describe the possibility of utilisation of the above mentioned products. Some examples of fibres from different terrestrial and sea grass and legume species are presented. In general a potential utilization of fibres from different non-conventional origins is discussed.

Prof. Dr. Stanisław Grundas
Bohdan Dobrzanski Institute of Agrophysics
Polish Academy of Sciences,
Poland

Dr. Andrzej Stępniewski
Bohdan Dobrzanski Institute of Agrophysics
Polish Academy of Sciences,
Poland

Section 1

General Information on Agrophysics

Chapter 1

Introduction to Scientific Discipline Agrophysics — History and Research Objects

B. Dobrzański, S. Grundas and A. Stępniewski

Additional information is available at the end of the chapter

http://dx.doi.org/10.5772/56982

1. Introduction

1.1. Agrophysics's definition and scope

From the very beginning of its existence the definition of Agrophysics as a science was described a number of times. In general, Agrophysics is a branch of natural and agricultural sciences which applies physics into agriculture. Therefore sometimes it is also called agricultural physics. It explores agricultural materials and processes to describe their physical properties in order to assure best quality of agricultural products or raw material for industry, taking into account the role of environment and other factors. As a field of science, Agrophysics is of interdisciplinary scope and it is closely related to Biophysics. It is however limited strictly to the agricultural environment, i.e., soil, plants and animals and also takes into account the knowledge of Agronomy and Agriculture Engineering.

Agrophysics deals with physical processes in the soil-plant-atmosphere system, taking into account various external factors (climate, impact of the machinery, pollution) and issues related to the growth, harvest, transport, storage and processing of agricultural materials.

Some examples of the wide scope of agrophysical investigation are: developing systems for monitoring and controlling the condition of soil (moisture, salinity etc.) and plant growth (maturity), evaluation of the soil's susceptibility to water and wind erosion, monitoring and diagnosis of soil biological activity, determination of pollution in agricultural products (fruits, vegetables etc.), the assessment of the technological value of grain, evaluation of quality of fruits and vegetables during their storage and changes of their nutrition value during storage.

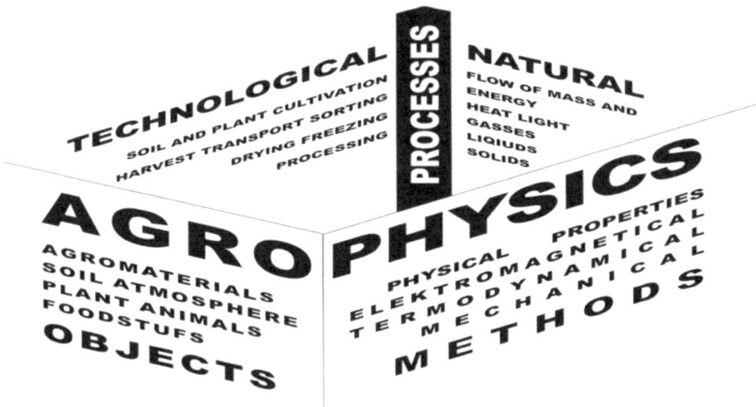

Figure 1. Graphical description of the definition of Agrophysics

The following definition of Agrophysics was recently included in the "Encyclopedia of Agrophysics":

"Agrophysics is a science that studies physical processes and properties affecting plant production. The fundaments (nutrients) and energy (light, heat) transport in the soil– plant–atmosphere and soil–plant–machine–agricultural products–foods continuums and way of their regulation to reach biomass of high quantity and quality with the sustainability to the environment. The knowledge of physical phenomena in agricultural environment allows increasing efficiency of use of water and chemicals in agriculture and decreasing biomass losses during harvest, transport, storage, and processing."

Agrophysics therefore aims at ecological use of agricultural ecosystem to assure best quality of agriculture products and the preservation of agricultural landscapes.

2. History

In the end of XIX century scientists recognized the need of application of physics to agriculture. The pioneer in this field of research was probably Franklin Hiram King. He was born in 1848 in Whitewater Wisconsin USA. He graduated from Cornell University and served as professor of natural sciences at River Falls State Normal School and after that as professor of agricultural physics at the University of Wisconsin. In 1888 the University of Wisconsin called him to the Chair of Agricultural Physics, the first of its kind in America. F.H. King was interested in a wide range of scientific problems but he made his major contributions to the applications of

http://dx.doi.org/10.5772/56982

physics to agriculture. Most attention was given to soil physics, e.g., he studied water-holding capacities, moisture requirements of plants, aeration, movement of water in soils, movement of groundwater, and the drafts of plows. During his last years in Madison he also began studies of soil fertility. He has been called the "father" of soil physics in the USA.

King published some books, the titles of which prove his scientific interest:

- Elementary lessons in physics of agriculture. F.H. King, 1894. Madison, WI.
- The soil: Its nature, relations and fundamental principles of management. Macmillan, 1895. New York.
- Irrigation and drainage. Macmillan, 1898. New York.
- Physics of agriculture. F.H. King, 1901. Madison, WI.
- Ventilation for dwellings, rural schools and stables. F.H. King, 1908. Madison, WI
- Farmers of forty centuries. F.H. King, 1911. Madison, WI.
- Soil management. Orange-Judd, 1914. New York.

F.H. King's observations and investigations could be nowadays regarded as a contribution to the so called sustainable agriculture. He died in 1911 in Madison, Wisconsin and his last two books were completed and published after his death.

Beginnings of application of physics in agriculture in Europe were in Soviet Russia. Agrophysical research was commenced in Russia in the middle of XX century. Russian physicist Abram Fieodorovich Ioffe (1880–1960) is regarded as the initiator of this branch of science. He studied measuring methods and used them in agriculture, biology and agrochemistry. He studied electromagnetism, radiology, crystals, high-impact physics, thermoelectricity and photoelectricity. He established research laboratories for radioactivity, superconductivity, and nuclear physics. In 1932 Ioffe organized the Institute of Physics and Agriculture in Leningrad (Sankt Petersburg at present) and became its first director. A.F. Ioffe published two fundamental works entitled: Physics and Agriculture (1955), and Physics for Agriculture (1959). His successors included: A.F. Chudnovsky (1910–1985), F.E. Koliasev (1898–1958);P.V. Vershinin (1909–1978); I.B. Revut (1909–1978), S.V. Nerpin (1915–1993), E.I. Ermakov (1929–2006), N.F. Bondarenko's (1928– 2003); I.S. Lisker; A.M. Globus, V.P. Yakushev and others. The research topics undertaken by the above mentioned scientists were: mathematical modeling of agricultural production, simulation of agrophysical systems and processes; interaction of biological objects with different physical fields (light, gravity, magnetic, electromagnetic, acoustic, electrostatic), information technologies of production management in arable farming and plant growing, agrophysical instrumentation, elaboration of vegetative systems in controlled climate, a new type of coordinate precision agriculture as the first step to the creation of the "electronic farmer".

The second scientific Institute whose main field of research is Agrophysics was established in Lublin, Poland. Created in 1968 by prof. Bohdan Dobrzański, the Institute of Agrophysics of the Polish Academy of Sciences was soon recognized as a leading research centre not only in

Eastern and Central Europe, but also in the global research. Numerous professors and younger researchers from the Institute frequently visited leading universities from Western Europe, Japan and the USA, bringing home the knowledge of recent trends and novel methods in the research field. In turn, the high quality of IA staff attracted a number of doctoral students and young researchers from Poland and abroad.This attractiveness, especially to young researchers made the Institute become also a leading educational centre at an advanced level in numerous research areas connected with environment, agriculture and food sector. The results of the research activities undertaken in the Institute were introduced into industry and gave basis for new products and technologies. The outstanding scientists of that period were: I. Dechnik (1929 - 2003), R. Walczak (1943 - 2003), M. Malicki (1939 - 2009), J. Stawiński (1942 - 2005), B. Szot (1933 – 2012), J. Gliński, W. Stępniewski, J. Lipiec, K. Konstankiewicz,

Figure 2. Coworkers of prof. B. Dobrzański from the Instite of Agrophysics at the Conference in Prague 1985 (from left; in the front row: R. Walczak, B. Szot, G. Skubisz, B. Dobrzański, J. Gliński; in the secondrow: M.Molenda, A. Pukos, W. Stępniewski, W. Woźniak, A. Kuczyński, K. Konstankiewicz, J. Lipiec)

The main fields of scientific activity of the Institute of Agrophysics PAS were as follows: investigation of physical and physical–chemical processes of mass and energy exchange in the soil – plant – atmosphere system, physical properties of agricultural materials and processes affecting plant production as well as processes related to gathering, transport and storage of agricultural materials.

The main feature of these studies was the elaboration of new theoretical and experimental research methods, developing of physical – mathematical models and their experimental verification, processing of data, taking into account their variabilityin time and space. Data bases and thematic maps created in this process can be used in practice for agricultural and environmental protection.

At present, there are more than 100 employees of interdisciplinary character. The scientific staff of the institute constitutes an interdisciplinary team of physicists, chemists, agronomists, horticulturists, biologists, engineers, geographers and mathematicians.

Agrophysics initially focused mainly on the study of soil and plant materials. Over time, this research area started to gradually expand, on the one hand including ever more elements of the soil-plant-atmosphere system, on the other hand, more and more focused on the process of food production and its quality - from the stage of agricultural production through the period of storage of agricultural products, their processing until the final product.

Agrophysics began also to be useful in the wider environment, being a part in the study of not only the degradation of soils, but also researching marshy land formation and greenhouse gas emissions. Currently Agrophysics concentrates on a number of agricultural specialties, it is used to interpret interactions, design, control and optimization of processes. It is also widely used in environmental protection, pedology, tillage and plant engineering, agriculture, agri-food technology, and others.

3. Agrophysics in other countries

Agrophysics was developed also in other countries, initially in Eastern European countries. In Czechoslovakia (Prague) R. Řezniček together with his coworkers (J. Blahovec, J. Pecen, P. Hnilica) conducted some important investigations at the Chair of Physics of Prague Agricultural University. They concentrated on physical properties of cereal grains and developed some interesting testing methods. In Czech M. Kutilek from Czech Technical University in Prague conducted advanced studies within soil science too. Also in Hungary (Gödöllö and Budapest) researchers carried out studies on physical properties of agricultural materials: soil, seed, grains, vegetables and fruits. In this context the following names should be mentioned: G. Sitkei, I. Husar, G. Várallyay, I. Farkas and others. Also German researchers joined international agrophysical society. The universities in Hohenheim (H-D. Kutzbach, E. Schlihting, K. Stahr), Bonn (K.-H Kromer), Kiel (R. Horn), Berlin (G. Wessolek) and in Potsdam (J. Helebrand) were engaged in application of physics in agriculture. The events called "Agrofizik Tagung" were organized on a regular base in Germany. These symposiums gathered scientists interested in this topic from Germany. Agrophysics was researched also in Spain. Spanish scientists were interested both in soil (J. Moreno, M. Aranda and D. de la Rosa) and fruit properties (M. Ruiz-Altisant, J. Caniavate). Belgium has also undertaken serious research activities to become a strong agrophysical centre. Catholic University in Louvain (J. DeBaerdemaker, B. Nicolai) and University of Ghent (M. De Boodt, D. Gabriëls) were main centers of agrophysical research there. In France agrophysics was developed in INRA Montfavet (S. Auber, P. Varoquaux), while in Italy the University of Torino (A. Ferero) was the centre of agrophysical studies. Same investigation were also conducted in Belarus and a leader there was I.I. Lishtvan. In Austria prof. W.E.H. Blum interested in some agrophysical aspects of soil sceinces. From Sovakia it is necessarry to mentione Jech.

Agrophysics was developed not only in Europe. A fundamental publication on this topic "Physical Properties of Agricultural Products" has been written by N.N. Mohsenin – scientist of Iranian origin who worked in the USA. Prof. A. Tabatabaeefar is a continuator of Mohsenin agrophysical research in Iran. There were also other scientists interested in agricultural physics i.e. S. Gunasekaran, O.R. Kunze, J.I. Ross, G. Brusevitz, F. McClure, Y.A. Pachepsky, S.A. Thompson, P.P. Chen and others. Some topics of applied physics in agriculture were undertaken in Japan (R. Hatano) and China (T. Ren,).

In Canada at the Guelph University physics in agriculture was practiced by number of researchers (W.K. Bilanski, R.L. Kushwaha). Also at University of Saskatchewan F. Sosulski. Agrophysics was developed even in New Zealand by C.J. Studman.

In Israel some aspects of agrophysics were studied by I. Shmulevich, E. Bresler, K. Peleg.

The above mentioned names and centres do not exhaust the list of places and persons who conduct their investigations in the field of applied physics in agriculture. These are only examples of people who cooperated with scientists from the Institute of Agrophysics PAS proving that agrophysics is a world-wide recognized science discipline.

4. Agrophysical conferences

Conferences on Agrophysics have always been an occasion for long discussionson all aspects of physical, physicochemical and biological processes of mass and energy exchange in soil-plant-atmosphere system and of plant production, as well as characteristics of agricultural products and materials, agrophysical measuring methods, soil degradation and remediation problems.

Since Agrophysics became a widely practiced discipline of science there were already a number of agrophysical conferences organized. They gathered scientists which main research interests focus on physics in agriculture. The conferences were organized as follows:

- Lublin, Poland 1976, B. Szot;
- Gödöllö, Hungary 1980, I. Husár;
- Prague, Czech Republic 1985, R. Řezniček;
- Rostock, Germany 1989, H.-J. Hellebrand;
- Bonn, Germany 1993, K.-H. Kromer;
- Lublin, Poland 1997, J. Gliński;
- Prague, Czech Republic 2001, J. Blahovec;
- Louven, Belgium 2004, J. De Baerdemaeker;
- Lublin, Poland 2005, R. Walczak;
- Lublin, Poland 2011, J. Horabik;

A number of scientific papers that were submitted and presented during the above mentioned conferences were published both in conference materials as well as in special issues of scientific journals.

5. Agrophysical journals

Institute of Agrophysic of the Polish Academy of Sciences is also the publisher of an outstanding scientific journal entitled "**International Agrophysics**". The editorial board states that: "the journal focuses on physical properties and processes affecting biomass production and processing. The main topics are: mass (water, air, plant nutrients) and energy (light, heat) transport in the soil-plant-atmosphere continuum, ways of their regulation in order to reach biomass of high quantity and quality. The description of new methods and devices for measurements of the physical properties of agro- and biomaterials are published. The journal is also open to wider aspects of environmental and agricultural physics". The present Editor-in-Chief is prof. J. Gliński.

International Agrophysics is indexed by Journal Citation Reports with 1,574 impact factor.

A second journal published by the Institute of Agrophysics is "Acta Agrophysica". Acta Agrophysica has been published since 1993. At the beginning it contained mainly monographies and dissertations which were published irregularly. From 2012 Acta Agrophysica is a quarterly. It publishes papers presenting the results of fundamental and applied studies from the field of application of physics for the solution of problems relating to the management and protection of the natural environment, sustainable agriculture, and food processing. Papers can be published both in Polish and English. The present Editor-in-Chief is prof. J. Horabik.

The third journal published by the Institute of Agrophysics is "Acta Agrophysica Monographiae" which publishes reviewed papers based on original research results as well as monographs pertaining to the field of agrophysics. The monographs are published in Polish or English. The present Editor-in-Chief is prof. J. Horabik.

The above mentioned journals are available in electronic versions on the internet.

6. Dictionaries and books, maps

In order to facilitate international collaboration and allow unification of terminology, a number of bilingual and multilingual dictionary of agrophysical terms (nomenclature) was prepared under editorial supervision of prof. Ryszard Dębicki and prof. Jan Gliński. The following languages were taken into consideration: English, Russian, French, Spanish, German.

"Atlas of the Redox Properties of Arable Soils in Poland" prepared by Ostrowski, J.; Stępniewska, Z.; Stępniewski, W.; Gliński, J. was published 1996. It contains a wide range of data on soils in Poland presented in the form of cartographic maps. The Atlas gives a comprehensive information on soils in Poland and can be regarded as a research tool for studying spatial characteristics of agriculture. Practical application of the database lays in the power to generate, for the first time in the world, the thematic maps on spatial differentiation of the redox soil properties throughout the country. A set of these maps has been presented in the published atlas.

The last but most fundamental publication on Agrophysics is the "Encyclopedia of Agrophysics" edited by J. Gliński, J. Horabik and J. Lipiec by Springer in 2011. This book provides an up-to-date information on the physical properties and processes affecting the quality of the environment and plant production. Encyclopedia of Agrophysics is a publication complementary to the Encyclopedia of Soil Science, (November 2007) which has been published in the field of Earth sciences series of Springer. The Encyclopedia presents a set of about 250 informative articles and ca 400 glossary terms covering all aspects of Agrophysics. It contains 450 illustrations on more than 1000 pages.

7. Data banks (soil probe bank)

One of the comprehensive accomplishment of the Institute of Agrophysics PAS is the bank of soil samples collected all over Poland. The collection consists of over a thousand profiles (3 levels) and their full characteristics. This is a unique set of samples which can enable scientists to monitor changes of agriculture environment in Poland.

8. Agrophysics — European centre of excellence

The Centre Of Excellence For Applied Physics In Sustainable Agriculture "AGROPHYSICS" was created on February 28-th. 2003, within the scheme of the 5-th. Framework Programme of the European Union. The project was realized for three years and ended on February 28^{th}2006. The main goal of the project was to support the research potential of the region and strengthen its integration with the European Research Area.

The main points of the action plan of the CoE were:

- to develop programmes aimed at significant technological applications, while preserving high standard of fundamental research,
- to concentrate on modern research fields with potential applications in industry,
- to expand the Institute's activities towards market-oriented research,
- to participate in European projects within European research programmes,
- to intensify activities in attracting funds from the State Committee for Scientific Research (KBN; now – the Ministry of Science and Higher Education) under research grants,
- to start co-operation with Polish industry and SMEs,

- to develop a scientific network with universities,
- to make the IA more attractive for students by combining post-graduate and doctoral studies with a contribution to research projects,
- to establish a Foundation for Development of Agrophysical Research in order to be able to obtain more funds for research.

The prolongation of the existence of the CoE and the consequence of this prolongation was a possibility for dynamic development of the structural base of the Institute. The full renovation of buildings as well as thorough exchange of research equipment was accomplished thanks to European funds.

9. Polish Society of Agrophysics

Polish Society of Agrophysics (PSA) was founded in 1996 in Lublin. The initiator and the first president of PSA was prof. Bogusław Szot (1933 - 2012). He has been managing the Society until his death in October 2012. During his presidency ten branches in major scientific centers in Poland were established, where agrophysical investigations are performed (Fig. 3.). The Society is very active and every year it organizes meetings, seminars and symposia. Today PSA comprises over 360 members, among whom over 150 hold the title of full professor.

Figure 3. Location of PSA branches in Poland

The main goal of the Society is to agglomerate researchers who conduct their research in the field of agrophysics. The Society has organized nine national and international conferences with over 1400 participants (115 of them from 29 countries outside Poland). Every conference has its own book of abstract, where the most recent results are presented. The most important papers are published in International Agrophysics and Acta Agrophysica – both journals edited at the Institute of Agrophysics PAS. Current information on the activities of PSA is distributed through its Information Bulletin.

10. Scientific Committee of Agrophysics of Polish Academy of Sciences

The Committee of Agrophysics of Polish Academy of Sciences was established in 1981.

The Scientific Committees of the Polish Academy of Sciences are self-governing, nationwide representation of various disciplines or groups as well as interdisciplinary scientific problems integrating scholars throughout Poland. The scientific committee includes the Members of the Polish Academy of Sciences of the relevant specialty (outstanding scientists), eminent researchers representing universities, institutes of the Polish Academy of Sciences and the scientific institutes and research departments, as well as representatives of other institutions, including economic and social organizations. The scientific committees of the Polish Academy of Sciences are the most representative group of experts in the discipline.

The main task of the Scientific Committee of Agrophysics is to promote research of physical and physicochemical properties of the natural environment with particular emphasis on the system: soil-plant-atmosphere-machine-crops, agricultural-machine-food products. At present the chairman of the Committee is prof. Bohdan Dobrzański Jr.

Actually there are three sections operating within the structure of the Committee:

- Section of Physics Application in Engineering of Agricultural Production and in Food Technology,
- Section for Physical Measurement Techniques for the Agricultural Environment Protection,
- Section of Physical Methods of Evaluation of the Quality of Agricultural Products.

Since the beginning of its existence the Committee's special attention was directed to the integration of the scientific community and its involvement in the process of creatively solving important problems in agricultural research. The Committee in its activities focused also on identifying the main areas of research and science policy in the agrophysics. In previous years the Committee also financed research projects in the field of agrophysics. It has developed extensive cooperation with other committees working in the field of agricultural sciences. It also promoted international cooperation as a partner institution in organization of international conferences.

11. Foundation for Development of Agrophysical Sciences (FRNA)

Foundation for Development of Agrophysical Sciences was established for the purpose of supporting and promoting the development of agrophysical research and activity of the Institute of Agrophysics PAS. The implementation of the main goals of the Foundation is achieved by publishing (Scientific Publishing House of FRNA) in the field of natural sciences and application of physics in agriculture. According to the up-to-date techniques, the Foundation proposed e-files i.e. electronic publications on CD and DVD instead traditionally printed books. The Foundation also sponsors various initiatives and initiates actions toward popularization of agrophysics as science.

12. Prototype apparatus and investigation methods

For many years a workshop for prototype apparatus production was being organized at the Institute of Agophysics PAS in Lublin. A number of pioneer measuring techniques were elaborated and used in innovative apparatus. Both construction and production of this equipment was realized at the Institute. The founder and the first manager of the workshop was M. Grochowicz, PhD.

Under this activity a measuring systems for determination of moisture content and salinity of soil and other porous materials has been elaborated. The measuring techniques are based on the Time-Domain Reflectometry (TDR) which is worldwide patented. In the last years an innovative silo for drying and safe storage of rapeseed was constructed. The silo allows automated post-harvest drying, cooling, and storage of rapeseeds in neutral gas atmosphere to assure best quality of raw material for oil production. It prevents the development of heat processes as well as moulds and fungi growth in stored seeds. Also some prototype measuring apparatus were elaborated in order to evaluate quality parameters of fruit tissue especially apple tissue.

All the above mentioned activities, institutions and persons prove the necessity and usability of Agrophysics as a branch of applied science, which can help to solve current problems in agriculture and which can benefit in progress of this part of economy.

Author details

B. Dobrzański, S. Grundas and A. Stępniewski

Institute of Agrophysics Polish Academy of Sciences, Lublin, Poland

References

[1] Dobrzanski, B. (1986). Możliwości rozwoju agrofizyki. Wykład z okazji nadania tytułu Doktora Honoris Causa ART w Olsztynie, Kaseta video, ART, Olsztyn.

[2] Dobrzanski, B. (1981). Badania w zakresie fizyki i fizykochemii gleb. Kosmos, , 2, 135-139.

[3] Dobrzanski, B, Dechnik, I, & Glinski, J. (1979). Rozwój badań agrofizycznych w Polsce. Post. Nauk Roln., 5.

[4] Dobrzanski, B, Glinski, J, & Szot, B. (1988). Agrophysical investigations in Poland at present and in future. Physical properties of agricultural materials and products. Hemisphere Publ. Corp., New York, , 873-976.

[5] Glinski, J, Horabik, J, & Lipiec, J. (2011). Encyclopedia of Agrophysics. Springer.

[6] Haman, J, Horabik, J, & Pukos, A. (1985). Mechanical investigations of agricultural materials in the Institute of Agrophysics. Zesz. Probl. Post. Nauk Roln., z. , 304, 9-16.

[7] Szot, B. Dobrzański jr B., (1995). Agrofizyka dla techniki rolniczej. Zesz. Probl. Post. Nauk Roln., z. , 424, 79-86.

[8] Tanner, C. B, & Simonson, R. W. (1993). Franklin Hiram King- Pioneer Scientist. Soil Science Society of America Journal , 57(1)

[9] http://www.agrophys.ru/

[10] http://www.ipan.lublin.pl/

[11] http://www.international-agrophysics.org/

[12] http://www.acta-agrophysica.org/

[13] http://www.acta-agrophysica-monographiae.org/

[14] http://www.komagrof.pan.pl/

Section 2

Physical Properties of Soil and Environment

Chapter 2

Aquametry in Agrophysics

Wojciech Skierucha, Agnieszka Szypłowska and Andrzej Wilczek

Additional information is available at the end of the chapter

http://dx.doi.org/10.5772/52505

1. Introduction

Aquametry is a part of metrology that uses the available measurement techniques in the measurement of water content in solid, liquid and heterogeneous materials. A similar branch of metrology, called hygrometry, deals with determination of the water vapour content in air and other gases (Kraszewski, 2001). The aim of agrophysics is to apply physical methods and techniques for studies of materials and processes which take place in agriculture. Possible test objects may therefore include soil, fruit, vegetables, intermediate and final products of the food industry, grain, oils, etc.

Historically, the primary application of aquametry in agrophysics is the measurement of soil water content. The available soil water content measurement techniques are evolving to follow the technological development in metrology. Soil is an inhomogeneous and complex medium in physical, chemical and biological aspects, which makes determination of soil water content a difficult technical and methodological problem. New moisture measurement methodologies and techniques, developed for the purpose of soil testing, in many cases have been later adapted to other agrophysical fields of interest, including storage of grain, conservation and quality testing of food products, transportation, climate change research and safety measures against floods and landslides.

Water is the basic biological solvent and an absolutely necessary component of every life form on Earth. The shortage of water is the main growth limiting factor for plants and other organisms in arid regions. Sufficient continuous supply of fresh water is obviously a fundamental necessity for human and animal life. Because water is a deficit resource in many parts of the world, the sustainable and harmonious development of societies needs reasonable and responsible water management policy. Continuous long-term monitoring of soil moisture on local and global scales, extremely significant for agriculture (including irrigation), environmental protection, meteorology, climatology, scientific research and water management, requires reliable measurement techniques. The development of remote and proximal soil moisture determination methods is therefore one of the most important aims of aquametry.

The practical agricultural applications of water content determination are not restricted to soil measurements. The storage of grain, seeds and other crops, as well as their processing and resultant food products, usually require control of moisture level to prevent spoilage and proliferation of pests, to preserve their quality and extend their shelf life, and to ensure appropriate technological processing conditions. Another fundamental application of aquametry is therefore the evaluation of the state and content of water in various materials of agricultural origin, including food products. Because of high variability of such materials, which may be liquids, solids, porous materials or heterogeneous mixtures of complicated structure and chemical composition, the development of appropriate aquametry techniques may become a challenging endeavour.

Due to the recent technological development in the fields of electronics, informatics, microwave techniques and mobile communication, there is a significant progress in indirect dielectric measurement methods, especially the broadband spectroscopic techniques. These methods enable fast, selective and non-destructive measurements using portable meters and sensors that can be applied in real time testing during production processes and monitoring of storage of various agricultural materials and food products. On the other hand, there are several significant obstacles hampering the development of effective dielectric techniques for determination of moisture content and related material properties. The main difficulties lie in the necessity of performing basic research to understand water-involving physical and chemical processes at micro- and macroscopic scales, mechanisms of polarization and influence of electromagnetic waves on studied materials, development of selective sensor techniques and achieving the required accuracy.

1.1. Physical fundamentals and definitions

The amount of water in a given body may be described by the mass or volume of water relative to the mass or volume of the whole moist object or just the appropriate dry mass (Kraszewski, 2001; Hillel, 2004). On a wet basis, the mass water content of a body is equal to the mass of the water, m_w, divided by the bulk mass of the moist body, m_b, which is the sum of the mass of the water and the mass of the dry material, m_d

$$\xi = \frac{m_w}{m_b}, \quad m_b = m_w + m_d \tag{1}$$

The mass water content on a dry basis is defined as the ratio of the mass of water to the mass of the dry material

$$\eta = \frac{m_w}{m_d} \tag{2}$$

Another useful quantity is the volumetric water content given by the relation

$$\theta_v = \frac{V_w}{V_b} \tag{3}$$

where V_w is the volume of water contained in a given body and V_b is the bulk volume of this body. Knowing the bulk density of the tested material, $\rho_b = m_b / V_b$, and the density of water, ρ_w, the volumetric water content may be expressed by the mass water content calculated on a wet basis as follows

$$\theta_v = \xi \frac{\rho_b}{\rho_w} \tag{4}$$

The total amount of water contained in a given body does not fully determine all of its moisture related properties. The structure and chemical composition of a given body can greatly influence the properties of contained water. The molecule of water, consisting of two hydrogen atoms bound to an atom of oxygen, exhibits polar structure. It may be described as a regular tetrahedron, with the oxygen atom in the centre and the hydrogen atoms with partial positive charges in two corners. Free corners of the tetrahedron are occupied by two electron orbitals. Effectively, though a water molecule is not electrically charged, it possesses an electrical dipole moment. Water molecules are thus enabled to form hydrogen bonds with each other and another compatible polar chemical groups.

Because of the unique properties of water molecules, generally it is possible to distinguish three states of water in a moist material (Hillel, 2004; Lewicki, 2004; Chen and Or, 2006):

1. Bound water – consisting of water molecules bound by hydrogen bonds to a macromolecule (so called structure water – its molecules are immobilised and become structural parts of a macromolecule) and of oriented water molecules forming hydration shells around ions, polar chemical groups and whole macromolecules (so called hydration water). Number of water molecules in a hydration shell, as well as their orientation, distortion and number of layers depend on the surface charge density of an ion or on a structure of a macromolecule in question. Furthermore, even non-polar compounds interact with water, affecting the distribution and orientation of surrounding water molecules – the effect is called hydrophobic hydration (Lewicki, 2004).

2. Capillary water – because water exhibits surface tension, in non-hydrophobic porous materials water can be held in the pores through a capillary pressure, defined as a difference of pressures above and below the surface of the water meniscus. It causes the effect of a capillary rise above the free water surface to the height defined by the surface tension, contact angle, capillary radius and the difference between the water and gas (air) densities. Water is more easily held in smaller pores than in large ones. Capillary water is at equilibrium with the films of bound water adsorbed on the surfaces.

3. Free water – movement of its molecules is not constrained by any kinds of bonds with other chemical constituents of a given material.

A physical thermodynamic quantity describing the state of water in a given body (liquid or solid) is called the water activity coefficient, a_w (Lewicki, 2004). It is defined as the ratio of the water vapour pressure in a given material, p, to the vapour pressure of pure free water, p_0, at the same temperature and total pressure

$$a_w = \frac{p}{p_0} \tag{5}$$

Water activity is especially useful in food industries, in processing technology and to assess food safety. The water activity is greatly affected by the amount of solute molecules and their interaction with water. For example, the more of the water molecules are bounded by the solute molecules, the lower the a_w is. The proliferation of microorganisms, causing spoilage of food, depends on the value of the water activity. For given microbial species, there is an inhibiting value of a_w, below which the proliferation is ceased. This is realised in many food preservation methods, such as freezing, salting and drying (Tapia et al., 2007).

To characterise the state of water in porous bodies such as soil, a quantity related to water activity, called water potential, ϕ , is defined as (Or and Wraith, 2002):

$$\phi = \phi_m + \phi_o + \phi_g + \phi_p = RT \ln \frac{p}{p_0} \tag{6}$$

Here R = 8.31 $JK^{-1}mol^{-1}$ is the gas constant and T is the thermodynamic temperature. The definition above pertains to the total water potential, which is the sum of several terms:

- matric potential, denoted ϕ_m, which depends on the capillary pressure and the adsorption of water on the surfaces of solid particles,
- osmotic pressure potential, ϕ_o, describing the influence of solutes on the state of water; this term is often neglected for porous bodies with no diffusion barriers,
- gravitational potential, ϕ_g, arising from the gravitational force being exerted on the water confined in a given porous body,
- pressure potential, ϕ_p, defined as the hydrostatic pressure exerted by unsupported water (i.e., saturating the soil) overlying a point of interest.

To express water potential in terms of pressure, one needs to multiply it by water density. Because pressure of soil water may be lower than the atmospheric, matric potential may possess negative values. Disregarding the negative sign, water potential is called tension or suction. Water potential governs the direction of the water flow, that is water tends to flow from areas with high water potential (or low suction) towards areas, in which the potential is lower (or the suction higher). For example, when considering the soil-plant-atmosphere continuum, the differences in water potential between soil and roots determines the absorption of water by plants.

http://dx.doi.org/10.5772/52505

2. Measurement methods and equipment

To properly and comprehensively characterise the influence of water contained in a given object on the properties of the said object, several parameters must be determined. In case of agricultural materials and food products, the most important of those parameters are: water content (by mass or by volume), water activity/potential and in case of soils also salinity, oxygenation and temperature (Malicki, 1999).Soil, being a porous three-phase body comprising of various mineral particles of different size and composition (clay, silt and sand particles) with soil water and air filling the pores, containing a countless number of various chemical compounds and organic matter, like microorganisms, flora and fauna species, is surely one of the most complex and variable medium in Nature. The accurate measurement of physical properties of such materials is thus a difficult challenge. Many of the modern aquametry techniques, appropriate for the soil, are also suitable for other materials like grain and food products, which are not as complex as soil. Therefore, the measurement methods and equipment described in this section will pertain mostly to soil water examinations.

From the agrophysical point of view, determining the status of water in soil and other porous materials is extremely important because each phenomenon and process taking place in such medium depends on its water-related properties (Blahovec, 2011). To achieve this goal, various methods and techniques have been developed, including sampling, non-destructive proximal and remote sensing, automatic monitoring systems, etc.

2.1. Common issues of water content measurement techniques

2.1.1. Selectivity

The measurement of physical quantities is rarely direct, that is the value of a quantity in question is usually inferred from another quantity or quantities determined during the measurement process. For example, soil water content may be determined from measurements of mass of the moist and dry sample, electrical resistance or capacity, neutron scattering, gamma-ray absorption or dielectric permittivity of the soil, depending on the applied measurement technique. The conversion function between the intermediary and the target quantities is called the calibration function. The critical issue in the selection of the measurement method is its selectivity, that is the lack of influence on the final result from factors other than the desired quantity. Good selectivity enables the usage of general calibrations, valid for many types of soil and independent of local conditions.

The physical quantity enabling selective determination of soil water content in several modern measurement techniques is dielectric permittivity. The relative dielectric permittivity of water for frequencies of the electric field below the relaxation frequency of 18 GHz is equal to about 80, while for the soil solid phase its value is of about 4 – 6. Therefore, the dielectric permittivity of soil is strongly dependent on the water content. The influence of the temperature in many cases may be neglected (Skierucha, 2009). The salinity of the soil may also interfere with the water content measurement, however its influence is negligible for the electric field frequencies of the order of 400 MHz and higher (Skierucha and Wilczek, 2010). The dielectric measurement methods, described in detail in section 2.3.2, are therefore highly selective and give reliable results for various field conditions.

Salinity of the soil may also be determined by electrical measurements. Salts dissolved in the soil water are the sources of ions, which enable the electric current conduction. Therefore, the soil electrical conductivity may be used as a salinity measure. Electrical conductivity is highly dependent not only on the concentration of ions, but also on temperature and water content. However, there are integrated dielectric sensors that enable simultaneous measurements of soil moisture, temperature and electrical conductivity.

2.1.2. Resolution and sampling

In case of inhomogeneous medium, the volume of the sample, on which the measurement is performed, may influence the final result. Because of the spatial and temporal variability of the soil, the measured values may differ even for points in proximity to each other.

Soil water content may be examined at a point, field, catchment, country or global scale. The spatial and temporal resolution and coverage of the measurements depend on the measurement and monitoring technique, selection of which is dependent on the purpose of the measurements. For example, satellite sensing provides global coverage with spatial resolution of many kilometres, while proximal techniques give soil moisture values close to a point scale, with the spatial and temporal coverage limited by the number of monitoring stations and the chosen time schedule for the measurements (Vereecken et al., 2008; Skierucha and Wilczek, 2010).

Several soil water content measurement techniques, like the thermogravimetric method described in section 2.2, require acquisition of traditional samples. The process is usually disruptive to the surroundings and may disturb the sample itself. The extraction, transport and processing of the sample may introduce errors, which may be reduced by increasing the number and volume of taken samples. Such measurement methods are usually laborious and time-consuming, which makes them unsuitable for monitoring applications.

2.1.3. Non-destructive measurements

There is a number of measurement techniques which do not require extraction of samples and provide means for non-destructive or even non-invasive testing.

Non-destructive measurement methods do not disturb the sample, enabling repetitive examinations of the same object. However, installation of probes and measurement stations in the soil may still be required. The advantage is that after the initial disturbance during the installation, the act of measurement do not affect the tested sample nor its surroundings. Such techniques are therefore suitable even for long-term monitoring purposes.

http://dx.doi.org/10.5772/52505

Name of the measurement method	Directly measured quantity, physical principle, soil property measured and references	Remarks
TDR (time domain reflectometry)	Velocity of propagation of the electromagnetic wave (step or needle pulse) along the metallic parallel or coaxial waveguide (TDR probe) fully inserted into the soil. It is very well correlated with the real part of the soil complex dielectric permittivity as well as the amount of water in soil (Topp et al., 1980; Noborio et al., 1999).	Commonly recognized alternative for the thermogravimetric method, instruments are still very expensive, usually no site calibration required.
	Attenuation of the electromagnetic wave during its travel in the TDR probe, which results mainly from the soil electrical conductivity dependent ion conduction. Signal attenuation is correlated with the soil bulk electrical conductivity and soil salinity (electrical conductivity of soil extract) (Malicki and Walczak, 1999; Robinson et al., 2003).	Not applicable for very saline soils and long probe rods, limited accuracy caused by the possible change of the TDR probe geometry.
FDR (frequency domain reflectometry)	Phase shift (dependent on soil bulk dielectric permittivity) and amplitude attenuation (dependent on soil salinity) of an electrical signal a probe inserted into the soil treated as a lossy capacitor. Measurement is done in a single frequency generated by the internal probe oscillator (50-150 MHz) (Veldkamp and O'Brien, 2000).	Requires soil site calibration, probes and meters are commercially available and cheaper than TDR instrumentation, low power consumption as compared to TDR technique.
Neutron scattering	Number slow neutrons that are produced from the collision of fast neutrons with hydrogen molecules in soil, which is linearly related to the soil volumetric water content. Fast neutron generator and the counter are installed in the vertical access tube for the measurements in different layers of soil (Evett and Steiner, 1995).	Requires soil site calibration ,precise but expensive, additional cost with special licensing, operator training, handling, radiation materials waste disposal, health hazard.
Tensiometry	Suction force or pressure exerted on a pressure transducer in a water filled tube connected with soil matrix by a porous cap. The measured physical quantity is a matrix potential of soil water, which is an basic element of the total potential of water in the soil (Mullins, 2001; Sisson et al., 2002).	Limited range of work (down to about -85 kPa), requires frequent servicing (air bubbles),in drought conditions water moves from the tensiometer to the soil.
Electrical resistance blocks	Electrical resistance, measured with an alternating current bridge (usually ≈ 1000 Hz) of electrodes encased in some type of porous material (gypsum, nylon fabric, fiberglass) that within about two days will reach a quasi-equilibrium state with the soil. This method determines soil water content and water potential as a function of electrical resistance (Spaans and Baker, 1992; Hillel, 2004).	Sensitive to soil salinity and temperature, requires soil specific calibration, very economic and field installations can work for several years, supplementary to tensiometers in the range up to -1500 kPa.

Table 1. Selection of the most popular non-destructive soil water status measurement methods, from (Skierucha, 2011)

On the other hand, non-invasive techniques do not introduce any disturbance of the tested material. Particularly, they do not use any probes which must be installed in the measurement site. Non-invasive methods include techniques such as airborne and satellite remote sensing (Jackson et al., 1996) or near infra-red reflectance spectroscopy (NIRS) (Cécillon et al., 2009). Remote techniques usually examine only the topsoil and require ground measurements for the proper calibration. However, they are irreplaceable for the monitoring of the soil properties on a global scale.

Non-destructive measurement methods have been adapted by agrophysics from various branches of science and industry. Therefore, the physical principles of those techniques vary greatly, from optical to electrical and nuclear methods. Despite the underlying measurement principle, the sensor device suitable for non-destructive and repetitive testing should possess several universal features, such as a data-logging option, a power supply enabling long-term operation without recharging and the communication capability for the remote configuration and transferring of the measurement results to the user. An overview of various non-destructive soil water measurement methods is presented in Table 1. The measuring principle of each technique is briefly described, along with the most important advantages and limitations.

Because the determination of soil water status requires measurement of several parameters, the construction of integrated sensors, which measure several soil properties at the same time and on the same sample volume, becomes increasingly popular. For example, TDR and FDR methods enable the development of the soil water content and soil salinity measurement device, with the possibility of an addition of a temperature sensor (Skierucha et al., 2006).

2.2. Direct methods of water content measurement

2.2.1. Thermogravimetric method

The traditional and standard water content measurement technique, called the thermogravimetric method, is based on a very simple concept of weighting a sample of a moist material, drying it and then weighting again. The difference between the mass of the moist and the dry material, equal to the mass of water evaporated from the original sample, divided by the mass of the dry material (assuming all of the water evaporated during the drying process), is an exact definition of mass water content on a dry basis (Equation 2). This quantity is also called gravimetric water content. Volumetric water content, when required, may be then easily calculated from the mass water content, according to relations presented in section 1.1.

However, the practical application of the thermogravimetric method is not so simple. First, a sample of a material under test needs to be collected. In case of soil testing, this process is invasive and disruptive to the soil profile and its surroundings. To achieve comparative results, the process of drying needs to be standardised. The temperature and time length of the drying should be adjusted to the specifics of a given material. For soils, a sample is dried by placing in an oven at temperature of 105°C for 24 hours (Hillel, 2004). However, this is still an arbitrary standard. It is difficult to completely dry a material containing microscopic pores and solid particles which easily adsorb water. Some soils, especially those containing much clay, may still hold some water even after the standard drying. On the other hand, some soils may contain many compounds which tend to decompose and evaporate along with water, while the sample is in the oven. The evaporation of other compounds beside water causes overestimation of the initial water content of the sample.

Another source of error in the moisture content measurement by the thermogravimetric method is the extraction of a sample and its transport to the laboratory, where usually the oven is located. Each disturbance of the sample during this process may cause errors. Furthermore, in this method the sample is irreversibly destroyed during the drying process, allowing only one measurement of a given sample.

There exist a method of drying alternative to the oven, in which the sample is placed in a container and impregnated with alcohol (Hillel, 2004). The alcohol is then burned off, what causes the evaporation of water. This method may be used in the field.

The thermogravimetric method, despite being destructive, laborious, time-consuming, prone to errors and completely impractical for monitoring purposes, is commonly regarded as a reference method. Most importantly, it is used as a calibration standard for other moisture measurement techniques.

2.2.2. Karl Fischer titration

One of the most important chemical method of water content measurement is the Karl Fischer titration(Isengard, 2001). It is regarded as a reference method for determining the moisture content of food products.

In the volumetric variation of this technique, a sample of tested material, i.e. the analyte, is placed in a titration cell along with the working medium and the titrant solution. The chemical components added to the analyte are: alcohol (ROH, usually methanol), sulphur dioxide (SO_2), a base (B, usually imidazole) and iodine (I_2). The overall reaction that occurs in the titration cell is as follows:

$$3B + ROH + SO_2 + I_2 + H_2O \rightarrow 3BH^+ + ROSO_3^- + 2I^- \quad (7)$$

In the net reaction, one mole of iodine is consumed per one mole of water from the analyte. The amount of water is therefore measured by the consumption of iodine. In the coulometric variation of this method, iodine is not added in the solution, but it is produced in the cell from iodide by anodic oxidation.

The end-point of the reaction is detected through electrochemical means. There are two platinum electrodes placed in the titration cell. In the so called bipotentiometric technique, a constant current is maintained on the electrodes and the voltage is monitored. In the biamperometric variation, the voltage is kept constant and the current is measured. When all the water is consumed, the redox reactions between iodine and iodide ions occur, what causes an abrupt rise of current (biamperometric technique) or a sudden drop of voltage (bipotentiometric technique). The determination of the end-point of the reaction therefore allows for the calculation of the total amount of water in the sample. The end-point is usually amended by a stop delay time correction, accounting for water that is held by the sample and not immediately available for the reaction.

The Karl Fischer titration is very accurate and can determine even extremely small amounts of water. However, the main disadvantage is that all the water in the sample should be made available for the reaction, which is sometimes difficult to achieve, for example in case of insoluble materials. This measurement method is also destructive, but fortunately it does not require large samples.

2.3. Indirect methods of moisture content measurement

Direct methods of moisture content determination in soils and biomaterials usually are time consuming, require laboratory equipment and are not practical for use in automatic systems to monitor environmental conditions and to control industrial technological processes. Due to spatial (from water volume fraction in soil micropores to water balance in continents) and temporal diversity (from milliseconds when analyzing water fluxes to days or weeks in weather prediction) of moisture content in analyzed objects, selectivity requirements and cost of the applied measurement equipment, the indirect methods of moisture content measurement require interdisciplinary approach that links deep knowledge about physical, chemical and biological processes in tested materials and engineering invention in designing appropriate sensors and meters.

The selection of moisture content measurement techniques presented below is not complete, it only covers representative or the most common ones.

2.3.1. Neutron scattering

One of the most accurate non-destructive water content measurement method involves detection of scattered neutrons emitted from a radioactive source placed within the soil (Hillel, 2004; Robinson et al., 2008). Although it requires installation of an access tube with the neutron source and the probe, which is invasive, the method allows for repetitive measurements of the site without further disturbances.

The neutron moisture meter commonly uses mixtures of radium and beryllium or americium and beryllium as the source of radiation. Radioactive elements emit into the soil gamma radiation and "fast neutrons", that is neutrons with high energy up to 15 MeV. The neutrons then collide with atomic nuclei present in the soil, scatter and lose energy, until they approach a typical energy of particles at a given temperature, that is of about 0.03 eV. Neutrons with such energy are called "thermalized" or "slow". Then they are finally absorbed by the atomic nuclei in soil. It happens that the neutrons are most effectively scattered through collisions with nuclei which mass is similar to their own, that is with protons, constituting the nuclei of hydrogen atoms. The most prevalent source of hydrogen atoms in soil is naturally water. The thermalized neutrons are then detected by the probe. The number of counts of the slow neutrons is approximately proportional to the volumetric water content in the soil.

With proper usage, this moisture content measurement method is very accurate. However, to test the water content of the top layer of the soil requires special precautions preventing escape of fast neutrons to the atmosphere. The main drawbacks of this technique are: the radiation hazard, expensive equipment, personnel training, low spatial resolution and the necessity of the site specific calibration.

Recently, a new remote sensing technique based on the scattering of the cosmic-ray neutrons has been proposed (Zreda et al., 2008). Cosmic rays, consisting mainly of protons, collide with atomic nuclei in atmosphere, creating cascades of secondary particles, including neutrons. Those neutrons may then penetrate soil and scatter, the process of which depends on the soil water content, as described above. Some of these neutrons may diffuse back to the atmosphere, where they may be detected by remote sensors placed several metres above the ground. The measurement results are integrated over large areas (approx. 670 m in diameter), therefore they may be used as an intermediate between on site and satellite sensing (Dorigo et al., 2011). This moisture sensing technique is non-invasive and does not require usage of any radioactive or otherwise hazardous elements. Furthermore, it is suitable for long-term environmental monitoring purposes.

2.3.2. Dielectric methods – electromagnetic aquametry

Dielectric properties or permittivity of agricultural products are of interest for several reasons. They include the sensing of moisture content in these products through its correlation with the dielectric properties of cereal grain and oilseed crops, the influence of permittivity on the dielectric heating of product at microwave or lower radio frequencies, and the potential use of dielectric permittivity for sensing quality factors other than moisture content (Nelson, 2005; Skierucha et al., 2012).

The subject of interest for electromagnetic aquametry is analysing solids of different form and structure, as well as liquids containing water, for identification of their properties when placed in electromagnetic fields of radio and microwave frequencies (attenuation, reflection, phase angle, shift of resonant frequency, etc.). The physical principle of the dielectric moisture content measurement methods is based on the high value of relative dielectric permittivity of free water (about 80 at room temperature) with respect to air (equal to 1) and other materials (for example, dry soil has relative dielectric permittivity of about 4 – 6).

Microwave aquametry, as a branch of electromagnetic aquametry (Kraszewski, 2005) of materials' dielectric properties, applies high measurement frequencies, where only dipole polarization of free and bound water particles is active. The measurement techniques of microwave aquametry provide information about free water content. The following advantages of microwave aquametry were obvious since the early experiments:

a. contrary to lower frequencies, the conductivity effects on material properties can be neglected,

b. penetration depth is much larger than that of infrared radiation and permits the probing of a significant volume of material being transported on a conveyor or in a pipe,

c. physical contact between the equipment and the material under test is not required, allowing on-line continuous and remote moisture sensing,

d. in contrast to infrared radiation, it is relatively insensitive to environmental conditions, thus dust and water vapour in industrial facilities do not affect the measurement,

e. water reacts specifically with certain frequencies in the microwave region (relaxation) allowing even small amounts of water to be detected.

f. contrary to chemical methods, it does not alter or contaminate the test material, thus the measurement is non-destructive.

Since heterogeneous systems have interfaces where the materials of different electrical properties contact each other, producing interfacial polarization that is due to the build-up of charge on the interfaces and causing current flow. Through introduction of external electrical field of variable frequency, it is possible to observe changes of the complex dielectric permittivity of the examined heterogenic material (electrical dispersion). The real part of the permittivity describes the ability of the material to polarize the internal electrical dipoles and charge carriers, and its imaginary part describes the energy loss of the electric field (dielectric loss and conductivity loss). These changes are characteristic for each analysed material because of its unique physical and chemical properties. The dielectric relaxation due to interfacial polarization provides information on the heterogeneous structure and the electrical properties of the constituent components.

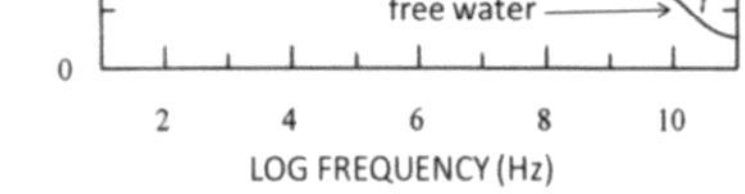

Figure 1. Idealized spectrum of the real part of the complex dielectric permittivity of cell suspensions and tissues. The step changes in dielectric permittivity are called dispersions and are due to the loss of particular polarization processes as frequency increases (from (Markx and Davey, 1999))

For most substances the electrical permittivity and conductivity are constant only for a limited range of frequencies. Within increasing frequency permittivity decreases, while the conductivity increases abruptly. These abrupt changes are called dispersions, and each of them represents a specific process of polarization. Biological materials are characterized by high dispersion, especially at low frequencies (Figure 1). It is caused by interfacial polarization on the surfaces between different materials, from which the cell is formed (Markx and Davey, 1999). The α-dispersion is due to the tangential flow of ions across cell surfaces, the β-dispersion results from the build-up of charge at cell membranes due to the Maxwell–Wagner effect, the δ-dispersion is produced by the rotation of macromolecular side-chains and bound water, and the γ-dispersion is due to the dipolar rotation of small molecules, particularly water. The low frequency polarization masks the bound water and free water dispersion effects. Therefore, the use of microwaves in the analysis of electromagnetic field interactions with soils and biomaterials is recommended.

2.3.2.1. Frequency domain sensors

The most common dielectric techniques for determination of moisture content in soils and biomaterials use capacitance sensors. They usually work in low frequencies up to 150 MHz. A sensor is in a form of a capacitor of parameters modified by dielectric permittivity and electrical conductivity of the surrounding material. The representative probes applied for the measurement of soil moisture and electrical conductivity are: Theta Probe or ECHO moisture sensor with performance and characteristics described broadly in literature (Li et al., 2005; Kizito et al., 2008). Capacitance sensors are not expensive and when equipped with wireless communication and scattered on large areas, they are especially useful in monitoring soil moisture for environmental and irrigation scheduling (Zhang et al., 2011), although they are not so accurate as TDR sensors (Evett et al., 2012).

Radio frequency and microwaves techniques include: reflection measurements with the use of an open-ended coaxial probe (Skierucha et al., 2004; Agilent, 2006), transmission measurements with the use of materials' samples placed inside transmission lines and microwave resonators (King, 2000).

The complex relative permittivity ε^* of a material can be expressed in the following complex form:

$$\varepsilon^* = \varepsilon' - j\varepsilon'' = \varepsilon' - j\left(\varepsilon_d'' + \frac{\sigma}{\varepsilon_0 \omega}\right) \tag{8}$$

The real part ε' is referred to as the dielectric constantand represents stored energy when the material is exposed to anelectric field, while the loss factor ε'', which is the imaginary part, influences energy absorption and attenuation, ε_d'' stands for the contribution due to dipoles rotation,σ $(\mathrm{S\ m^{-1}})$ is the ionic conductivity, ω $(\mathrm{rad\ s^{-1}})$ is the angular frequency and ε_0 is the permittivity of free space or vacuum (8.854 10^{-12} F $\mathrm{m^{-1}}$), and $j=\sqrt{-1}$. Mechanisms that contribute to the dielectric loss in heterogeneous mixtures include polar, electronic, atomic and Maxwell–Wagner responses. Aquametry measurements at RF and microwave frequencies are of practical importance and they are currently used for applications in food processing (Venkatesh and Raghavan, 2004), food treatment (Marra et al., 2009) and quality determination of biomaterials and food products (Wang, 2003; Sosa-Morales et al., 2010).

2.3.2.2. TDR technique

Time domain reflectometry (TDR) is a fast, accurate, and safe technique. The basic principle of time domain reflectometry (TDR) is the same as in radars. The system sends an electromagnetic pulse along the waveguide, which reflects on the mismatch impedance. This technique assumes that the material is homogeneous in the vicinity of the waveguide forming a TDR sensor.

The TDR probe consists of two waveguides connected together: a coaxial one, called the feeder, and a parallel one, called the sensor, made of two or three parallel metal rods inserted into the measured medium (Figure 2).

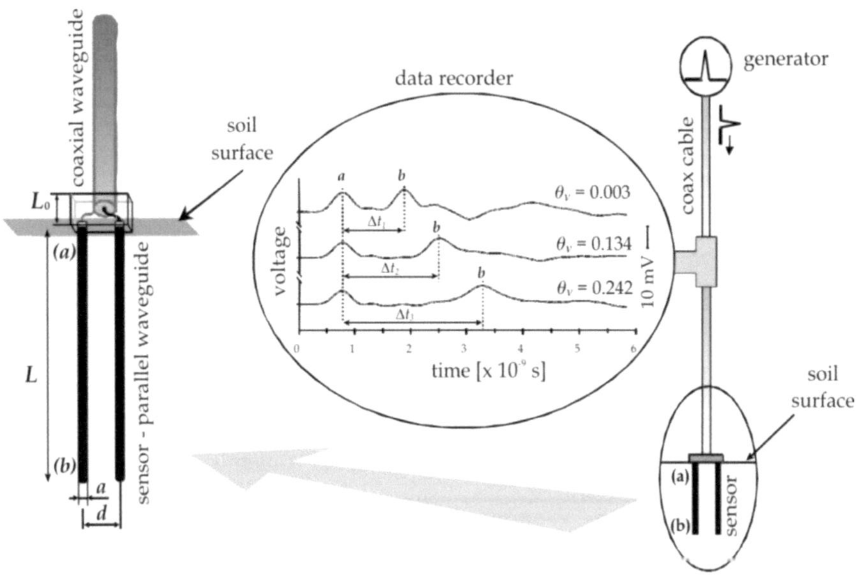

Figure 2. Hardware setup for simultaneous measurement of soil water content and electrical conductivity using Time Domain Reflectometry method, from (Skierucha and Malicki, 2004)

The initial needle pulse or step pulse travels from the generator by the feeder towards the sensor. The recorder registers this pulse as it passes a T-connector. There is a rapid change in geometry of the electromagnetic wave travel path between the feeder and the sensor. At this point, some energy of the pulse is reflected back to the generator, and the remaining pulse is traveling along the parallel waveguide to be reflected completely from the rods ending. The successive reflections are recorded for calculation of the time distance between the two reflections (a) and (b). Three reflectograms (voltage as a function of time at a chosen point in the feeder) are presented in Figure 2. They represent cases when the sensor was placed in dry, wet and water saturated soil. The time distance, Δt, necessary for the pulse to cover the distance equal to the double length of metal rods in the measured medium, increases with the soil dielectric constant, thus with water content. The reason for that is the change of electromagnetic propagation velocity $v$in media of different dielectric constants, according to Equation (9)

http://dx.doi.org/10.5772/52505

$$v \approx \frac{c}{\sqrt{\varepsilon(\theta_v)}} = \frac{c}{n} = \frac{2L}{\Delta t} \tag{9}$$

where c is a velocity of light in free space, $\varepsilon(\theta_v)$ is the real part of the complex dielectric permittivity dependent on its volumetric water content, n is the medium refractive index; L is the length of TDR probe rods inserted into the soil.

Also, the amplitude of the pulse at the point (b) decreases with the increase of soil bulk (or apparent) electrical conductivity, EC_b $(\mathrm{S\ m^{-1}})$, according to Equation (10) (Dasberg and Dalton, 1985)

$$EC_b \approx \frac{\sqrt{\varepsilon(\theta_v)}}{120\pi L} \ln\left(\frac{U_{in}}{U_{out}}\right) \tag{10}$$

where U_{in} and U_{out} are the amplitude of the pulse before and after attenuation caused by the pulse travel twice a distance of the probe length, L. The value of EC_b is a strong indicator of the ionic concentration in soils, i.e. its salinity (Malicki and Walczak, 1999; Friedman, 2005). The TDR determined dielectric constant can be utilized to determine the volumetric water content, θ_v, on the base of empirical calibration (Topp et al., 1980; Malicki and Skierucha, 1989; Malicki et al., 1996) or theoretical models (Roth et al., 1990; Or and Wraith, 1999) of the sensor in the multiphase medium that includes the fraction of bound water beside free water, solid phase and air.

TDR sensors are more precise than capacitance sensors due to higher frequency range of work (about 1 GHz) (Robinson et al., 2008), which minimizes the influence of salinity of the tested material. Although the calibration equations $\theta_v(\varepsilon_b)$, such as Equation (11), are universal for majority of mineral soils giving the mean measurement accuracy ±2% of the measured value of θ_v, the presence of bound water (in materials of large specific surface area, like clay) and variable density of tested material can significantly increase the TDR determined volume fraction of water. Accounting for these effects, the decrease of the moisture measurement error by TDR technique requires material specific calibrations.

2.3.3. Material specific calibration

Moisture content of porous materials is difficult to monitor accurately because of the heterogeneity of pore space, bulk density and structure. There are several types of commercial moisture probes available, including those that employ time domain reflectometry and frequency domain reflectometry, calibrated in the field or laboratory conditions. However, while a moisture sensor performance can be expressed in the sense of measurement accuracy, predicting the final accuracy in terms of moisture content dramatically depends on the properties of the material to be monitored and the particular physical and chemical mechanisms in the process being examined.

The reference values used for calibration in the moisture content measurement of porous materials are the ones taken from thermogravimetric method described in part 2.2.

Calibration equation relating the dielectric constant to the soil moisture content are necessary. The $\theta_v(\varepsilon_b)$ equation generally is provided by a manufacturer. However, in some cases site-specific calibration may be needed. For example, field calibration may be necessary in fine-textured soils.

Time domain reflectometry (TDR) is becoming a widely used method to determine volumetric soil water content, θ_v, from measured apparent (effective, bulk) relative dielectric constant (permittivity), ε_b, using the empirical Topp-Davis-Annan (Topp et al., 1980) calibration equation:

$$\theta_v = -5.3\times10^{-2} + 2.92\times10^{-2}\varepsilon_b - 5.5\times10^{-4}\varepsilon_b^2 + 4.3\times10^{-6}\varepsilon_b^3 \tag{11}$$

Dirksen et al. (Dirksen and Dasberg, 1993) showed that this equation is not adequate for all soils. Other studies showed that bulk density, and thus also porosity, substantially affects the relation between dielectric constant and water content. Two equivalent, empirical, normalized conversion functions were found (Malicki et al., 1996), one accounting for soil bulk density and the other for soil porosity. Each of them reduced the root mean square error of the dielectric TDR determinations of moisture to 0.03, regardless of the materials bulk density and porosity.

2.4. Water potential

Total water potential in a material (e.g. soil), described by Equation (6), depends on several elements, two of which are of the most practical importance: matric potential ϕ_m and osmotic potential ϕ_o. The relation between volumetric water content and water potential of the soil is called soil-water characteristic curve. Its shape differs for clay, silt and sandy soil, as water retention characteristics differ for these soils. Knowing soil water characteristic curve and water potential value, one can calculate water content in this material. The typical measurement methods of water potential of soils and biomaterials use tensiometers and psychrometers as the sensing elements.

2.4.1. Tensiometry

A tensiometer is a device that measures how hard the plant is working to extract water from the soil. It directly measures the physical force that the root system must overcome in order to access water held in the soil (also known as matric potential ϕ_m). It is built from sealed water-filled tube, equipped with a porous tip installed in the ground to the desired root zone (Or and Wraith, 2002). When the matric potential of the soil is lower (more negative) than the equivalent pressure inside the tensiometer cup, water moves from the tensiometer along a potential energy gradient to the soil through the saturated porous cup, thereby creating suction sensed by the gauge. Water flow into the soil continues until equilibrium is reached and the suction inside the tensiometer equals the soil matric potential. When the soil is wetted, flow may occur in the reverse direction, i.e., soil water enters the tensiometer until a new equilibrium is attained.

2.4.2. *Thermocouple psychrometry*

According to equation (6), under equilibrium conditions the material's water potential is equal to the potential of water vapor in the surrounding air, which is measured by a psychrometers. A special construction of a psychrometer, called a thermocouple or Peltier psychrometer, has been developed to be used for applications in soil and biomaterials for the measurement of the sum of matric and osmotic potentials. The construction of a typical thermocouple psychrometers sensor and its connection to a readout device as well as a sample measurement device output are presented in Figure 3a (Skierucha, 2005).

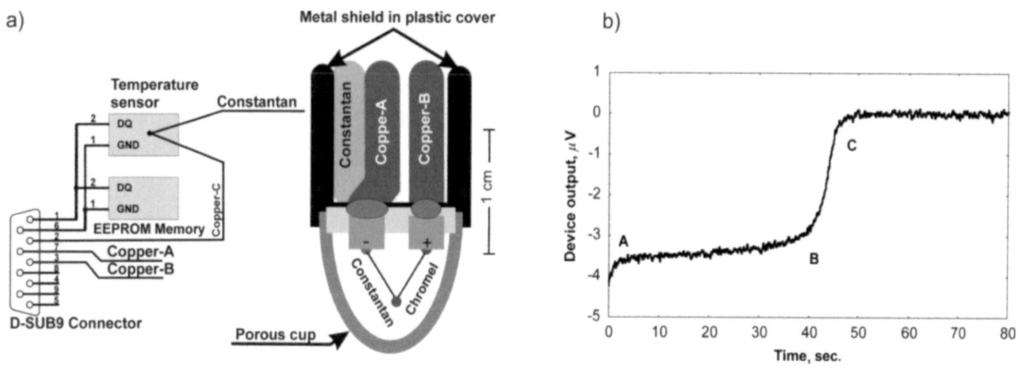

Figure 3. a) Peltier psychrometer sensor with porous ceramic thermocouple shield (Andraski and Scanlon, 2002) and the meter connector with installed electronics for cold junction compensation and storage of the sensor individual parameters; b) recorded output of the thermocouple psychrometer

The measurement cycle consists of steps that are controlled by the microcontroller and electronic circuitry of the meter. These steps may vary according to the water potential to be measured, temperature, required accuracy and time interval of measurements. Optimization of measurement process with respect to different restrictions may lead to various measurement procedures. The measurement sequence taking place to determine psychrometric water potential value is as follows:

1. The constantan/chromel thermocouple should not have water condensed on the fine wires. This is assured by passing the warming current of an appropriate value through the junction and then, after stopping the current, the junction should attain the temperature equilibrium with the air space surrounding it. Also the air space must be in the temperature and vapor equilibrium with the measured sample.

2. The Peltier cooling current is passing through the constantan/chromel thermocouple junction. The magnitude and duration of cooling current must be sufficient to cool the junction below the dew point temperature of the equilibrated air. When the tempera-

ture of the junction is below the dew point, water condenses on the junction from the surrounding air. The Peltier current is discontinued and the thermocouple output voltage starts to be monitored (Figure 3b, point A). During the evaporation of water condensed on the thermocouple junction, its temperature does not change rapidly (points A and B in Figure 3b). This temperature — the wet bulb depression temperature, depends on relative humidity of the air surrounding the sensor. The wet bulb depression lasts until all water evaporates from the junction and the thermocouple temperature returns to the ambient (points B and C in Figure 3b).

For soils, water potential measurements in field condition cannot be acceptably performed by one method in the full range of variability, i.e. from 0 MPa to the wilting point water potential –1.5 MPa, and below. Tensiometers work from 0 to about –0.09 MPa and respond only to soil matrix potential. Porous gypsum blocks are available in the full range of interest but they respond also to soil salinity and therefore need site-specific calibration. Psychrometric sensors seem to be the ideal solution because they measure the humidity of air that remains in equilibrium with a sample of material containing moisture and they respond to total soil water potential. However in contrary to tensiometers, they work in the range from about –0.3 to –6 MPa. Therefore, there is no reliable sensor covering the range of soil water potential from –0.09 to –0.3 MPa in the field conditions. The thermocouple psychrometry is reliable method of water potential measurement (Savage and Cass, 1984; Andraski and Scanlon, 2002), provided that proper precautions are applied to sensors. This include careful cleaning, handling and calibration of the sensors that are susceptible to acid environment. Special attention should be paid to eliminate the temperature gradients in the sensor during measurements. The current state of the thermocouple psychrometry is presented in (Andraski and Scanlon, 2002).The respective sensors are commercially available but the complexity of measurements and rigid temperature conditions forced on the measurement process make this method not convenient.

2.5. Other methods of water content measurement

There is a number of other moisture measurement methods based on several different physical principles (Hillel, 2004; Robinson et al., 2008; Vereecken et al., 2008).

One of the earliest method of soil moisture measurement, proposed at the end of the 19th century, bases on soil electrical conductivity. However, the electrical conductivity of soil depends not only on water content, but also on soil salinity, texture, composition and temperature (Robinson et al., 2008). On the other hand, when moisture content is known, soil electrical conductivity may be effectively used as a salinity measure. Despite this selectivity issue, soil moisture sensor devices, based on the electrical resistance measurements, are being used to evaluate the status of water in the soil.

Electrical resistance blocks are porous bodies comprising of gypsum, nylon or fiberglass, and containing two electrodes, may be used to evaluate soil moisture by measuring electrical resistance of soil water filling the pores. The soil water fills the pores in order to achieve the equilibrium in water tension between the soil and the block. Therefore, this method may be actually better suited for measurements of water potential than the water content.

Thermal properties of a material may also be used to determine its water content. Soil volumetric heat capacity (in units J m^{-3} K^{-1}), defined as the amount of energy needed to increase the temperature of a unit volume of soil by a degree, depends on volumetric water content, as well as other factors such as porosity and heat capacity of the solid phase. If those additional quantities are known, soil water content may be calculated from the calorimetric measurements of soil heat capacity. Soil thermal conductivity, defined as the ability to conduct heat (in units W m^{-1} K^{-1}), depends on the moisture as well. A guarded hot-plate method, in which the sample is placed between the heating and cooling plates and its thermal conductivity is measured, may therefore be used as a water content determination technique (Robinson et al., 2008).

A gamma-ray absorption moisture meter has also been developed, which possesses better spatial resolution than a neutron moisture meter. It consists of a gamma-ray source unit placed in the soil, usually containing radioactive caesium, and a detector, placed in the soil at some distance from the source. The amount of radiation detected by the probe depends on the attenuation coefficient and the distance from the source. It happens that the absorption of the gamma radiation by the soil depends on the moisture content. This method, applicable mostly in laboratory conditions, is cumbersome, presents radiation hazard and therefore is not very popular.

Other techniques of moisture measurement, applicable mostly to food products and agricultural materials, include direct methods such as infrared, halogen and microwave drying (similar in principle to oven drying, or thermogravimetric method for soils), desiccation by water transfer, distillation, chemical methods based on calcium carbide or calcium hydride reactions with water, methods combining evaporation with Karl Fischer titration or diphosphorus pentoxide method (Isengard, 2001). Among the indirect water detection techniques are optical methods such as polarimetry and refractometry, and near infrared (NIR) technique. Other parts of the electromagnetic spectrum are utilised by a low-resolution nuclear magnetic resonance (NMR) technique (based on the influence of a radio frequency pulse on a nuclear spin of hydrogen nuclei placed in a constant magnetic field) or a microwave cavity resonator method. The methods listed above require product-specific calibrations.

3. Aquametry applications in agrophysics

3.1. Soil quality

The appearance of large information banks on soil properties in Europe (Great Britain, France, Holland, Denmark etc.) was provoked by necessity of increasing of agricultural production economics, i.e. the commercial price of soil as the production medium. Protection of soil as element of natural environment was the reason of initiating in 1980 year of national program of soil mapping in Norway. The projects of soil monitoring were limited to the region of Western Europe and it was evident that later projects should have been integrated in the frame of the whole Europe (Montanarella, 2002). It was realized that soil performs the multitude of functions including supporting plant and animal productivity, maintaining or enhancing water and air quality and supporting human health and habitation, which all define the soil quality (Nortcliff, 2002). Soil quality is usually considered to comprise the following components: physical (texture, dry bulk density, porosity, aggregate strength and stability, soil compaction and crusting, etc.), chemical (pH, salinity, aeration status, organic matter content, cation exchange capacity, status of plant nutrients, concentration of toxic elements, etc.) and biological (populations of micro-, meso- and macroorganisms, respiration rate or other indicators of microbial activity, etc.). It is evident that almost all mentioned physico-chemical-biological parameters of soil depend on its moisture. Water is not only the medium, which is necessary for biological changes in evolution of flora and fauna, but it is also the transport medium of heat and energy in the soil (Heitman and Horton, 2011).

Agriculture has both positive and negative influence on environment. Its primary function is meeting the growing demand for food. Agriculture creates habitats not only for humans but also for wildlife and plays an important role in sequestering carbon, managing watersheds and preserving biodiversity. However, agriculture degrades natural resources by causing soil erosion, introducing unrecoverable hydrological changes, contributing in groundwater depletion, agrochemical pollution, loss of biodiversity, reducing carbon sequestration from deforestation and carbon dioxide emissions from forest fires (Doran, 2002).

Physical conditions in agriculture and environment can be defined as physical properties and processes involved in mutual relation between the processes of food and fibre production and the impact of these processes on natural agro-environment. They include topography, surface water and groundwater distributions, heat-temperature distributions, wind direction changes and intensity. There are no universal soil quality indicators developed. It is evident, that they should include water content and/or water potential of the soil as the fundamental parameters. A soil physical parameter defined as the slope of the soil water retention curve at its inflection point (Dexter, 2004) can be used as an index of soil physical quality that enables different soils and the effects of different management treatments and conditions to be compared directly.

3.2. Quality of food materials and products

Moisture content (or water activity) affects food quality, i.e. texture, taste, appearance and stability of foods depends on the amount of water they contain. A knowledge of the moisture content is often necessary to predict the behavior of foods during processing, e.g. mixing, drying, transportation, flow through a pipe or packaging, storage stability or shelf-life (Bell, 2007; Roudaut, 2007).

The tendency of microorganisms to grow in foods depends on their water content. For this reason many foods are dried below some critical moisture content. The cost of many foods depends on the amount of water they contain - water is an inexpensive ingredient, and manufacturers often try to incorporate as much of it as possible in a food, without exceeding some maximum legal requirement. Also, there are legal limits to the maximum or minimum amount of water that must be present in individual types of food (Tapia et al., 2007).

It is therefore important for food scientists to be able to reliably measure moisture contents. A number of analytical techniques have been developed for this purpose, which vary in their accuracy, cost, speed, sensitivity, specificity, ease of operation, etc. The choice of an analytical procedure for a particular application depends on the nature of the food being analysed and the reason the information is needed.

3.3. Environmental monitoring

The need for monitoring physical conditions in agriculture and environment is increasing because of increasing pressure on natural resources, sustainability, exhaustion of nonrenewable resources and climate change. Advances in sensor technology, computers, and communication devices results in great amounts of temporal and spatial information that should be processed in real-time (or near real-time) to produce unambiguous information for the decision making stage. There are two areas of development in the field of monitoring moisture content of the soil upper layer: ground monitoring stations connected into global networks covering the area of river basins, continents or the whole world (Dorigo et al., 2011) and satellite monitoring systems designed especially for the purpose of monitoring water on the Earth, like SMOS (Soil Moisture Ocean Salinity) (Kerr, 2007). Ground monitoring stations tend to use small and wirelessly connected moisture content sensors, while satellite systems process great amount of data on the base of models that must be verified and validated with the use of data from ground monitoring stations.

Integrated, robust, low-cost, and preferably real-time sensing systems are needed for monitoring physical conditions in agriculture and environment. Commercial products have become available for some sensor types. Others are currently under development, especially from the view of climate change (Seneviratne et al., 2010) and precision agriculture (Wang et al., 2006).

3.3.1. Ground monitoring systems with automated data acquisition and processing

The technological progress in material science, electronics, telecommunication and informatics effects in the development of new sensing devices that can be adopted in examining objects of agricultural and environmental studies. They include TDR and FDR probes for the simultaneous measurement of soil moisture, electrical conductivity and temperature (Skierucha et al., 2006). The sensing devices include sensors and transducers, where the former detects the signal or stimulus and the latter converts input energy of one form into output energy of another form. An example of the sensor is a thermistor giving the change of resistance as the function of temperature. Such a sensor associated with electrical circuitry forms an instrument, also called a transducer, that converts thermal energy into electrical energy.

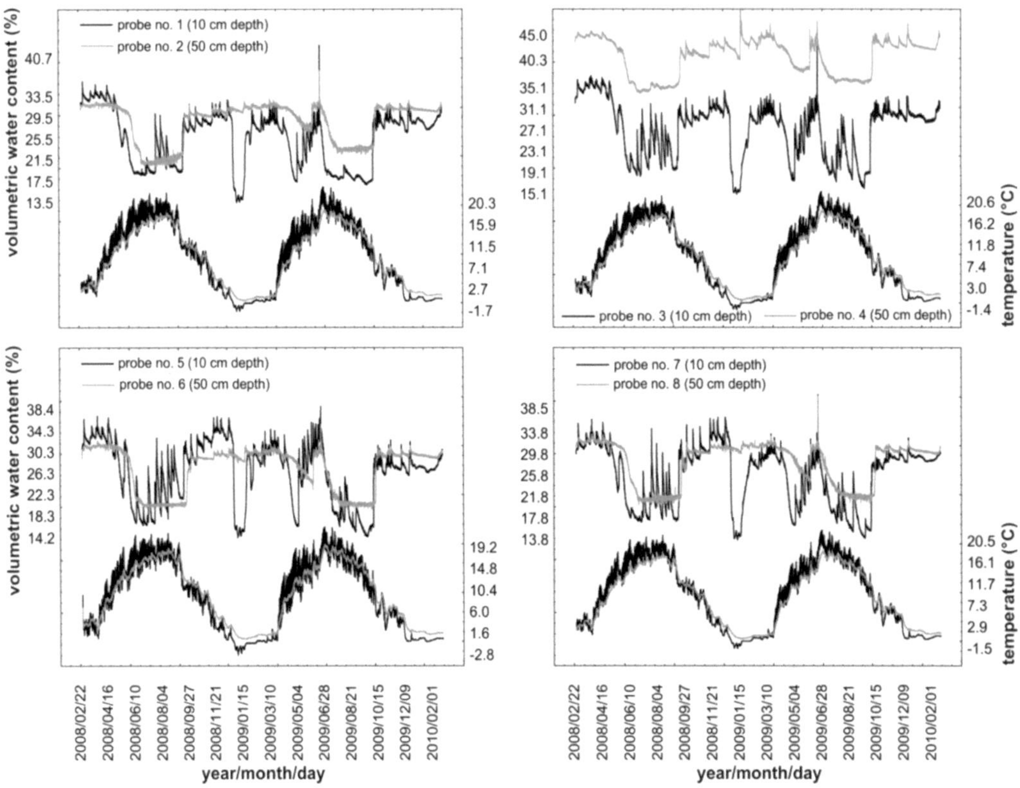

Figure 4. Temporal variability of soil moisture and temperature in P4 localization in Polesie National Park during the period 03.2008 – 02.2010

Another important element of a ground monitoring system is a data acquisition and processing unit, which monitors the output signal of the transducer and processes the resulting data into a form that can be understood by the end user. The basic features of this unit include user friendly interfaces for the operator, large storage memory, physical communication interfaces preferably with serial transmission from the instrument to the operator's notebook. Telemetry with the application of wireless networks is becoming popular especially for distant ground monitoring systems (Wang et al., 2006).

Monitoring stations must meet strong requirements concerning power consumption. The hardware designers should use low power electronic circuits and apply sleep mode operations whenever possible. Also, charging the internal battery may be accomplished with a solar panel.

Figure 4 presents a sample graph of time variability of the values of moisture and temperature of the soil in the P4 measurement point (rendzina soil) in the Polesie National Park in eastern Poland collected by monitoring stations of soil moisture, temperature and electrical conductivity (not presented in Figure 4).

The measurement data collected by the described system are uploaded to and distributed by the International Soil Moisture Network (Dorigo et al., 2011).The ISMN network enables supplementation of the soil moisture data at given locations with other physical parameters (so called metadata).

3.3.2. Remote sensing

Remote sensing of water in agricultural and environmental applications means the acquisition of relevant information about the condition and state of the land surface by sensors that are not in direct physical contact with it. The data are received mainly in the form of electromagnetic waves reflected from the land surface either in passive mode – when the source of energy is the sun and/or the Earth, or in active mode – when the source energy is artificially generated. The analyzed signal reflected from the land surface is composed of different wavelengths over the electromagnetic spectrum (Huete, 2004). The most important regions of electromagnetic spectrum for environmental remote sensing are listed in Table 2.

Spectral Region	Wavelength	Application
Ultraviolet (UV)	0.003 to 0.4 μm	Air pollutants
Visible (VIS)	0.4 to 0.7 μm	Pigments, chlorophyll, iron
Near infrared (NIR)	0.7 to 1.3 μm	Canopy structure, biomass
Middle infrared (MIR)	1.3 to 3.0 μm	Leaf moisture, wood, litter
Thermal infrared (TIR)	3 to 14 μm	Drought, plant stress
Microwave	0.3 to 300 cm	Soil moisture, roughness

Table 2. Regions of electromagnetic spectrum used in environmental monitoring (Huete, 2004)

Today a large number of satellite sensors observe the Earth at wavelengths ranging from visible to microwave, at spatial resolutions ranging from sub-meters to kilometers and temporal frequencies ranging from minutes to weeks or months (Artiola et al., 2004).

The remote sensed data provide information about ecosystem stability, land degradation and desertification (Huete, 2004), carbon cycling (Rosenqvist et al., 2003), soil moisture (Montanarella, 2002; Schmugge et al., 2002; Kerr, 2007), erosion and sediment yield, plant and weeds cover (Thorp and Tian, 2004).

Water in the soil influences the agricultural productivity as well as the weather and climate. Repeating weather disturbances caused by excessive amount of water or its enduring lack impose the necessity of monitoring water content of soil upper layers and deeper in soil profiles. There is a direct feedback between soil moisture and relative humidity of air. Weather prediction on the base of atmospheric parameters including barometric pressure, temperature and air humidity will be more accurate after including soil parameters, like moisture in soil profiles and temperature distribution in the soil. Although water in the soil has a minor contribution of the water balance in the continents, it greatly influences the global water balance (Seneviratne et al., 2010). Therefore, to increase weather prediction and protect people from weather cataclysms, it is necessary to collect and process data about soil moisture from ground and satellite measurements (Dorigo et al., 2011).

Monitoring of soil temperature, which is one of the most important physical parameter apart from water content or water matrix potential, is not a technical problem. There are various temperature sensors available including electronic ones that enable automatic measurement.

4. Summary

Aquametry in agrophysics integrates a number of interdisciplinary research and application issues, including: the state of water in soil and biomaterials, construction of the sensors and physical principles of applied measurement techniques, accuracy and representativeness. There is no universal recipe to determine the amount of water in a sample of material, because the objects of interest in agrophysics differ in scale, observation time, texture, temperature, required accuracy, etc. Therefore, there are so many measurement principles and techniques, each optimized to the object of interest, required measurement conditions, temporal and dimensional scales, and the function of the final information.

The current aquametry tools reflect actual state of technology development and they will continuously change. The information about water in the term of its quantity and quality is crucial for sustainable development, environmental protection and food production since water is not only the basic ingredient of food, but also is a vital element of our habitat. The received and processed data increase our knowledge for the benefit of social, political and economic sustainable development, security as well as for better understanding the nature.

Author details

Wojciech Skierucha*, Agnieszka Szypłowska and Andrzej Wilczek

*Address all correspondence to: w.skierucha@ipan.lublin.pl

Institute of Agrophysics, Polish Academy of Sciences, Lublin, Poland

References

[1] Kraszewski A. Microwave aquametry: an effective tool for nondestructive moisture sensing. Subsurface Sensing Technologies and Applications 2001;2(4):347–362.

[2] Hillel D. Water content and potential. In: Hillel D. (ed.) Introduction to Environmental Soil Physics. Burlington, USA: Academic Press; 2004. p93–125.

[3] Lewicki PP. Water as the determinant of food engineering properties. A review. Journal of Food Engineering 2004;61(4) 483–495.

[4] Chen Y, Or D. Effects of Maxwell-Wagner polarization on soil complex dielectric permittivity under variable temperature and electrical conductivity. Water Resources 2006;42(6) W06424.

[5] Tapia MS, Alzamora SM, Chirife J. Effects of water activity (aw) on microbial stability: as a hurdle in food preservation. In: Barbosa-Cánovas GV, Fontana AJJ, Schmidt SJ, Labuza TP. (eds.) Water activity in foods. Fundamentals and applications. Ames, USA: Blackwell Publishing and the Institute of Food Technologists; 2007. p239–271.

[6] Or D, Wraith JM. Soil Water Content and Water Potential Relationships. In: Warrick AW. (ed.) Soil physics companion. Boca Raton, Florida, USA: CRC Press LLC; 2002. p49–84.

[7] Malicki MA. Methodical questions of monitoring of water status in selected biological materials (in Polish). Lublin: Institute of Agrophysics PAS; 1999.

[8] Blahovec J. Water in forming agricultural products. In: Glinski J, Horabik J, Lipiec J. (eds.) Encyclopedia of Agrophysics. Dordrecht, The Netherlands: Springer; 2011. p971–973.

[9] Skierucha W. Temperature dependence of time domain reflectometry-measured soil dielectric permittivity. Journal of Plant Nutrition and Soil Science 2009;172(2) 186–193.

[10] Skierucha W, Wilczek A. A FDR sensor for measuring complex soil dielectric permittivity in the 10–500 MHz frequency range. Sensors (Basel, Switzerland) 2010;10(4) 3314–3329.

[11] Vereecken H, Huisman JA, Bogena H, Vanderborght J, Vrugt JA, Hopmans JW. On the value of soil moisture measurements in vadose zone hydrology: A review. Water Resources Research 2008;44 W00D06.

[12] Jackson TJ, Schmugge J, Engman ET. Remote sensing applications to hydrology: soil moisture. Hydrological Sciences Journal 1996;41(4) 517–530.

[13] Cécillon L, Barthès BG, Gomez C, Ertlen D, Genot V, Hedde M, Stevens A, Brun JJ. Assessment and monitoring of soil quality using near-infrared reflectance spectroscopy (NIRS). European Journal of Soil Science 2009;60(5) 770–784.

[14] Topp GC, Davis JL, Annan AP. Electromagnetic determination of soil water content: measurements in coaxial transmission lines. Water Resources Research 1980;16(3) 574–582.

[15] Noborio K, Horton R, Tan CS. Time domain reflectometry probe for simultaneous measurement of soil matric potential and water content. Soil Science Society of America Journal 1999;63(6) 1500–1505.

[16] Malicki MA, Walczak RT. Evaluating soil salinity status from bulk electrical conductivity and permittivity. European Journal of Soil Science 1999;50 505–514.

[17] Robinson DA, Jones SB, Wraith JM, Or D, Friedman SP. A review of advances in dielectric and electrical conductivity measurement in soils using time domain reflectometry. Vadose Zone Journal 2003;2(4) 444–475.

[18] Veldkamp E, O'Brien JJ. Calibration of a frequency domain reflectometry sensor for humid tropical soils of volcanic origin. Soil Science Society of America Journal 2000;64(5) 1549–1553.

[19] Evett SR, Steiner JL. Precision of neutron scattering and capacitance type soil water content gauges from field calibration. Soil Science Society of America Journal 1995;59(4) 961–968.

[20] Mullins CE. Matric Potential. In: Smith KA, Mullins CE. (eds.) Soil and Environmental Analysis: Physical Methods. New York: Marcel Dekker; 2001. p65–93.

[21] Sisson JB, Gee GW, Hubbell JM, Bratton WL, Ritter JC, Ward AL, Caldwell TG. Advances in tensiometry for long-term monitoring of soil water pressures. Vadose Zone Journal 2002;1 310–315.

[22] Spaans EJA, Baker JM. Calibration of Watermark soil moisture sensors for soil matric potential and temperature. Plant and Soil 1992;143 213–217.

[23] Skierucha W. Nondestructive measurements in soil. In: Gliński J, Horabik J, Lipiec J. (eds.) Encyclopedia of Agrophysics. Dordrecht, The Netherlands: Springer; 2011. p513–516.

[24] Skierucha W, Wilczek A, Walczak RT. Recent software improvements in moisture (TDR method), matric pressure, electrical conductivity and temperature meters of porous media. International Agrophysics 2006;20 229–235.

[25] Isengard H-D. Water content, one of the most important properties of food. Food Control 2001;12 395–400.

[26] Robinson DA, Campbell CS, Hopmans JW, Hornbuckle BK, Jones SB, Knight R, Ogden F, Selker J, Wendroth O. Soil moisture measurement for ecological and hydrological watershed-scale observatories: a review. Vadose Zone Journal 2008;7(1) 358–389.

[27] Zreda M, Desilets D, Ferré TPA, Scott RL. Measuring soil moisture content non-invasively at intermediate spatial scale using cosmic-ray neutrons. Geophysical Research Letters 2008;35 L21402.

[28] Dorigo WA, Wagner W, Hohensinn R, Hahn S, Paulik C, Xaver A, Gruber A, Drusch M, Macklenburg S, van Oevelen P, Robock A, Jackson T. The International Soil Moisture Network: a data hosting facility for global in situ soil moisture measurements. Hydrology and Earth System Sciences 2011;15(5) 1675–1698.

[29] Nelson SO. Dielectric spectroscopy in agriculture. Journal of Non-Crystalline Solids 2005;351(33-36) 2940–2944.

[30] Skierucha W, Wilczek A, Szypłowska A. Dielectric spectroscopy in agrophysics. International Agrophysics 2012;26 1–11.

[31] Kraszewski A. Recent developments in electromagnetic aquametry. In: Kupfer K. (ed.) Electromagnetic Aquametry. Berlin Heidelberg: Springer-Verlag; 2005. p1–14.

[32] Markx GH, Davey CL. The dielectric properties of biological cells at radiofrequencies: applications in biotechnology. Enzyme and Microbial Technology 1999;25(3-5) 161–171.

[33] Li X, Lei T, Wang W, Xu Q, Zhao J. Capacitance sensors for measuring suspended sediment concentration. Catena 2005;60(3) 227–237.

[34] Kizito F, Campbell CS, Campbell GS, Cobos DR, Teare BL, Carter B, Hopmans JW. Frequency, electrical conductivity and temperature analysis of a low-cost capacitance soil moisture sensor. Journal of Hydrology 2008;352(3-4) 367–378.

[35] Zhang R, Guo J, Zhang L, Zhang Y, Wang L, Wang Q. A calibration method of detecting soil water content based on the information-sharing in wireless sensor network. Computers and Electronics in Agriculture 2011;76(2) 161–168.

[36] Evett SR, Schwartz RC, Casanova JJ, Heng LK. Soil water sensing for water balance, ET and WUE. Agricultural Water Management 2012;104 1–9.

[37] Skierucha W, Walczak RT, Wilczek A. Comparison of Open-Ended Coax and TDR sensors for the measurement of soil dielectric permittivity in microwave frequencies. International Agrophysics 2004;18(3) 355–362.

[38] Agilent. Agilent 85070E Dielectric Probe Kit - Technical Overview. 2006. p1–12.

[39] King RJ. On-line industrial applications of microwave moisture sensors. Sensors Update 2000;7(1) 109–170.

[40] Venkatesh MS, Raghavan GSV. An overview of microwave processing and dielectric properties of agri-food materials. Biosystems Engineering 2004;88(1) 1–18.

[41] Marra F, Zhang L, Lyng JG. Radio frequency treatment of foods: Review of recent advances. Journal of Food Engineering 2009;91(4) 497–508.

[42] Wang S. Dielectric properties of fruits and insect pests as related to radio frequency and microwave treatments. Biosystems Engineering 2003;85(2) 201–212.

[43] Sosa-Morales ME, Valerio-Junco L, Lopez-Malo A, García HS. Dielectric properties of foods: Reported data in the 21st Century and their potential applications. LWT - Food Science and Technology 2010;43(8) 1169–1179.

[44] Skierucha W, Malicki MA. TDR method for the measurement of water content and salinity of porous media. Lublin: Institute of Agrophysics PAS; 2004.

[45] Dasberg S, Dalton FN. Time Domain Reflectometry Field Measurements of Soil Water Content and Electrical Conductivity. Soil Science Society of America Journal 1985;49(2) 293–297.

[46] Friedman SP. Soil properties influencing apparent electrical conductivity: a review. Computers and Electronics in Agriculture 2005;46(1-3) 45–70.

[47] Malicki MA, Skierucha W. A manually controlled TDR soil moisture meter operating with 300 ps rise-time needle pulse. Irrigation Science 1989;10(2) 153–163.

[48] Malicki MA, Plagge R, Roth CH. Improving the calibration of dielectric TDR soil moisture determination taking into account the solid soil. European Journal of Soil Science 1996;47(3) 357–366.

[49] Roth K, Schulin R, Fluhler H, Attinger W. Calibration of time domain reflectometry for water content measurement using a composite dielectric approach. Water Resources Research 1990;26(10) 2267–2273.

[50] Or D, Wraith JM. Temperature effects on soil bulk dielectric permittivity measured by time domain refiectometry: a physical model. Water Resources Research 1999;35(2) 371–383.

[51] Dirksen C, Dasberg S. Improved calibration of time domain reflectometry soil water content measurements. Soil Science Society of America Journal 1993;57(3) 660–667.

[52] Skierucha W. Design and performance of psychrometric soil water potential meter. Sensors and Actuators A: Physical 2005;118(1) 86–91.

[53] Andraski BJ, Scanlon BR. Thermocouple psychrometry. In: Dane JH, Topp GC. (eds.) Methods of Soil Analysis. Part 4. Physical Methods. Madison, WI, USA: Soil Science Society of America, Inc.; 2002. p609–642.

[54] Savage MJ, Cass A. Measurement of water potential using in situ thermocouple hygrometers. Advances in Agronomy 1984;37 73–126.

[55] Montanarella L. Monitoring the change. Institute of Environment & Sustainability, Joint Research Centre - Project MOSES, Ispra, Italy; 2002. p1–20.

[56] Nortcliff S. Standardisation of soil quality attributes. Agriculture, Ecosystems & Environment 2002;88(2) 161–168.

[57] Heitman JL, Horton R. Coupled heat and water transfer in soil. In: Glinski J, Horabik J, Lipiec J. (eds.) Encyclopedia of Agrophysics. Dordrecht, The Netherlands: Springer; 2011. p155–162.

[58] Doran JW. Soil health and global sustainability: translating science into practice. Agriculture, Ecosystems and Environment 2002;88 119–127.

[59] Dexter AR. Soil physical quality. Part I. Theory, effects of soil texture, density, and organic matter, and effects on root growth. Geoderma 2004;120 201–214.

[60] Bell LN. Moisture effects on food's chemical stability. In: Barbosa-Cánovas GV, Fontana AJJ, Schmidt SJ, Labuza TP. (eds.) Water activity in foods. Fundamentals and applications. Ames, USA: Blackwell Publishing and the Institute of Food Technologists; 2007. p173–198.

[61] Roudaut G. Water activity and physical stability. In: Barbosa-Cánovas GV, Fontana AJJ, Schmidt SJ, Labuza TP. (eds.) Water activity in foods. Fundamentals and applications. Ames, USA: Blackwell Publishing and the Institute of Food Technologists; 2007. p199–213.

[62] Kerr YH. Soil moisture from space: Where are we? Hydrogeology Journal 2007;15(1) 117–20.

[63] Seneviratne SI, Corti T, Davin EL, Hirschi M, Jaeger EB, Lehner I, Orlowsky B, Teuling AJ. Investigating soil moisture-climate interactions in a changing climate: A review 2010;99(3-4) 125–161.

[64] Wang N, Zhang N, Wang M. Wireless sensors in agriculture and food industry — recent development and future perspective. Computers and Electronics in Agriculture 2006;50(1) 1–14.

[65] Huete AR. Remote sensing for environmental monitoring. In: Artiola JF, Pepper IL, Brusseau ML. (eds.) Environmental Monitoring and Characterization. Elsevier Science; 2004. p183–206.

[66] Artiola JF, Pepper IL, Brusseau ML. Monitoring and characterization of the environment. In: Artiola JF, Pepper IL, Brusseau ML. (eds.) Environmental Monitoring and Characterization. Elsevier Science; 2004. p1–9.

[67] Rosenqvist Å, Milne A, Lucas R, Imhoff M, Dobson C. A review of remote sensing technology in support of the Kyoto Protocol. Environmental Science & Policy 2003;6 441–455.

[68] Schmugge TJ, Kustas WP, Ritchie JC, Jackson TJ. Remote sensing in hydrology. Advances in Water Resources 2002;25(8-12) 1367–1385.

[69] Thorp KR, Tian LF. A review on remote sensing of weeds in agriculture. Precision Agriculture 2004;5 477–508.

Chapter 3

Time Stability of Soil Water Content

Wei Hu, Lindsay K. Tallon, Asim Biswas and
Bing Cheng Si

Additional information is available at the end of the chapter

http://dx.doi.org/10.5772/52469

1. Introduction

Soil water is a key variable controlling water and energy fluxes in soils (Vereecken et al., 2007). It is necessary for plant and vegetation growth and development. Research has indicated that soil water content (*SWC*) varies both in space and time. Variations in both space and time present a substantial challenge for applications such as precision agriculture and soil water management.

Since the contribution of Vachaud et al. (Vachaud et al., 1985), a large body of research has indicated the presence of time stability of *SWC* (Biswas and Si, 2011c; Comegna and Basile, 1994; Grayson and Western, 1998; Hu et al., 2009; Martínez-Fernández and Ceballos; 2005; Mohanty and Skaggs, 2001), which means that the spatial pattern of *SWC* does not change with time at a certain probability. According to this concept, if a field is repeatedly surveyed for *SWC*, there is a high probability that a location with certain wetness characteristics (i. e., wet, dry, intermediate) will maintain those characteristics on subsequent occasions. Time stability has also been extended to describe the characteristics of *SWC* at point scales. A location will be regarded as time stable provided it can estimate the average *SWC* of an area.

Time stability of *SWC* has been observed at a large variety of scales ranging from plot (Pachepsky et al., 2005) to region (Martínez-Fernández and Ceballos, 2003) and related studies cover a range of investigated areas, sampling schemes, sampling depths, investigation periods, and land uses (Biswas and Si; 2011c; Brocca et al., 2009; Cosh et al., 2008; Hu et al., 2010a;Tallon and Si, 2004; Vachaud et al., 1985). As a result, a variety of methods have been developed to evaluate time stability of *SWC*, each with its own advantages and disadvantages. Time stability is usually used to characterize time persistence of the spatial pattern of *SWC* between measurement occasions, either at the measurement scale or at different scales (Biswas and Si, 2011c; Cosh et al., 2006; Kachanoski and De Jong, 1988; Vachaud et al., 1985).

On the other hand, *SWC* at time stable locations can be used to estimate the spatial average *SWC* of a given area (Grayson and Western, 1998). Therefore, quick and accurate estimation of soil water content at a field or large areas may be possible with only one representative location. Areal estimation of *SWC* from point source data has the potential to substantially reduce both the capital and labor costs involved in estimating average *SWC*, making the method appealing to a wide range of disciplines.

Time stability of *SWC* is controlled by various factors including soil, topography, vegetation, and climate (Brocca et al., 2009; Gómez-Plaza et al., 2001; Grayson and Western, 1998; Hu et al, 2010a; Tallon and Si, 2004). Information on the controls of time stability provides an essential insight into the mechanisms of soil water movement and storage. In addition, it is also important to identify the time stable location for average *SWC* estimation *a priori*.

The goal of this chapter is to provide a comprehensive review on the study of time stability of *SWC*. In doing so, we have introduced the concept, methodology, application, and the controlling factors step by step.

2. Concept

Soil water content varies with time and space. However, if a field is repeatedly surveyed for *SWC*, there is a high probability that a location with a certain relative wetness condition (i. e., wet, dry, and intermediate) at a given time will remain at the same relative wetness condition at other times. This phenomenon was first explained as time stability by Vachaud et al. (1985). This can be defined as the "time invariant association between spatial location and classical statistical parametric values" of *SWC*, most often the mean.

Time stability of *SWC* can be divided into two types, one is to describe the overall similarity of the spatial pattern between measurements, and the other is to describe the time-invariance of the relative *SWC* of a given location.

Repeated survey indicated that some locations are always wetter or drier than the average *SWC* of an area, resulting in the preservation of their ranks. Hence, the spatial pattern of *SWC* measured at one time will be similar to those measured on subsequent occasions. Statistically, this kind of time stability indicates that the orders, or ranks, of *SWC* at different locations do not change over time at some probability. If all locations were to maintain their rank on subsequent measurement occasions, then the spatial pattern of *SWC* will be identical. From this point, Chen (2006) augured that the term of "rank stability" or "order stability" should be better than "time stability" because "the stability is the order or rank of a soil property at different spatial points that does not change at some probability".

The most widely used method to describe time stability of *SWC* spatial patterns is Spearman's rank correlation analysis (Vachaud et al., 1985), while Pearson correlation analysis has also been widely used (Cosh et al., 2004). This is because Pearson correlation measures strictly linear relationships, while Spearman Rank correlation also measures nonlinear correlation between SWC values measured at two different occasions. Time stability of the spatial pattern of *SWC* is scale dependent because of the interaction between measurement locations (Kachanoski and De Jong, 1988). Scale dependence of the time stability was also investigated by different methods including spatial coherency analysis (Kachanoski and De Jong, 1988), wavelet coherency analysis (Biswas and Si, 2011c), and multivariate empirical mode decomposition (Rehman and Mandic, 2010). A detailed introduction of these methods is given in Section 3.

For time stability at point scales, *SWC* at each location is usually scaled in terms of the field average (Vachaud et al., 1985). This indicates that at a given location, the change in scaled *SWC* exhibits little change over time. Many indices are available to examine the degree of time stability at point scales. These indices include standard deviation of relative difference (Mohanty and Skaggs, 2001; Schneider et al., 2008; Vachaud et al., 1985), root mean square error (Bosch et al., 2006; Jacobs et al., 2004), width of the 90% empirical tolerance interval of relative water content (Guber et al., 2008), chi-squared statistic (Guber et al., 2008), root-mean-squared differences (Guber et al., 2008), temporal coefficient of variability (Starr, 2005), and mean absolute bias error (Hu et al., 2010a, 2010b). One of the applications of time stability at point scales is to identify catchment average soil moisture monitoring (CASMM) site (Grayson and Western, 1998) or benchmark location (Tallon and Si, 2004) for average *SWC* evaluation. Detailed introduction of these indices and applications is given in Section 3.

Generally, these two types of time stability are correlated. If the spatial pattern is time stable, there is a larger possibility of time stability at point scales. On the other hand, if more points are time stable, there is a larger possibility of time stability of spatial pattern. However, these two types of time stability are also distinguished for the following three aspects: (1) time stability of spatial pattern is used to evaluate the time stability of an area, while time stability of relative *SWC* is used to evaluate the time stability of a point. Therefore, time instability of spatial pattern does not mean the absence of time stable points (Grayson and Western, 1998; Schneider, 2008). On the other hand, if no points are time stable, it does not mean that spatial pattern is time unstable for any two measurement occasions; (2) they have different evaluation criteria as mentioned above, and (3) they have different applications. Time stability of spatial pattern is used to qualitatively describe the similarity of spatial distribution of *SWC* between different measurement occasions for the better understanding of soil water related process and influencing factors (Lin; 2006), while time stable locations can be selected to estimate the average *SWC* for an area (upscaling) by directly using the soil water content of the time stable location (Brocca et al., 2009; Cosh et al., 2008; Grayson and Western, 1998; Martínez-Fernández and Ceballos, 2005) or by considering the offset between the mean value and that of the time stable location (Grayson and Western, 1998; Hu et al., 2010b; Starks et al., 2008).

3. Methodology

There are different methods for analyzing time stability. We have documented the methods in two sections: (1) time stability of spatial patterns and (2) time stability of points.

3.1. Time stability of spatial patterns

3.1.1. *Pearson correlation analysis*

Pearson correlation coefficient is sensitive only to a linear relationship between two variables. A Pearson correlation coefficient, $r_{j,j'}$, between two spatial series of soil water content measured at time *j* and *j'* can be defined by

$$r_{j,j'} = \frac{\sum_i (SWC_j(i) - \langle SWC_j \rangle)(SWC_{j'}(i) - \langle SWC_{j'} \rangle)}{\sqrt{\sum_i (SWC_j(i) - \langle SWC_j \rangle)^2} \sqrt{\sum_i (SWC_{j'}(i) - \langle SWC_{j'} \rangle)^2}} \tag{1}$$

where $SWC_j(i)$ and $\langle SWC_j \rangle$ are soil water content at location *i* and spatial average at a given time *j*, respectively. $SWC_{j'}(i)$ and $\langle SWC_{j'} \rangle$ are soil water content at location *i* and spatial average at another time j', respectively. The resulting coefficients refer to the correlation of the spatial patterns of *SWC* from one time to another. It is expected that closely correlated patterns have a $r_{j,j'}$ near one, while uncorrelated patterns are indicated by $r_{j,j'}$ values near zero (Cosh et al., 2004). Student's *t* is generally used to test the significance of Pearson correlation coefficient. It can be implemented by some statistical software, such as Excel, SPSS, MathCad, Matlab, and SAS.

3.1.2. *Spearman's rank correlation analysis*

Spearman's rank correlation coefficient is a non-parametric measure of statistical dependence between two variables. It assesses how well the relationship between two variables can be described using a monotonic function. It is the Pearson correlation between the ranks of one series and the ranks of another series. Because ranking linearize some of the nonlinear relationships, it is sensitive to nonlinear relationships. Let R_{ij} be the rank of soil water content $SWC_j(i)$ and R'_{ij} the rank of $SWC_{j'}(i)$. The Spearman rank correlation coefficient, r_s, is calculated by

$$r_s = 1 - \frac{6\sum_{i=1}^{n}\left(R_{ij} - R'_{ij}\right)^2}{n(n^2 - 1)} \tag{2}$$

where *n* is the number of observations. A value of r_s=1 corresponds to identity of rank for any sites, or perfect time stability between time *j* and *j'*. The closer r_s is to 1, the more stable the spatial pattern will be. Student's *t* can be used to test the significance of r_s (Zar, 1972). Software such as Excel, SPSS, MathCad, Matlab, and SAS can implement this analysis.

3.1.3. *Average spatial coefficient of determination and temporal coefficient of variability*

If perfectly time stable soil water content pattern exists, the ratio of $SWC_j(i)$ to $\langle SWC_j \rangle$, can be a time independent scaling factor, $r(i)$, which is expressed as (Starr, 2005)

$$\frac{SWC_j(i)}{\langle SWC_j \rangle} = r(i) \tag{3}$$

Complete time stability is usually disturbed by a series of processes and therefore Eq. (3) needs an extra term to fit a more general situation (Starr, 2005)

$$\frac{SWC_j(i)}{\langle SWC_j \rangle} = r(i) + e_j(i) = s_j(i) \tag{4}$$

where $s_j(i)$ is a scaled factor depending on both time and location, and $e_j(i)$ is an additional term that accounts for random measurement errors, random sampling variability, and any true deviations from time stability. Assuming that $e_j(i)$ and $r(i)$ are independent, then the variance of $s_j(i)$ over space at any time, $\gamma_i^2[s_j(i)]$, can be written as

$$\gamma_i^2\left[s_j(i)\right] = \gamma_i^2\left[r(i)\right] + \gamma_i^2\left[e_j(i)\right] \tag{5}$$

where $\gamma_i^2[r(i)]$ and $\gamma_i^2[e_j(i)]$ are the variances of $r(i)$ and $e_j(i)$, respectively over space. The $\gamma_i^2[s_j(i)]$ can be calculated by averaging the calculated variances for each observation. Then, the average spatial coefficient of determination, R_s^2, which describes the proportion of the total variance explained by the time stability model of Eq. (3) can be written as

$$R_s^2 = \frac{\gamma_i^2\left[r(i)\right]}{\left\langle \gamma_i^2\left[s_j(i)\right]\right\rangle} \tag{6}$$

where $\langle \gamma_i^2[s_j(i)] \rangle$ is the mean value of $\gamma_i^2[s_j(i)]$ over time. Given $e_j(i)$ with an average value of zero over time, $r(i)$ can be approximated by time average of $s_j(i)$, $\langle s_j(i) \rangle$, i. e.,

$$r(i) = \left\langle s_j(i) \right\rangle \tag{7}$$

The value of R_s^2 approaches one for a complete spatial pattern described by Eq. (3) and zero for a situation where no variances can be explained by the time stability model. Therefore, greater R_s^2 means stronger time stability of spatial pattern. However, Starr (2005) pointed out it is not suitable for the situation where no spatial variability of *SWC* and completely time stability exists.

To deal with this issue, temporal coefficient of variability, CV_t, was developed to describe the degree of time stability (Starr, 2005) and can be written as

$$CV_t = 100\frac{1}{n}\sum_{i=1}^{n}\frac{\sigma_t\left[s_j(i)\right]}{\left\langle s_j(i)\right\rangle} \tag{8}$$

where $\sigma_t[s_j(i)]$ is the standard deviation of $s_j(i)$ over time, $\langle s_j(i)\rangle$ is the time average of $s_j(i)$. The smaller CV_t value indicates stronger time stability. The CV_t approaching zero indicates a perfect time stable pattern.

3.1.4. Empirical orthogonal function method

Empirical orthogonal function (EOF) analysis is a decomposition of a data set in terms of orthogonal basis functions which are determined from the data. It is the same as performing a principal components analysis on the data, except that the EOF method finds both time series and spatial patterns. The basic principle of EOF method is to partition a series into time-invariant spatial patterns (EOFs) of *SWC* and coefficients (ECs) which vary temporally but is constant spatially (Perry and Niemann, 2007). The original spatial series of *SWC* can be obtained by taking the sum of product of EOFs by ECs. A limited number of EOFs that present significant spatial variation of *SWC* can be selected and used to identify the dominant factors determining the spatial pattern.

First, the spatial anomalies of *SWC* are computed. The spatial anomaly at location i and time j, $z_j(i)$, can be calculated as

$$z_j(i) = SWC_j(i) - \frac{1}{n}\sum_{i=1}^{n} SWC_j(i) \tag{9}$$

Considering the spatial anomalies, the spatial covariance $v_{j,j'}$ at time j and j' can be calculated as

$$v_{j,j'} = \frac{1}{n}\sum_{i=1}^{n} z_j(i) z_{j'}(i) \tag{10}$$

where $z_{j'}(i)$ is the spatial anomaly at the same location *i* but time j'.

To consider all the measurement times, the matrix of spatial anomalies, *Z*, and its spatial covariance, *V*, can be written as

$$Z = \begin{pmatrix} z_{11} & \cdots & z_{1m} \\ \vdots & \ddots & \vdots \\ z_{n1} & \cdots & z_{nm} \end{pmatrix} \tag{11}$$

and

$$V = \frac{1}{n} Z^T Z \tag{12}$$

respectively, where *m* is the number of sampling times.

The next step is to diagonalize the spatial covariance matrix *V* by finding its eigenvectors, *E*, and eigenvalues, *L*. Mathematically, they should satisfy the equation

$$VE = LE \tag{13}$$

where *E* is an *m*×*m* matrix that contains the eigenvectors as columns

$$E = \begin{pmatrix} e_{11} & \cdots & e_{1m} \\ \vdots & \ddots & \vdots \\ e_{m1} & \cdots & e_{mm} \end{pmatrix} \tag{14}$$

and *L* is an *m*×*m* matrix that contains the associated eigenvalues along the diagonal and zeros at off-diagonals

$$L = \begin{pmatrix} l_{11} & \cdots & 0 \\ \vdots & \ddots & \vdots \\ 0 & \cdots & l_{mm} \end{pmatrix} \tag{15}$$

The eigenvectors in *E* represent the weights applied to each component in *V* to diagonalized *V*. This transformation is a rotation of the original axes in multi-dimensional space, where each dimension corresponds to a sampling time. Axes are orthogonal to each other and explain different amount of covariance in the spatial anomaly dataset. The eigenvalues contained in *L* represent the variance that occurs in the direction of each newly identified axis. After diagonalization of V, E and L are arranged accordingly to keep the eigenvalues in L sorted in a descending order. Therefore, the first axis explains the most covariance in the spatial anomaly dataset, and the second axis explains the second most covariance and so on. Each axis represents a direction in the multidimensional space. The total variance of the spatial anomaly data is the sum of the diagonal values in L and is equal to the total variance of the original spatial anomalies. Therefore, the portion of the variance, P_j, that the jth axis can explain is

$$P_j = \frac{l_{jj}}{\sum_{k=1}^{m} l_{kk}} \tag{16}$$

The relative importance of each axis in explaining the variability of spatial anomalies can be defined by the relative magnitude of the associated eigenvalue in L.

Each spatial anomaly can be described in terms of the new variable axis. A matrix F containing the coordinates of the spatial anomalies on the new axis can be obtained by projecting the spatial anomalies onto the rotated axis. Mathematically, this operation is

$$F = ZE \tag{17}$$

Each column in F is EOF, and corresponding column in E is expansion coefficient (EC). From the aspect of time stability application, usually a limited number of EOFs that explains significant amount of the spatial variability of the original SWC series is selected to present the underlying time-invariable spatial pattern. North et al. (1982) indicated a set of selection criterion. An EOF is considered significant provided the lower confidence limit of its eigenvalue is larger than the upper confidence limit of the eigenvalue of the next most important EOF (North et al., 1982).

3.1.5. Spectral coherency analysis

Spectral coherency is used to measure the similarity between two spatial series in a frequency domain, which can be converted to a spatial scale. This method assumes the spatial series to be stationary and linear. Therefore, it can be used to find the scale specific information on time stability of spatial pattern between different measurements.

Spectral coherency analysis involves calculation of the power spectrum of a variable $V_t^2(f_K)$ which estimates the observed variance as a function of spatial scale.

$$V_t^2(f_K) = (2h+1)^{-1} N^{-1} \sum_{l=-h}^{h} \left[\left(\sum_{j=1}^{N} SWC_t(j)\cos(2\pi f_{K+1} j) \right)^2 + \left(\sum_{j=1}^{N} SWC_t(j)\sin(2\pi f_{K+1} j) \right)^2 \right] \tag{18}$$

where $f_K = K / N$, K = 0, 1, 2, …, N/2 cycles h^{-1}, and h is the smoothing coefficient determining the degree of averaging of adjacent independent frequencies and the degrees of freedom for the individual spectral variance estimates. The value of h cannot be too large because power spectrum of a variable may incorporate more bias from smoothing of the spectra, although larger value of h will result in smaller variance of the estimate (Gómez-Plaza et al., 2000).

The covariance between $SWC_j(i)$ and $SWC_{j'}(i)$ can be estimated as a function of spatial scale by calculating sample cross spectrum $V_{j,j'}(f_K)$

$$V_{j,j'}\left(f_K\right) = (2h+1)^{-1} N^{-1} \sum_{l=-h}^{h} F_j\left(f_{K+1}\right)^* F_{j'}\left(f_{K+1}\right) \tag{19}$$

where the asterisk (*) is a complex conjugate of

$$F_j\left(f_K\right) = \sum_{i=0}^{N-1} SWC_j(i)\left[\cos\left(2\pi f_K i\right) - i\sin\left(2\pi f_K i\right)\right] \tag{20}$$

$$F_{j'}\left(f_K\right) = \sum_{i=0}^{N-1} SWC_{j'}(i)\left[\cos\left(2\pi f_K i\right) - i\sin\left(2\pi f_K i\right)\right] \tag{21}$$

where $f_K = K / N$, K = 0, 1, 2, …, N/2 cycles h^{-1}, and $i = (-1)^{0.5}$.

Spectral coherency function $R^2_{j,j'}(f_K)$ can then be calculated in a similar manner to calculate coefficient of determination (Kachanoski and De Jong, 1988)

$$R_{j,j'}{}^2\left(f_K\right) = \frac{\left|V_{j,j'}\left(f_K\right)\right|^2}{V_j^2\left(f_K\right) V_{j'}^2\left(f_K\right)} \tag{22}$$

The coherency function $R_{j,j'}(f_K)$ estimates the proportion of the spatial variance of $SWC_j(i)$ which can be explained by the spatial variance of $SWC_{j'}(i)$, as a function of spatial scale. Thus, spectral coherency is useful to measure time stability of soil water content as a function of spatial scale.

The significance test method for spectral coherency consist of parametric methods based on an assumed theoretical distribution and nonparametric method such as reshuffling, bootstrapping, and bagging method. We encourage readers to see Si (2008) for a more detailed account on significance test.

3.1.6. *Wavelet coherency analysis*

Wavelet analysis, differing with spectral analysis, can identify localized features of soil processes (Si, 2008). It is suitable to reveal the scale and location specific time-persistence of spatial pattern of *SWC* between sampling occasions assuming the soil water content system is linear (Biswas and Si, 2011c). Wavelet coherency of two spatial series can describe the time stability of spatial pattern. It requires the calculation of wavelet coefficients for each of the two data series and associated cross-wavelet spectrum. Many publications on the introduction of wavelet analysis (Farge, 1992; Kumar and Foufoula-Georgiou, 1993, 1997) and wavelet coherency (Biswas and Si, 2011c; Grinsted et al., 2004; Si, 2008; Si and Zeleke, 2005) can be found. Here, we will present the basic procedure to calculate wavelet coherency.

First, wavelet coefficient, $W_i^Y(s)$, is calculated with the continuous wavelet transform (CWT) for a *SWC* series of length n (Y_i, $i = 1, 2, \ldots, n$) with equal incremental distance δx. This can be defined as the convolution of Y_i with the scaled and normalized wavelet using fast Fourier transform (Si and Zeleke, 2005; Torrence and Compo, 1998):

$$W_i^Y(s) = \sqrt{\frac{\delta x}{s}} \sum_{j=1}^{n} Y_j \psi\left[(j-i)\frac{\delta x}{s}\right] \tag{23}$$

where $\psi[]$ is the mother wavelet function and s is the scale. The mother wavelet functions include Morlet, Mexican hat, Harr, and others (Si, 2008). Depending on the purpose, different mother wavelet functions can be selected. The Morlet wavelets allow us to detect both location dependent amplitude and phase for different frequencies in the spatial series (Torrence and Compo, 1998), which can be written as

$$\psi[\eta] = \pi^{-1/4} e^{i\omega\eta - 0.5\eta^2} \tag{24}$$

where ω is dimensionless frequency and η is dimensionlessspace. The wavelet is stretched in space (x) by varying its scale (s), so that $\eta = s/x$. The Morlet wavelet (with ω =6) is a good choice for feature extraction purpose like identifying the scales and locations, since it provides a good balance between space and frequency localization.

The wavelet coefficients $W_i^Y(s)$ can be expressed as $a + ib$ where a and b are the real and imaginary components of $W_i^Y(s)$. For the polar form of complex numbers, $W_i^Y(s) = |W_i^Y(s)|(cos\theta + isin\theta)$, where $\theta = arctan\frac{b}{a}$ is called the phase or argument of $W_i^Y(s)$. The wavelet power spectrum is defined as $|W_i^Y(s)|^2$ and the local phase is defined as the complex argument of $W_i^Y(s)$.

After calculating the wavelet spectra $W_i^Y(s)$ and $W_i^Z(s)$ corresponding to two *SWC* spatial series Y and Z, respectively for two different times, cross wavelet power spectrum $W_i^{YZ}(s)$ at scale s and location i can be defined as

http://dx.doi.org/10.5772/52469

$$W_i^{YZ}(s) = \left[W_i^Y(s)\right]\overline{\left[W_i^Z(s)\right]} \tag{25}$$

and the wavelet coherency of two spatial series can be written as

$$\rho_i^2(s) = \frac{\left|\overleftrightarrow{W_i^{YZ}(s)}\right|^2}{\left|\overleftrightarrow{W_i^{Y}(s)}\right|^2 \left|\overleftrightarrow{W_i^{Z}(s)}\right|^2} \tag{26}$$

where $\overleftrightarrow{(\cdot)}$ is a smoothing operator, which can be written as

$$\overleftrightarrow{W} = SM_{scale}\left[SM_{space}(W)\right] \tag{27}$$

where SM_{space} denotes smoothing along the wavelet scale axis and SM_{scale} smoothing in spatial distance. The following smoothing function is the normalized real Morlet wavelet and has a similar footprint as the Morlet wavelet

$$\frac{1}{s\sqrt{2\pi}}\exp\left(-\frac{\tau^2}{2s^2}\right) \tag{28}$$

where τ denotes location. Therefore, the smoothing along locations can be written as

$$SM_{space}(W,s,\tau) = \sum_{k=1}^{m}\left\{W(s,\tau)\frac{1}{s\sqrt{2\pi}}\exp\left[-\frac{(\tau - x_k)^2}{2s^2}\right]\right\}\Big|_s \tag{29}$$

The Fourier transform of Equation (28) is $\exp(-2s^2\omega^2)$, where ω is the frequency. Eq. (29) can be implemented using Fast Fourier Transform (FFT) and Inverse FFT (IFFT) based on convolution theorem and is written as

$$SM_{space}(W,s,x) = IFFT\left\{FFT\left[W(s,\tau)\right]\left[\exp\left(-2s^2\omega^2\right)\right]\right\} \tag{30}$$

The smoothing along scales can be written as:

$$SM_{scale}\left(W,s_k,x\right)=\frac{1}{2p+1}\sum_{j=k-p}^{k+p}\left[SM_{space}\left(W,s_j,x\right)\Pi\left(0.6s_j\right)\right]\Big|_x \tag{31}$$

where P is the number of terms on each symmetrical half of the window, and Π is the rectangle function. The factor of 0.6 is the empirically determined scale decorrelation length for the Morlet wavelet (Si and Zeleke, 2005; Torrence and Compo, 1998).

There are many methods available to test the statistical significance of wavelet coherency. The Monte Carlo simulation or reshuffling method is among the suggested ones. A detailed description of these methods can be found in Si (2008). Matlab codes for calculating wavelet coherency are available at URL: http://www.pol.ac.uk/home/research/waveletcoherence/ (Grinsted et al., 2004).

3.1.7. Multivariate empirical mode decomposition

Empirical mode decomposition (EMD), extracts oscillations from the soil water content series into a finite and often small number of intrinsic mode functions (IMFs) according to the energy associated with different space scales (Huang et al., 1998; Huang and Wu, 2008). Unlike spectral and wavelet methods, EMD does not call for any assumption of the data and works directly in the spatial domain with the basis completely derived from the data. Therefore, it is intuitive, direct, *a posteriori* and adaptive (Huang et al., 1998; Huang and Wu, 2008). The locality and adaptivity of EMD can deal with different types of spatial series including non-stationary and nonlinear (Huang et al., 1998). Therefore, EMD has a great potential to find the underlying scale of spatial series of soil moisture without imposing any mathematical assumption on the measured data (Biswas and Si, 2011b).

Multivariate empirical mode decomposition (MEMD) is the multivariate extensions of standard EMD (Huang et al., 1998). An important step for MEMD is the computation of the local mean because local extrema are not well defined directly for multivariate spatial data. Moreover, the notion of "oscillatory model" defining an IMF is rather confusing for multivariate spatial data. To deal with these problems, Rehman and Mandic (2010) produced a multiple m-dimensional envelopes by taking projections of multiple inputs along different directions in an m-dimensional space.

Assuming $\{v(s)\}_{s=1}^{S}=\{\upsilon_1(s),\ \upsilon_2(s),\ \ldots,\ \upsilon_m(s)\}$ being the m spatial data sets of soil water content as a function of space (s), and $x^{\theta_k}=\{x_1^k,\ x_2^k,\ \ldots,\ x_m^k\}$ denoting a set of direction vectors along the directions given by angles $\theta^k=\{\theta_1^k,\ \theta_2^k,\ \ldots,\ \theta_{m-1}^k\}$ on a m-1-dimensional sphere (k is the number of direction used to calculate the projections and envelope curves). Then, IMFs of the m spatial data sets can be obtained by MEMD as follows:

1. Choose a suitable point set for sampling on an m-1-dimensional sphere. This can be done by sampling unit hyperspheres (m-spheres) based on both uniform angular sampling methods and quasi-Monte Carlo-based low-discrepancy sequences.

2. Calculate a projection, denoted by $p^{\theta_k}(t)\}_{s=1}^{S}$, of the spatial data sets $\{v(s)\}_{s=1}^{S}$ (S is the number of sampling point for each time) along the direction vector x^{θ_k}, for all k (the whole set of direction vectors), giving $p^{\theta_k}(s)\}_{k=1}^{K}$ as the set of projections.

3. Find the spatial instants $s_i^{\theta_k}$ corresponding to the maxima of the set of projected data sets $p^{\theta_k}(s)\}_{k=1}^{K}$.

4. Interpolate $\left[s_i^{\theta_k}, v\left(s_i^{\theta_k}\right)\right]$ to obtain multivariate envelope curves $e^{\theta_k}(s)\}_{k=1}^{K}$ for all the data sets considered.

5. For a set of K direction vectors, the mean $m(s)$ of the envelope curves is calculated as:

$$m(s)=\frac{1}{K}\sum_{k=1}^{K} e^{\theta_k}(s) \tag{32}$$

6. Extract the "detail" $d(s)$ using $d(s)=v(s)-m(s)$. If the "detail" $d(s)$ fulfills the stoppage criterion for a multivariate IMF, apply the above procedure to $v(s)-d(s)$, otherwise apply it to $d(s)$. The stoppage criterion for multivariate IMFs is similar to that proposed by reference (Huang et al., 2003).

MEMD has the ability to align "common scales" present within multivariate data. Each common scale is manifested in the common oscillatory modes in all the variates within an *m*-variate IMF. After MEMD analysis, scale and location specific time stability of *SWC* can be easily identified by comparing and calculating spearman's rank correlation coefficient between the IMFs with the same numerical numbers for different measurements using Eq. (2). The exact scale for each IMF can be obtained from the instantaneous frequencies by Hilbert transformation with IMF (Huang et al., 1998). The instantaneous frequencies of soil water content can be converted to period (1/frequency), which was further converted to the spatial scale after multiplying the period with the sampling interval. MEMD can be completed using Matlab program such as that written by Rehman and Mandic (2009) (http://www.commsp.ee.ic.ac uk/~mandic/research/emd.htm). This method requires that soil water content measurement should be obtained in intervals with equidistance.

3.2. Time stability at points

Time stability at point scales are usually obtained in terms of average *SWC* estimation quality. Direct and indirect methods can be used to estimate average *SWC* by the measurement at time stable locations. The direct method estimates average *SWC* directly by measuring *SWC* at a time-stable location (Brocca et al., 2009; Cosh et al., 2008; Grayson and Western, 1998; Pachepsky et al., 2005). The indirect method estimates average *SWC* by considering the offset between the average *SWC* and the measurement value at a time-stable location (Grayson and Western, 1998; Starks et al., 2006). All the indices are listed in Table 1. Each index has its own advantages and disadvantages, and these are discussed in the application (Section 4.2). For all indices, smaller value of time stability index at a given location indicates stronger time stability and also higher quality of average *SWC* estimation by that location.

Index	Formula and explanation	Average SWC estimation Method†	Reference
Standard deviation of relative difference, $SDRD(i)$	where $SDRD(i)=\sqrt{\frac{1}{m-1}\sum_{j=1}^{m}(\delta_j(i)-\langle\delta_i\rangle)^2}$ and $\delta_j(i)=\frac{SWC_j(i)-\langle SWC_j\rangle}{\langle SWC_j\rangle}$	Direct and indirect	Vachaud et al., 1985
Root mean square error, $RMSE(i)$		Direct	Jacobs et al., 2004
Standard deviation of relative SWC, $\sigma(\beta_i)$ ‡	$\langle\delta_i\rangle=\frac{1}{m}\sum_{j=1}^{m}\delta_j(i)$ where $RMSE(i)=(\langle\delta_i\rangle^2+SDRD(i)^2)^{1/2}$ and $\sigma(\beta_i)=\sqrt{\frac{1}{m-1}\sum_{j=1}^{m}(\beta_j(i)-\langle\beta_i\rangle)^2}$	Indirect	Pachepsky et al., 2005
Width of the 90% empirical tolerance interval of relative water content, $T(i)$	where $\beta(i)_{P=0.95}$ and $\beta(i)_{P=0.05}$ are the relative SWC values at cumulative probability of 95% and 5%, respectively.	Indirect	Guber et al., 2008
Chi-squared statistic, $\chi^2(i)$	where σ_j is the standard deviation of SWC at observation time j.	Indirect	Guber et al., 2008
Root-mean-squared differences, $D(i)$		Indirect	Guber et al., 2008
Mean absolute bias error, $MABE(i)$		Indirect	Hu et al., 2010a, 2010b

†Direct refers to estimating average *SWC* directly by measuring *SWC* at a time-stable location and indirect refers to estimating average *SWC* by considering the offset between the mean value and the measurement value at a time-stable location.

‡ $\sigma(\beta_i)=SDRD(i)$.

Table 1. Time stability indices at point scales and their formulas

4. Application

4.1. Time stabilityof spatial pattern

Information on the spatial pattern of soil water content with time is important to understand the changes in hydrological processes and in making predictions with physical and statistical models. In this section, we have documented the use of each method introduced above to characterize the similarity of the spatial pattern of *SWC*.

Spearman's rank correlation analysis is the most widely used method to examine the similarity of the spatial patterns of *SWC*. Numerous studies have used this method and have confirmed the presence of time stability of the spatial pattern (Biswas and Si, 2011c; Hu et al., 2009, 2010b; Martínez-Fernández and Ceballos, 2003; Vachaud et al., 1985; Vivoni et al., 2008). In the first of this kind, Vachaud et al. (1985) measured soil water storage of 0 to 1.0 m depth at 17 neutron access tubes located at 10 m intervals for 24 occasions over a time span of two and a half years. According to the cumulative probability function for the two extreme conditions, i. e., the wettest on 7 September 1982, and the driest on 17 August 1983, many locations maintained their ranks (Figure 1). For example, location #1 always had the maximum soil water storage, location #4 had the minimum soil water storage irrespective of the environmental conditions (e. g., wet or dry). Spearman's rank correlation coefficients of the soil water storage between measurements were also calculated to examine the degree of time stability of the spatial pattern. As an example, the correlation between seven measurements representing the range of the water storages recorded over the entire observation period is presented in Table 2. All rank correlation coefficients were highly significant at the 0.1% two-tailed test, indicating strong time stability of the spatial pattern of soil water storage. On the other hand, time instability of spatial pattern between measurements was also reported (Comegna and Basile, 1994; Mohanty and Skaggs, 2001). Although it is very popular among researchers, Spearman's rank correlation analysis should be viewed only as a statistical tool measuring the degree of concordance between two rankings (Vachaud et al., 1985). It may be questionable if differences between measured values are smaller than experimental uncertainties. This may be the case in situations where either the probability density function is very uniform or the experimental determinations are very crude.

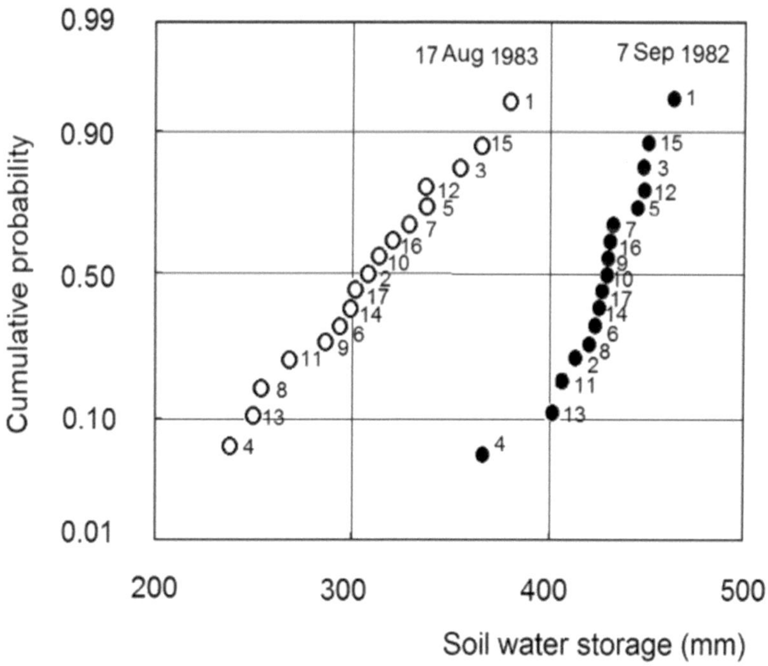

Figure 1. Cumulative probability function of soil water storage at 0-1.0 m measured in the most dry, and the most wet conditions. Numbers refer to measuring locations. (Reproduced from Vachaud et al. (1985)).

Dates	7 Sep 1982	28 Oct 1981	2 Dec 1982	29 Jul 1981	18 Aug 1981	16 Jul 1982	25 Aug 1981
Average storage (mm)	428.1	425.77	424.77	382.06	354.72	347.23	344.46
(1)	1						
(2)	0.953	1					
(3)	0.941	0.953	1				
(4)	0.988	0.961	0.946	1			
(5)	0.953	0.922	0.882	0.968	1		
(6)	0.863	0.789	0.843	0.836	0.882	1	
(7)	0.860	0.824	0.824	0.794	0.863	0.939	1

Table 2. Matrix of Spearman's rank correlation coefficient between series of soil water storage measurements obtained on the 17 locations on seven dates. (Reproduced from Vachaud et al. (1985))

As an alternative to Spearman's rank correlation analysis, Pearson correlation coefficients were also used to describe time stability of the spatial pattern (Cosh et al., 2004). It was observed that the Pearson correlation analysis revealed a pattern similar to that of the Spearman's rank correlation analysis (Cosh et al., 2004). Note that the tendency of soil moisture sensors to occasionally report erroneous measurements may negatively affect the value of Spearman's rank correlation coefficient, while the Pearson correlation coefficient is less sensitive to this problem. This is because the total number of measurements will not affect Pearson correlation coefficient as it affects the rank correlation coefficient (Cosh et al., 2006, 2008). However, both Spearman's rank correlation and Pearson correlation analysis are only suitable to describe how much spatial pattern at one measurement time can be preserved at another time (Rolston et al., 1991). If we want to get the information on the similarity of spatial pattern over multiple measurement occasions, mean value of correlation coefficients can be used (Hu et al., 2009).

With the aim to characterize the degree of time stability of spatial pattern among multiple data series, Starr (2005) suggested two indices, i. e., average spatial coefficient of determination (Rs^2) and temporal coefficient of variability (CV_t). The authors reported that 26 to 76% (with mean of 47%) of the observed variability was explained by the time stable pattern. The reported CV_t value ranged from 11% to 18% (with mean of 16%) for different sampling transects. On the other hand, the Rs^2 may not be a good measure of how time stable a *SWC* pattern is at a field that is uniform and completely time stable. This is because the Rs^2 will be zero in this case (Starr, 2005).

Perry and Niemann (2007) used the EOF method to characterize the time stable spatial pattern of *SWC*. They identified two dominant spatial patterns of soil water variability at Tarrawarra catchment (Figure 2). The first spatial pattern was more important for medium wet periods and was associated with lateral redistribution of soil water. The second spatial pattern served to modify the first spatial pattern at some days which was associated to evapotranspiration. The importance of each spatial pattern on each measurement occasion can be reflected by the

EC values (Figures 2c and 2d). There was a large decrease in EC1 weight during the dry season, which agrees well with the finding of Western et al. (1999) that soil moisture patterns on moderately wet occasions show a strong topographic influence. The EOF method can also be used to predict the *SWC* distribution for unobserved times if the average *SWC* and the temporal varying coefficient (ECs), which reflect the relative importance of different spatial patterns can be obtained. Average *SWC* can be estimated by many methods such as linear regression (Western et al., 1999), a dynamic multiple linear regression of soil moisture against topographic attributes (Wilson et al., 2005), and by measurement from representative location (Grayson and Western, 1998). Examples of EOF method can also be found in other study (Ibrahim and Huggins, 2011; Jawson and Niemann, 2007; Joshi and Mohanty, 2010; Korres et al., 2010; Yoo and Kim, 2004). Different numbers, usually one to four, of significant EOFs were obtained to explain most of *SWC* variations (usually more than 70%). The time-invariant spatial patterns were usually related to topography-related factors (Yoo and Kim, 2004), sand content (Jawson and Niemann, 2007), mixed influences of rainfall, topography, and soil texture (Joshi and Mohanty, 2010), elevation and wetness index under wet conditions and soil properties (ECa and bulk density) in dry conditions (Jawson and Niemann, 2007).

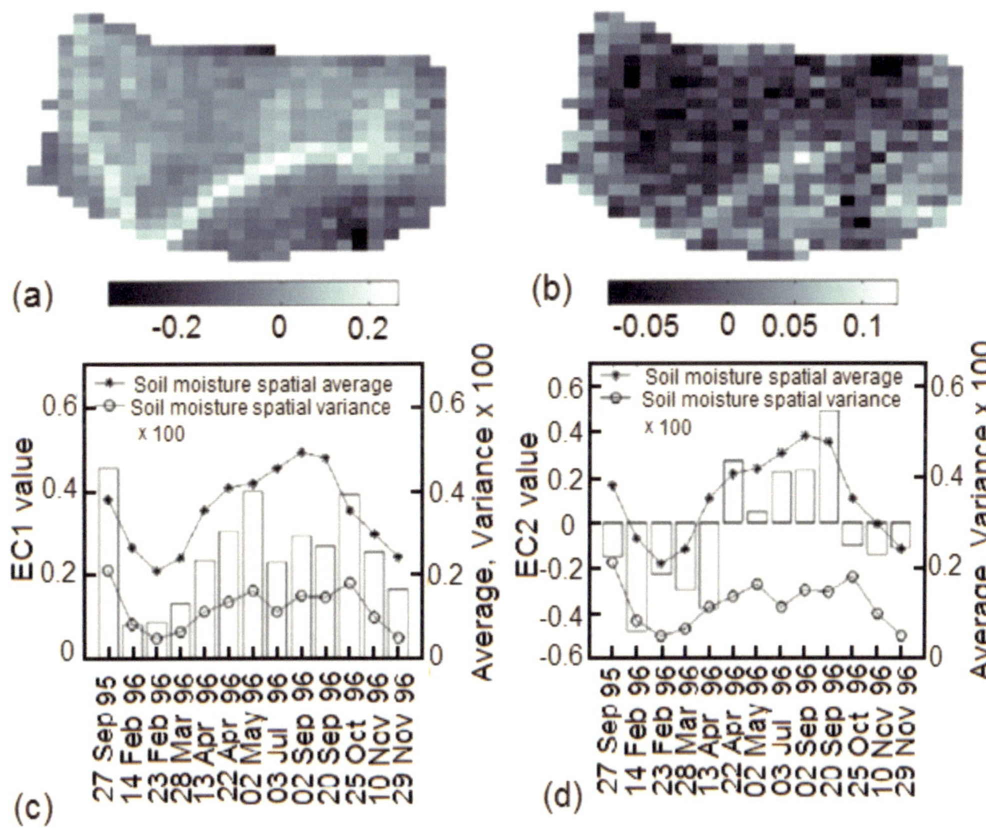

Figure 2. a) EOF1 and (b) EOF2 computed from the spatial anomalies of soil moisture. The bars in (c) and (d) are EC1 and EC2, respectively. The spatial average and variance of the soil moisture on each sampling date are also shown in (c) and (d) for comparison. (Reproduced from Perry and Niemann (2007))

Time stability of soil water content is usually scale dependent because of the interactions of soil moisture between measurement locations (Kachanoski and De Jong, 1988). The first effort to reveal the scale specific time stability was made by Kachanoski and De Jong (1988) who used spectral coherency to explore the time stability of soil water storage in a rolling landscape (Figure 3). They observed that for both wet and drying periods, the spectral coherency was high at large scales (>40 m). For smaller scales (<40 m), however, the spectral coherency was high for the drying period but greatly weakened during the alternating wetting and drying period. Therefore, their results clearly showed the loss of time stability at small scales (<40 m). Gómez-Plaza et al. (2000) also used the spectral coherency to explore the scale dependence of time stability in burned and unburned areas in a semi-arid environment. They clearly observed that soil moisture patterns were maintained in all cases at large scales (transect scale), while soil moisture tended to become time unstable, depending on the range of scale and section of transect considered at medium scales.

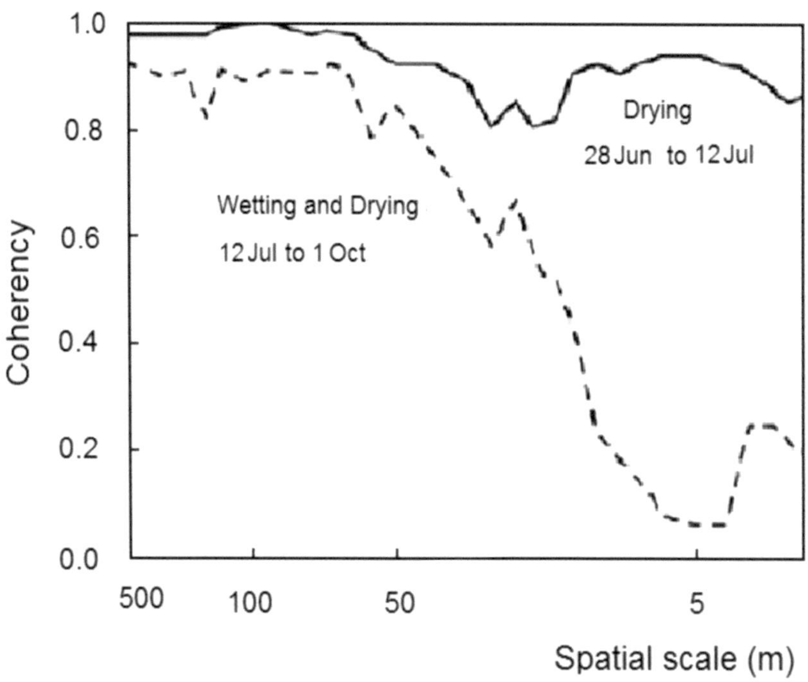

Figure 3. Coherency spectrum for soil water storage for the drying (June 28-July 12) and recharge (July 12-October 1) period. (Reproduced from Kachanoski and De Jong (1988))

Spectral coherency was verified to be robust to look for the scale dependency of time stability. Its multi-variate extension is also feasible (Si, 2008), although no publications have been found on this point. However, spatial coherency analysis assumes the stationarity in the measured data series over space, hence may not capture localized features of time stability (Biswas and Si, 2011a).

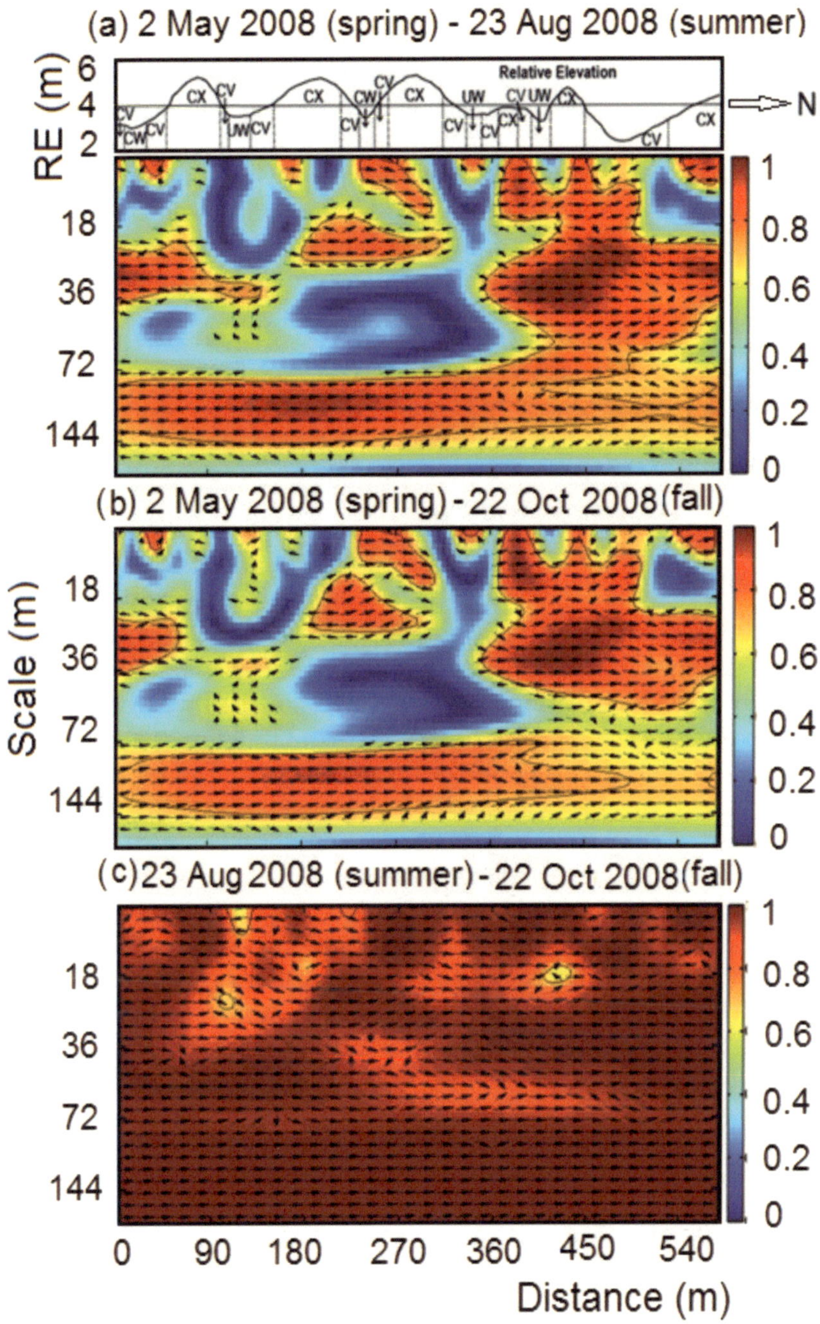

Figure 4. The inter-season wavelet coherencies between the soil water storage measured during (a) spring (2 May 2008) and summer (23 August 2008), (b) spring (2 May 2008) and fall (22 October 2008), and (c) summer (23 August 2008) and fall (22 October 2008) for the whole soil profile. Cross sectional view of the transect with landform elements is shown at the top. The X-axis indicates distance along the transect; the Y-axis indicates the scale in meter; the solid black line indicates 5% significance level; the color bar indicates strength of correlation, and the direction of arrow indicates the phase relationship or type of correlation. (Reproduced from Biswas and Si (2011c)).

Wavelet coherency can deal with the non-stationary series, it is therefore suitable to identify both scale and location specific time stability (Biswas and Si, 2011a). An excellent application of wavelet coherency in time stability analysis was made by Biswas and Si (2011a). They identified scales and locations of time stability between the spatial patterns of soil water storage in a hummocky landscape in central Saskatchewan, Canada. As Figure 4 shows, strong time stability at all scales and locations existed between summer and fall, when the vegetation was present in field. The spatial pattern in spring was different from that of summer or fall. The large-scale (> 72 m) time stability was present over the whole transect. The medium-scale spatial patterns were time stable in large depressions that had fewer landform elements. The small-scale time stability was mostly random.

Wavelet coherency can also be used to identify the correlation between different variables at different scales and positions (Lakshmi et al., 2004; Si and Zeleke, 2005). One limitation of wavelet coherency analysis is that it is based on the assumption of linearity of systems. Moreover, the application of wavelet coherency on time stability analysis was only confined to two spatial series. Extension of this method to multiple spatial series is expected to understand the scale and location specific time stability when soil water content from different seasons are considered together.

MEMD is a data-driven method that can deal with both non-stationary and nonlinear series. Recently, Hu et al. (unpublished data) (2012a) applied MEMD to time stability analysis using soil water storage measurements from 20 occasions in a hummocky landscape in central Saskatchewan, Canada. At first, multiple spatial series of soil water storage were separated into different IMFs. The IMFs for the 0-20 cm layer on a selection of measurement occasions is given in Figure 5. Each IMF corresponds to a specific scale common to all the spatial series considered. Spearman's rank correlation analysis was then conducted for each scale. According to their study, the dominant scale of variations of soil water storage was 104-128 m (IMF5), for 0-20 cm, 20-40 cm, 40-60 cm, and 60-80 cm, and 315-412 m (IMF7) for 80-100 cm, 100-120 cm, and 120-140 cm. Time stability generally increased with scale and it was the strongest at the dominant scale of variation. At different scales, the time stability generally ranked in the order of intra-season>inter-annual>inter-season. Therefore, the study of Hu et al. (2012a) verified that scale specific time stability of soil moisture can be identified using MEMD method.

According to the study reviewed above, time stability analysis is developing from two spatial series to multiple series, from one scale to multiple scales. With the advance of methodology, a deeper understanding of spatial patterns of soil moisture can be expected. Future work should make full use of various methods, especially those most recently developed, such as EOF, wavelet coherency, and MEMD to gain greater insight into the spatial pattern dynamics at different scales.

4.2. Identify time stable locations for average soil water estimation

One of the most important applications of time stability is to identify a representative location for average soil water content estimation for a given area. Identification of representative locations cover different land uses, different climate zones, and places with various topographical attributes (Gao et al., 2011).

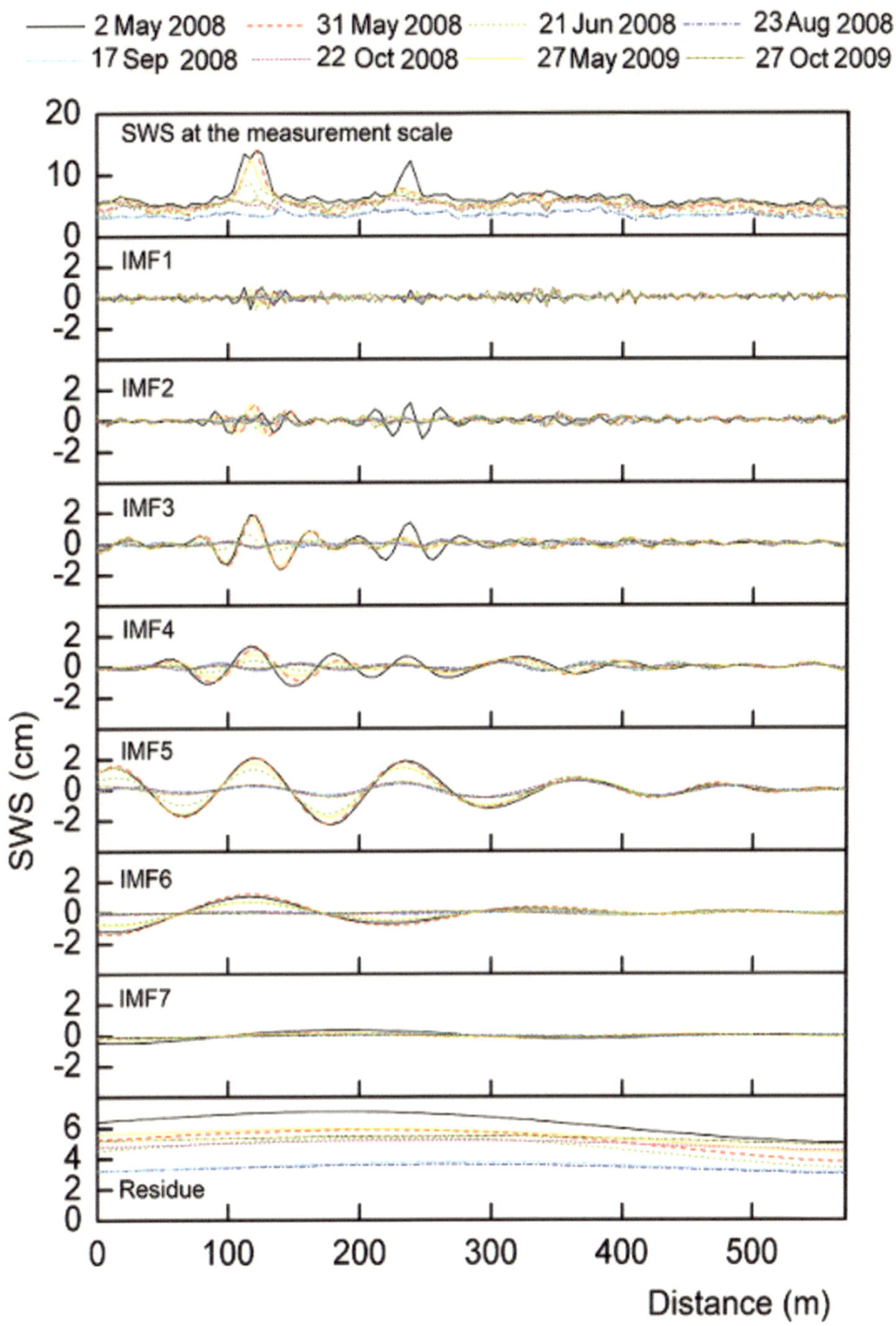

Figure 5. Intrinsic mode functions (IMFs) and residues based on multivariate empirical mode decomposition using soil water storage series (top) of the surface layer (0-20 cm). The horizontal axis is the distance between a sampling location and the origin of the transect. The vertical solid bar in Y-axis shows the scale of IMFs. (Reproduced from Hu et al. (2012a)).

Various time stability indices for use at the point scale have been used to identify representative locations. Most studies have been aimed at finding locations where soil water content can directly represent the mean of an area (Brocca et al., 2009; Cosh et al., 2008; Grayson and Western, 1998; Martínez-Fernández and Ceballos; 2005). In this way, *SWC* at the time stable location should approximate the average *SWC* of the study area. By extension, point scale time stability representing the areal average also means that the mean relative difference of soil water content, $\langle \delta_i \rangle$, at the stable location *i* should be close to zero at all times. Practically, it is difficult to find a location where the mean relative difference is the close to zero at all times. In this situation, the time stability index *SDRD* is widely used to judge the persistence of the stable location. Therefore, the mean relative difference and associated *SDRD* should be considered together. Successful application of this method can be found in many publications (Brocca et al., 2009; Cosh et al., 2006, 2008; Gómez-Plaza et al., 2000; Grayson and Western, 1998; Tallon and Si, 2004). In the study of Grayson and Western (1998), although the overall spatial pattern of soil moisture was time unstable, some locations were time stable and can represent the average *SWC* of a landscape. In the Tarrawarra watershed (Figure 6), the mean relative differences for locations 1, 4, 12, and 13 were close to zero, and their *SDRD* were also relatively small, so these locations were time stable and could be used as representative location for average *SWC* estimation. In the study of Tallon and Si (2004) which was conducted on a chernozemic soil on the Canadian prairies (75 km northeast of Saskatoon, SK, Canada) time stability of spatial patterns of soil moisture for different depths (30, 60, 90, 120 and 160 cm) were observed. With the exception of one location that occurred at two separate depths, the locations representative of average *SWC* differed with soil depths.

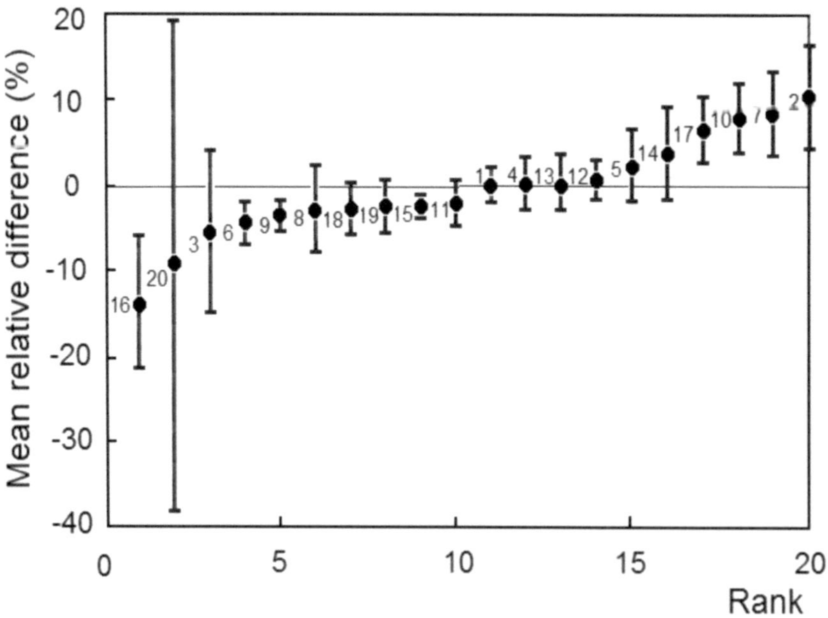

Figure 6. Ranked mean relative differences of soil water content for 0-60 cm measured by neutron probe in the Tarrawarra watershed. Also shown is one standard deviation error and site numbers (Reproduced from Grayson and Western (1998)).

Although mean relative difference and *SDRD* have been used successfully in the past in identifying representative locations for average *SWC*, the question becomes a matter of which index is more important when the location with mean relative difference closest to zero disagrees with that with the minimum *SDRD*. Most researchers believed that the *SDRD* was more important (Brocca et al., 2009; Cosh et al., 2006, 2008; Gómez-Plaza et al., 2000; Grayson and Western, 1998), while mean relative difference was also emphasized (Martínez-Fernández and Ceballos, 2003).

Instead of considering two indices simultaneously, one index, *RMSE*, combines both indices into one, thus making the identification of time stable location more objective. By this index, drier locations were usually found to have lower *RMSE* values than the wetter points (Cosh et al., 2004, 2006). This suggests that improved consistency will result from the selection of slightly drier sampling locations (Cosh et al., 2004). Operationally, it facilitates the selection of representative locations. However, it may confuse the concept of time stability, because for locations with large absolute values of relative difference (i. e. extremely dry or wet conditions) but very small *SDRD*, they may have large value of *RMSE* although they are completely time stable.

Another method for estimating average soil water content with time stable location involves a consideration of the constant offset (mean relative difference) for time stable location (Grayson and Western, 1998). Then $\langle SWC_j \rangle$ can be estimated by

$$\beta_j(i) = \frac{SWC_j(i)}{\langle SWC_j \rangle} \tag{33}$$

In this situation, time stable locations are usually defined by the value of *SDRD* (Grayson and Western, 1998). However, there are two limitations in using *SDRD* as a criterion. First, the value of the *SDRD* may be affected to some extent by the magnitude of the mean relative difference; second, the value of the *SDRD* cannot be directly related to the estimation error of the mean value (Hu et al., 2010b).

Indices that can consider the estimation error are needed for identifying representative time stable locations. One example was conducted by Guber et al. (2008). They applied $\chi^2(i)$ and $D(i)$ to rank the utility of sampling locations to estimate average *SWC* by considering the estimation error. Another alternative is the mean absolute bias error (*MABE*), which was developed directly to relate to the relative error of mean estimation (Hu et al., 2010a, 2010b). Although *MABE* was positively correlated to *SDRD* in some cases (Gao et al., 2011; Hu et al., 2010a, 2010b), the lowest value of *SDRD* did not coincide with the lowest value of *MABE* (Figure 7). Therefore, *MABE* outperformed *SDRD* in terms of representative location identification for indirect estimation of average *SWC* (Hu et al., 2010b).

The degree of representativeness of time stable locations may also change with season or year due to the change of the relative importance of different factors affecting time stability (Guber et al., 2008; Martínez-Fernández and Ceballos, 2003). Schneider et al. (2008) showed that the selected representative locations were appropriate to predict average soil water content of the sites for multiple years, although the time stability characteristics of some locations varied between years. An interesting study conducted by Guber et al. (2008) indicated that the utility of a location to estimate the estimation accuracy of field average *SWC* was not influenced by measurement frequency.

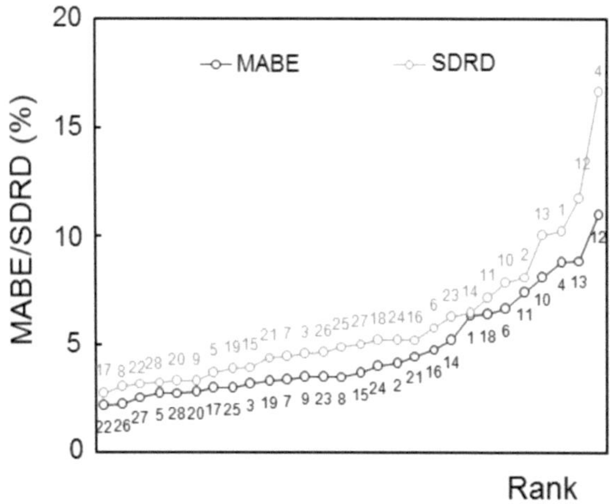

Figure 7. Rank ordered mean absolute bias error (*MABE*) and standard deviation of relative differences (*SDRD*) for the soil water storage of 0-1.0 m measured with neutron probe from October, 2004 to September, 2006. Also shown is location number (Reproduced from Hu et al. (2010b))

Many indices have been available to identify the representative location for average *SWC* estimation. However, few works have been conducted to compare their performances with a few exceptions (Gao et al., 2011; Hu et al., 2010a, 2010b, 2012b). *MABE* was reported to be a better index than *SDRD* in terms of average soil water storage estimation by the indirect method (Hu et al., 2010a, 2010b). More recently, with the soil water storage datasets from a transect in a Canadian prairie area and a watershed on the Chinese Loess Plateau, Hu et al. (2012b) evaluated the performance of different time stability indices in terms of the average soil water storage estimation quality judged by different goodness-of-fit indices. Their results showed that *MABE*, χ^2, *D*, and CV_t outperformed *SDRD* and *T*. If root mean squared deviations (*RMSD*) and Nash-Sutcliffe coefficient of efficiency (*NSCE*) were used as goodness-of-fit indices, *D* was the best time stability index. If absolute mean difference (*BIAS*) and absolute bias relative to mean (*RBIAS*) were adopted, *MABE* was the best time stability index. Their results also showed that average soil water storage estimation by the indirect method was more accurate than that by the direct method. It is necessary to compare their performances under other different hydrological backgrounds to obtain a more suitable index for identifying representative location. Furthermore, all these methods still need an intensive measurement of soil water content over time before a representative location can be identified. According to Martínez-Fernández and Ceballos (2005) about one year (a complete seasonal cycle) of measurements is needed to determine the representative location. Therefore, there is an urgent need to look for the possible definable features of the terrain

http://dx.doi.org/10.5772/52469

and soils that could be used to determine the representative locations *a priori* (Martínez-FernándezandCeballos, 2005).

5. Controlling factors

Time stability of soil water content is controlled by factors that consistently influence soil moisture distribution in a similar way at all times. Most studies focus on factors influencing the underlying spatial pattern of soil moisture, and relatively few studies are associated with factors influencing the degree of time stability at point scales. We will review the controlling factors of the two types of time stability below.

5.1. Controlling factors of time stability of spatial pattern

Controlling factors of underlying spatial pattern are usually defined by comparing spatial patterns of time stability and various factors (Hu et al., 2010a; Perry and Niemann, 2007; Schneideret al., 2008). Related studies referred to different climate conditions, from semi-arid (Hu et al., 2010a) to humid environments (Perry and Niemann, 2007) and different land uses, such as rangeland (Mohanty and Skaggs, 2001), grass (Jacobs et al., 2004), forest (Lin, 2006), and agricultural land (Starr et al., 2005).

Soil properties such as soil texture are usually found to influence the spatial pattern of time stability. For example, Vachaud et al. (1985) found that the scaled factor of soil water content were significantly correlated to the silt plus clay contents on a field in St. Martin d'Heres, France. Mohanty and Skaggs (2001) observed that the sandy loam soil produced the best time-stable soil moisture pattern and silty loam soil produced an intermediate level of time stability in the Little Washita watershed of the southern part of the Great Plains of the US. Jacobs et al. (2004) pointed out that lower sand content may produce more time stable patterns in the Iowa. Grant et al. (2004) found a significant correlation between clay content, coarse fragment and the rank of mean relative difference of soil water storage in the Reynolds Mountain East catchment in the Owyhee Mountains. Starr (2005) found that coarser particle size classes were generally drier than adjacent finer particle size class soil on a potato farm located near Houlton, Maine. Hu et al. (2010a) recognized that sand and silt content can explain about 41.4% - 65.5 % of the spatial pattern of mean relative difference of soil water content. Except for soil texture, other soil properties such as organic matter content (Hu et al., 2009), bulk density (Jacobs et al., 2004), and soil thickness (Zhu and Lin, 2011) have also been recognized as the controlling factors for time stability of spatial pattern. However, soil properties may also have no influences on the spatial pattern of time stability. For example, Tallon and Si (2004) reported that clay content showed the least amount of control of spatial patterns on a chernozemic soil on the Canadian prairies (75 km northeast of Saskatoon, SK, Canada). On ungrazed sites, Schneider et al. (2008) found that neither bulk density, organic carbon nor sand and clay content could explain the time-stable characteristics of the sampling points. Comegna and Basile (1994) and Starr (2005) both observed the absence of the controlling influence of soil texture and they attribute it to the homogeneous distribution of soil texture in their study areas.

Topographic characteristics have also been recognized to influence the time stability of spatial patterns. Gómez-Plaza et al. (2000) showed that when the factors affecting soil moisture were limited to topographical position or local topography at the transect scale, spatial patterns of soil moisture presented time stability. Joshi et al. (2011) observed that fields with flat topography (LW21) showed the worst time stability features compared to the fields having gently rolling topography (LW03 and LW13) in Oklahoma. The influence of topography on the time stability of spatial pattern may be scale dependent. For example, although the Pearson correlation coefficient between soil water storage and elevation was very small, Biswas and Si (2011a) identified that strong scale-specific correlations existed between soil water storage and elevation at different scales using wavelet coherency, which contributed to the time stability of soil water storage at different scales. Kachanoski and De Jong (1988) attributed to the loss of time stability of soil water storage at scale smaller than 40 m to the role of spatial pattern of surface curvature.

Usually, soil properties and topographical properties jointly control the time stability of soil moisture spatial pattern. According to Joshi et al. (2011), soil properties (i. e., percent silt, percent sand, and soil texture) and topography (elevation and slope) jointly affected the spatiotemporal evolution and time stability of soil moisture at both point and footprint scales in the Little Washita watershed, Oklahoma and in the Walnut Creek watershed, Iowa. Zhu and Lin (2011) also observed that both soil and terrain controlled soil moisture variation at different seasons and soil depths at farm scale.

The relative role of soil and topographic properties in controlling the spatial variability and time stable pattern of soil moisture is usually related to the dominant hydrological process taking place for different soil water conditions between wet and dry periods. The SWC patterns existing between wet and dry periods are referred to as preferred states (Graysonet al., 1997; Western et al., 1999). When dry conditions dominate, evapotranspiration is greater than precipitation, and local controls dominate mainly by vertical movement of soil water with no connection between adjacent points. In this situation, differences in vegetation and soil properties such as texture would be responsible for the spatial pattern of soil moisture (Graysonet al., 1997). During wet states, evapotranspiration is less than precipitation and non-local controls govern soil moisture distribution. These controls are related to upslope topography and include catchment area, aspect, depth and soil profile curvature (Gómez-Plaza et al., 2001). Western at al. (1999) attributed these differences in local and non-local controls to the high degree of organization during wet periods that consist of connected bands of high soil water content in drainage lines. Soil water in these drainage lines is laterally redistributed by both overland and subsurface flow. Using EOF analysis, Perry and Niemann (2007) found that underlying spatial patterns of soil moisture were controlled by local soil properties in wet and dry conditions and topographic characteristics during intermediate conditions. Their results agreed with the findings of Western et al. (1999) that soil moisture patterns on moderately wet dates show a strong topographic influence, one that seems to diminish on both dry and very wet days. Note that soil porosity, rather than topography, will dominate the spatial distribution of soil moisture under very wet conditions (Western et al., 1999). Non-local control can also be found in the study of Biswas and Si (2012). In their study, the coefficient of determination between soil water storage and elevation increased more than eightfold after shifting the spatial series of soil water storage by a length equal to that of the existing slope. Although local and non-local control was widely recognized, it is still difficult to make a very clear limit between them, most likely due to the possible influences of topography on soil properties

Because of the different controls in different soil water conditions, time stability of soil water content tends to be stronger under the similar soil water conditions than those with quite different soil water conditions. For example, Martinez-Fernández and Ceballos (2003) observed that the periods with the lowest time stability coincided with situations involving the transition from dry to wet. Gao et al. (2011) revealed that the time stability of root zone soil water (0–60 cm) was higher in either dry or wet season than that including both, and soil water exhibited very low time stability during the transition period from dry to wet.

In addition to soil and topography, other factors such as vegetation (Pachepsky et al., 2005; Schneideret al., 2008), soil depth (Huet al., 2009) can also influence the time stability of soil moisture. For example, Gómez- Plaza et al. (2000) found that the existence of vegetation could destroy the time stability of spatial patterns of soil moisture. Mohanty and Skaggs (2001) found varying degrees of time stability depending on vegetation cover as well as topography. Generally, time stability of spatial pattern is strongest in deep soil layers than in shallow layers (Cassel et al., 2000). However, stronger time stability of spaital pattern at shallower depth (20 cm) than deeper soil depths was also observed (Hu et al., 2009, 2010a).

Although many factors have been revealed to influence the time stability of spatial pattern, no single, or even combined, factor can explain all time stable patterns. This suggests that some influencing factors may still remain undiscovered. In addition, random measurement errors also contribute to some variability of the time stable pattern (Starr, 2005). According to the study of Starr (2005), random measurement error can contribute 20% of the variability of underlying spatial pattern. Therefore, to better understand the spatio-temporal variability and time stability of spatial pattern of soil moisture, controlling factors should be made clearer when more accurate measurement of soil water content should also be expected.

5.2. Controlling factors of time stability at the point scale

Revealing the controlling factors of time stability at the point scale is crucial in identifying time stable locations for average *SWC* estimation. Soil and topography are usually found to influence the time stability of soil water content at point scales. Brocca et al. (2009) observed that the time stable positions were linked to topographic characteristics, primarily the upslope drainage area but also elevation and slope. Cosh et al. (2008) found that almost 60% of the *SDRD* variability can be credited to the soil type parameters. Soil parameters as major controlling factors is a finding that is in agreement with those of Hu et al. (2010a) who found that soil texture can significantly ($P<0.05$) affect the stability of soil water content. Gao et al. (2011) observed that time-stable locations corresponded to relatively high clay contents, mild slopes and planar surfaces, which agreed with the previous studies, where both soil type and topography influenced time stability at point scales (Jacobs et al., 2004; Joshi et al., 2011; Mohanty and Skaggs, 2011). However, poor relationships between time stable locations and soil and topographic properties were also found (Tallon and Si, 2004). Land use was observed to have no influences on time stability of soil water content at point scales (Hu et al., 2010a; Joshi et al., 2011).

Time stability at point scales was usually associated with the soil water conditions of that location. For example, Martinez-Fernández and Ceballos (2005) found that time stability as characterized by *SDRD* was usually stronger in dry locations than wet locations for all depths, and they attribute it to the predominance of the sandy fraction with the resultant weak ability to retain water in dry locations. Similar results were obtained in other studies (Bosch et al., 2006; Jacobs et al., 2004; Martínez-Fernández and Ceballos, 2003). However, the relationship of time stability and soil water condition is likely influenced by the time stability index used. In the LYMQ watershed on the Loess Plateau in China, Hu et al. (2010a) also found stronger time stability in drier locations using *SDRD*, while they observed stronger time stability in wetter locations when using *MABE*.

Time stability at point scales depends on different periods and soil depths. Generally, stronger time stability was found in wetter periods (Guber et al., 2008; Zhao et al., 2010; Zhou et al., 2007). Zhou et al. (2007) believed that the lateral redistribution of soil moisture in the wet/dry transition period and uneven evapotranspiration and root water uptake in drying process was the reasons for the relatively higher time stability during wetting periods. Guber et al. (2008) related the weaker time stability in the May–June period to the active vegetative growth of corn. Zhao et al. (2010) attributed the high temporal stability under wet conditions to an enhanced capillary movement of water from the subsoil to the topsoil. Time stability at point scale was usually found to be stronger at deeper depths observed by different time stability indices (Guberet al., 2008; Hu et al., 2010a, 2010b; Starks et al., 2006). This was generally in agreement with the relationship of time stability of spatial pattern with depth (Cassel et al., 2000).

Due to the complex influences of various factors on the time stability at the point scale, future work should be devoted to quantitatively define the contribution of various factors on time stability of soil water content. Models of the relationship of time stability and environmental factors should also be developed for the purpose of identification of time stable positions with known soil and topographic information. In addition, because of the different relationships between factors and different time stability indices (Hu et al., 2010a), attention should also be paid to the selection of time stability index for average *SWC* estimation in a field.

6. Conclusion

We reviewed the studies on time stability of soil water content, including its concept, methodology, application, and controlling factors. The following conclusions can be drawn: (1) Two types of time stability of soil water content can be classified; one is to describe the over-

http://dx.doi.org/10.5772/52469

all similarity of soil water content spatial patterns between different occasions, and the other is to describe the time invariance of the relative soil water content with time at point scales; (2) Time stability can be used to analyze the time persistence of spatial pattern of soil water content. Spearman's rank correlation, Pearson correlation analysis, average spatial coefficient of determination, and temporal coefficient of variability can be used to characterize the time stability of spatial pattern for two spatial series. Empirical orthogonal function can be used to extract the same spatial patterns of soil moisture for multiple spatial series. Spectral coherency, wavelet coherency, and multivariate empirical mode decomposition can be used to identify the scale specific time stability of a spatial pattern. However, the spectral coherency method assumes spatial series are stationary and linear, wavelet methods assume the spatial series to be linear, and both of them are only applicable to the case of two spatial series. For multivariate empirical mode decomposition, it can deal with non-stationary and nonlinear systems with multiple spatial series; (3) Time stability can be used to identify the most time-stable locations for average *SWC* evaluation for a given field. Standard deviation of relative difference, root mean square error, standard deviation of soil water content, width of the 90% empirical tolerance interval of relative water content, and mean absolute bias error can be used to identify the most time-stable location. However, the performance of these indices should be compared in different environmental backgrounds to look for the most suitable index; (4) Time stability of soil water content can be influenced by many factors, including soil, topography, vegetation, and climate. It also shows non-local control and local control, depending on the soil water conditions. Knowledge of the controlling factors for time stability of spatial pattern and that at point scale is important for understanding the soil water processes, soil water management, and identification of time-stable locations for average *SWC* evaluation. Future work should quantify the influences of different factors on time stability, and make deeper understanding of nonlocal and local control on time stability of soil water content; (5) Time stability of soil water content is also depth-wise and scale specific. Usually, time stability of soil moisture increased with depth and scale. Time stability analysis can not only facilitate our understanding to soil water related processes but also greatly reduce the sampling work for soil water content in fields.

Acknowledgements

The project was funded by Natural Science and Engineering council (NSERC) of Canada.

Author details

Wei Hu[1], Lindsay K. Tallon[1], Asim Biswas[2] and Bing Cheng Si[1]

*Address all correspondence to: bing.si@usask.ca

1 University of Saskatchewan, Department of Soil Science, Saskatoon, Canada

2 Department of Natural Resource Sciences, McGill University, Ste-Anne-de-Bellevue, Quebec, Canada

References

[1] Biswas, A., & Si, B. C. (2011a). Identifying Scale-Specific Controls of Soil Water Storage in a Hummocky Landscape Using Wavelet Coherency. Geoderma, 165, 50-59.

[2] Biswas, A., & Si, B. C. (2011b). Revealing the Controls of Soil Water Storage at Different Scales in a Hummocky Landscape. Soil sci. soc. am. j., 75, 1295-1306.

[3] Biswas, A., & Si, B. C. (2011c). Scales and Locations of Time Stability of Soil Water Storage in a Hummocky Landscape. J. hydrol., 408, 100-112.

[4] Biswas, A., & Si, B. C. (2012). Identifying Effects of Local and Nonlocal Factors of Soil Water Storage Using Cyclical Correlation Analysis. Hydrol. Process., doi: 10.1002/hyp.8459.

[5] Bosch, D. D., Lakshmi, V., Jackson, T. J., Choi. M., & Jacobs, J. M. (2006). Large Scale Measurements of Soil Moisture for Validation of Remotely Sensed Data: Georgia Soil Moisture Experiment of 2003. J. hydrol., 323, 120-137.

[6] Brocca, L., Melone, F., Moramarco, T., & Morbidelli, R. (2009). Soil Moisture Temporal Stability over Experimental Areas in Central Italy. Geoderma, 148, 364-374.

[7] Cassel, D. K., Wendroth, O., & Nielsen, D. R. (2000). Assessing Spatial Variability in an Agricultural Experiment Station Field: Opportunities Arising from Spatial Dependence. Agron. j., 92, 706-714.

[8] Chen, Y. (2006). Letter to the Editor on Rank Stability or Temporal Stability. Soil sci. soc. am. j., 70, 306.

[9] Comegna, V., & Basile, A. (1994). Temporal Stability of Spatial Patterns of Soil Water Storage in a Cultivated Vesuvian Soil. Geoderma, 62, 299-310.

[10] Cosh, M. H., Jackson, T. J., Bindlish, R., & Prueger, J. H. (2004). Watershed Scale Temporal and Spatial Stability of Soil Moisture and Its Role in Validating Satellite Estimates. Remote sens. environ., 92, 427-435.

[11] Cosh, M. H., Jackson, T. J., Moran, S., & Bindlish, R. (2008). Temporal Persistence and Stability of Surface Soil Moisture in a Semi-Arid Watershed. Remote sens. environ., 112, 304-313.

[12] Cosh, M. H., Jackson, T. J., Starks, P., & Heathman, G. (2006). Temporal Stability of Surface Soil Moisture in the Little Washita River Watershed and Its Applications in Satellite Soil Moisture Product Validation. J. hydrol., 323, 168-177.

[13] Farge, M. (1992). Wavelet transforms and their applications to turbulence. Annu. rev. of fluid mech., 24, 395-457.

[14] Gao, X. D., Wu, P. T., Zhao, X. N., Shi, Y. G., & Wang, J. W. (2011). Estimating Spatial Mean Soil Water Contents of Sloping Jujube Orchards Using Temporal Stability. Agr. water manage., 102, 66-73.

[15] Gómez- Plaza, A., Alvarez-Rogel, J., Albaladejo, J., & Castillo, V. M. (2000). Spatial Patterns and Temporal Stability of Soil Moisture across a Range of Scales in a Semi-Arid Environment. Hydrol. process., 14, 1261-1277.

[16] Gómez-Plaza, A., Martínez-Mena, M., Albaladejo, J., & Castillo, V. M. (2001). Factors Regulating Spatial Distribution of Soil Water Content in Small Semi-Arid Catchments. J. hydrol., 253, 211-226.

[17] Grant, L., Seyfried, M., & McNamara, J. (2004). Spatial Variation and Temporal Stability of Soil Water in a Snow-Dominated, Mountain Catchment. Hydrol. process., 18, 3493-3511.

[18] Grayson, R. B., & Western, A. W. (1998). Towards Areal Estimation of Soil Water Content from Point Measurements: Time and Space Stability of Mean Response. J. hydrol., 207, 68-82.

[19] Grayson, R. B., Western, A. W., Chiew, F. H. S., & Blöschl, G. (1997). Preferred States in Spatial Soil Moisture Patterns: Local and Nonlocal Controls. Water resour. res., 33, 2897-2908.

[20] Grinsted, A. J., Moore, C., & Jevrejeva, S. (2004). Application of the Cross Wavelet Transform and Wavelet Coherence to Geophysical Time Series. Nonlinear proc. geoph., 11, 561-566.

[21] Guber, A. K., Gish, T. J., Pachepsky, Y. A., van Genuchten, M. T., Daughtry, C. S. T., Nicholson, T. J., & Cady, R. E. (2008). Temporal Stability in Soil Water Content Patterns across Agricultural Fields. Catena, 73, 125-133.

[22] Hu, W., Biswas, A., & Si, B. C. (2012a). Application of Multivariate Empirical Mode Decomposition for Revealing Scale Specific Time Stability of Soil Water Storage. J. hydrol., submitted.

[23] Hu, W., Shao, M. A., Han, F. P., Reichardt, K., & Tan, J. (2010a). Watershed Scale Temporal Stability of Soil Water Content. Geoderma, 158, 181-198.

[24] Hu, W., Shao, M. A., & Reichardt, K. (2010b). Using a New Criterion to Identify Sites for Mean Soil Water Storage Evaluation. Soil sci. soc. am. j., 74, 762-773.

[25] Hu, W., Shao, M. A., Wang, Q. J., & Reichardt, K. (2009). Time Stability of Soil Water Storage Measured by Neutron Probe and the Effects of Calibration Procedures in a Small Watershed. Catena, 79, 72-82.

[26] Hu, W., Tallon, L. K., & Si, B. C. (2012b). Evaluation of Time Stability Indices for Soil Water Storage Upscaling. J. hydrol., accepted.

[27] Huang, N. E., Shen, Z., Long, S. R., Wu, M. C. L., Shih, H. H., Zheng, Q., Yen, N. C., Tung, C. C., & Liu, H. H. (1998). The Empirical Mode Decomposition and the Hilbert

Spectrum for Nonlinear and Non-Stationary Time Series Analysis. Proc. r. soc. a, 454, 903-995.

[28] Huang, N. E., Wu, M. C. L., Long, S. R., Shen, S. S. P., Qu, W., Gloersen, P., & Fan, K. L. (2003). A Confidence Limit for the Empirical Mode Decomposition and Hilbert Spectral Analysis. Proc. r. soc. a, 459, 2317-2345.

[29] Huang, N. E., & Wu, Z. (2008). A Review on Hilbert-Huang Transform: Method and Its Applications to Geophysical Studies. Rev. geophys, 46, RG2006, doi: 10.1029/2007RG000228.

[30] Ibrahim, H. M., & Huggins, D. R. (2011). Spatio-Temporal Patterns of Soil Water Storage under Dryland Agriculture at the Watershed Scale. J. Hydrol., 404, 186-197.

[31] Jacobs, J. M., Mohanty, B. P., Hsu, E. C., & Miller, D. (2004). SMEX02: Field Scale Variability Time Stability and Similarity of Soil Moisture. Remote sens. environ., 92, 436-446.

[32] Jawson, S. D., & Niemann, J. D. (2007). Spatial Patterns from EOF Analysis of Soil Moisture at a Large Scale and Their Dependence on Soil, Land-Use, and Topographic Properties. Adv. water resour., 30, 366-381.

[33] Joshi. C., & Mohanty, B. P. (2010). Physical Controls of Near-Surface Soil Moisture across Varying Spatial Scales in an Agricultural Landscape During SMEX02. Water resour. res., 46, W12503, doi: 10.1029/2010WR009152.

[34] Joshi. C., Mohanty, B. P., Jacobs, J. M., & Ines, A. V. M. (2011). Spatiotemporal Analyses of Soil Moisture from Point to Footprint Scale in Two Different Hydroclimatic Regions. Water resour. res., 47, W01508, doi:10.1029/2009WR009002.

[35] Kachanoski, R. G., & De Jong, E. (1988). Scale Dependence and the Temporal Persistence of Spatial Patterns of Soil Water Storage. Water resour. res., 24, 85-91.

[36] Korres, W., Koyama, C. N., Fiener, P., & Schneider, K. (2010). Analysis of Surface Soil Moisture Patterns in Agricultural Landscapes Using Empirical Orthogonal Functions. Hydrol. earth syst. sci., 14, 751-764.

[37] Kumar, P., & Foufoula-Georgiou, E. (1993). A Multicomponent Decomposition of Spatial Rainfall Fields: 1. Segregation of Large- and Small-Scale Features Using Wavelet Transforms. Water resour. res., 29, 2515-2532.

[38] Kumar, P., & Foufoula-Georgiou, E. (1997). Wavelet Analysis of Geophysical Applications. Rev. geoph., 35, 385-412.

[39] Lakshmi, V., Piechota, T., Narayan, U., & Tang, C. L. (2004). Soil Moisture as an Indicator of Weather Extremes. Geophys. res. lett., 31, L11401, doi:10.1029/2004GL019930.

[40] Lin, H. (2006).Temporal Stability of Soil Moisture Spatial Pattern and Subsurface Preferential Flow Pathways in the Shale Hills Catchment. Vadose zone j., 5, 317-340.

[41] Martínez-Fernández, J., & Ceballos, A. (2003). Temporal Stability of Soil Moisture in a Large-Field Experiment in Spain. Soil sci. soc. am. j., 67, 1647-1656.

[42] Martínez-Fernández, J., & Ceballos, A. (2005). Mean Soil Moisture Estimation Using Temporal Stability Analysis. J. hydrol., 312, 28-38.

[43] Mohanty, B. P., & Skaggs, T. H. (2001). Spatio-Temporal Evolution and Time-Stable Characteristics of Soil Moisture within Remote Sensing Footprints with Varying Soil Slope and Vegetation. Adv. water resour., 24, 1051-1067.

[44] North, G. R., Bell, T. L., Cahalan, R. F., & Moeng, F. J. (1982). Sampling errors in the estimation of empirical orthogonal functions. Mon. weather rev., 110, 699-706.

[45] Pachepsky, Y. A., Guber, A. K., & Jacques, D. (2005). Temporal Persistence in Vertical Distributions of Soil Moisture Contents. Soil sci. soc. am. j., 69, 347-352.

[46] Perry, M. A., & Niemann, J. D. (2007). Analysis and Estimation of Soil Moisture at the Catchment Scale Using Eofs. J. hydrol., 334, 388-404.

[47] Rehman, N., & Mandic, D. P. (2009). http://www.commsp.ee.ic.ac.uk/~mandic/research/emd.htm.

[48] Rehman, N., & Mandic, D. P. (2010). Multivariate Empirical Mode Decomposition. Proc. r. soc. a, 466, 1291-1302.

[49] Rolston, D. E., Biggar, J. W., & Nightingale, H. I. (1991). Temporal Persistence of Spatial Soil-Water Patterns under Trickle Irrigation. Irrig. sci., 12, 181-186.

[50] Schneider, K,, Huisman, J. A., Breuer, L., Zhao, Y., & Frede, H. G. (2008). Temporal Stability of Soil Moisture in Various Semi-Arid Steppe Ecosystems and Its Application in Remote Sensing. J. hydrol., 359, 16-29.

[51] Si, B. C. (2008). Spatial Scaling Analyses of Soil Physical Properties: A Review of Spectral and Wavelet Methods, Vadose zone j., 7, 547-562.

[52] Si, B. C., & Zeleke, T. B. (2005). Wavelet Coherency Analysis to Relate Saturated Hydraulic Properties to Soil Physical Properties. Water resour. res., 41, W11424, doi: 10.1029/2005WR004118.

[53] Starks, P. J., Heathman, G. C., Jackson, T. J., & Cosh, M. H. (2006). Temporal Stability of Soil Moisture Profile. J. hydrol., 324, 400-411.

[54] Starr, G. C. (2005). Assessing Temporal Stability and Spatial Variability of Soil Water Patterns with Implications for Precision Water Management. Agr. water manage., 72, 223-243.

[55] Tallon, L. K., & Si, B. C. (2004). Representative Soil Water Benchmarking for Environmental Monitoring. J. environ. inform., 4, 581-590.

[56] Torrence, C., & Compo, G. P. (1998). A Practical Guide to Wavelet Analysis. Bull. am. meteor. soc., 79, 61-78.

[57] Vachaud, G., Passerat De Silans, A., Balabanis, P., & Vauclin, M. (1985). Temporal Stability of Spatially Measured Soil Water Probability Density Function. Soil sci. soc. Am. J., 49, 822-828.

[58] Vereecken, H., Kamai, T., Harter, T., Kasteel, R., Hopmans, J., & Vanderborght, J. (2007). Explaining Soil Moisture Variability as a Function of Mean Soil Moisture: A Stochastic Unsaturated Flow Perspective. Geophys. res. lett., 34, DOI: 10.1029/2007GL031813.

[59] Vivoni, E. R., Gebremichael, M., Watts, C. J., Bindlish, R., & Jackson, T. J. (2008). Comparison of Ground-Based and Remotely-Sensed Surface Soil Moisture Estimates over Complex Terrain During SMEX04. Remote sens. environ., 112, 314-325.

[60] Western, A., Grayson, R., Blöschl, G., Willgoose, G., & McMahon, T. (1999). Observed Spatial Organization of Soil Moisture and Its Relation to Terrain Indices. Water resour. res., 35, 797-810.

[61] Wilson, D. J., Western, A. W., & Grayson, R. B. (2005). A Terrain and Data-Based Method for Generating the Spatial Distribution of Soil Moisture. Adv. water resour., 28, 43-54.

[62] Yoo, C., & Kim, S. (2004). EOF Analysis of Surface Soil Moisture Field Variability. Adv. water resour., 27, 831-842.

[63] Zar, J. H. (1972). Significance Testing of the Spearman Rank Correlation Coefficient. J. am. stat. assoc., 67, 578-580.

[64] Zhao, Y., Peth, S., Wang, X. Y., Lin, H., & Horn, R. (2010). Controls of Surface Soil Moisture Spatial Patterns and Their Temporal Stability in a Semi-Arid Steppe. Hydrol. process., 24, 2507-2519.

[65] Zhou, X., Lin, H., & Zhu, Q. (2007). Temporal Stability of Soil Moisture Spatial Variability at Two Scales and Its Implication for Optimal Field Monitoring. Hydrol. earth syst. sci. discuss, 4, 1185-1214.

[66] Zhu, Q., & Lin, H. (2011). Influences of Soil, Terrain, and Crop Growth on Soil Moisture Variation from Transect to Farm Scales. Geoderma, 163, 45-54.

Chapter 4

Model Averaging for Semivariogram Model Parameters

Asim Biswas and Bing Cheng Si

Additional information is available at the end of the chapter

http://dx.doi.org/10.5772/52339

1. Introduction

Soil varies considerably from location to location (Nielsen et al., 1973). Knowledge on soil spatial variability is important in ecological modelling (Burrough, 1983; Corwin et al., 2006), environmental prediction (Trangmar et al., 1985), precision agriculture (Goderya, 1998), soil quality assessment (Heuvelink and Pebesma, 1999; McBratney et al., 2000), and natural resources management. Adequate understanding of the variability in soil properties as a function of space and time is necessary for developing logical, empirical and physical models of soil and landscape processes (Burrough, 1993; Foussereau et al., 1993; Wilding et al., 1994). Geostatistics, a widely used approach, has been used to identify the spatial structure in the variability of soil attributes (Vieira, 2000; Carvalho et al., 2002; Vieira et al., 2002). Semivariance function characterizes the spatial continuity between points. When the semivariance is plotted against the lag distance or separation distance between points, the plot is called semivariogram (Fig. 1; McBratney and Webster, 1986; Isaaks and Srivastava, 1989). The structure of the semivariogram is explained by three properties; the nugget, the sill and the range (Fig. 1). These spatial structures of semivariogram help in identifying autocorrelation and replicating samples, revealing dominant pattern in data series, identifying major ongoing processes (Si et al., 2007), designing experiments (Fagroud and van Meirvenne, 2002) and monitoring networks (Prakash and Singh, 2000), selecting proper data analysis method and interpreting data (Lambert et al., 2004), and assessing simulation and uncertainty analysis in a better way (Papritz and Dubois, 1999). The semivariogram structures also help to quantify spatial dependence between observations. Modelling of observed semivariance values helps in predicting the spatial distribution of attribute values (Goovaerts, 1998). The spatial distribution of attribute values is very important in separating random noise in semivariance, and interpolation and mapping analysis such as kriging (Deutsch and Journel, 1998; Nielsen and Wendorth, 2003).

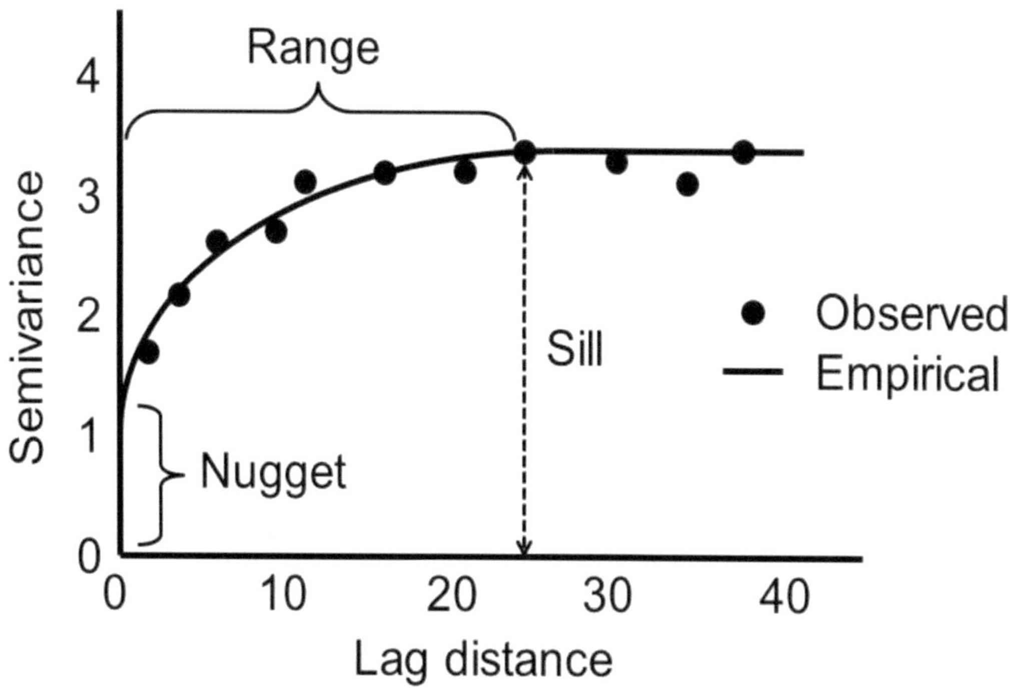

Figure 1. A typical example of semivariogram showing different components.

The choice of theoretical models and its fitting procedure is very important to get a better prediction of unsampled locations (McBratney and Webster, 1986). Spherical model (Burgess and Webster, 1980; Vieira et al., 1981; Van Kuilenburg et al., 1982; Vauclin et al., 1983; Trangmar, 1985), exponential model (David, 1977), Gaussian model (McBratney and Webster, 1986) and linear plateau model (Burgess and Webster, 1980; Hajrasuliha et al., 1980; Vauclin et al., 1983) are important among the most commonly used semivariogram models in the field of soil science. The maximum likelihood method (Cressie, 1991) or least square regression (Vieira et al., 1981; Yost et al., 1982; Trangmar et al., 1985) including the weighted least square methods (Cressie, 1985) optimize the parameter value by minimizing the deviations of model prediction from the experimental semivariances. Small sum of deviations between the model and the experimental values indicate superior performance of the models. Small and comparable sum of the deviations indicates comparable performance of the models. Therefore, selection of model is prerequisite for better prediction (Burnham and Anderson, 2002). In reality, there can be several models to choose from. Selection of good models requires balancing of goodness of fit and complexity, which is generally, determined using a likelihood ratio approach leading to a chi-square test. The Akaike Information Criterion (AIC) values can be calculated from maximum likelihood function (Akaike, 1973), which is used to evaluate the performance of the models (Webster and McBratney, 1989). The smallest AIC value indicates the best model, but the values of parameters such as nugget, sill and range can be different from model to model, even though they have the same physical meaning (Trangmar, 1985). Different value of a parameter for different models clearly indicates the uncertainty associated with the parameter, which makes the interpretation difficult. Therefore, interpretation of a particular spatial process from the parameter values of a particular model will be biased. In addition the parameter value from individual models may have large uncertainty. This uncertainty is inherent to the model selection and is well beyond just the issue of determining best model(s). Model averaging is a technique designed to help in accounting the uncertainty associated with the models and their parameters.

The model averaging is not a new concept in statistics. People started combining and averaging models since 1970s for different purposes (Hoeting et al., 1999). During averaging models are assigned with weights based on their performance. The AIC values are used to calculate the weights for the models (Burnham and Anderson, 2002). Before the recent development of computational power, the averaging procedure ignored the uncertainty associated with models. Recently, the use of model averaging is increased to reduce the uncertainty associated with the model selection. Information on the use of model averaging procedure in soil science is scarce (Webster and McBratney, 1989) and there is no information of reducing uncertainty associated with the semivariogram model parameters from averaging of commonly used semivariogram model parameters. Therefore, the objective of this paper is to reduce the uncertainty associated with semivariogram model parameters through averaging. The weighted average of the parameter values of commonly used models can be a better way of describing the spatial processes.

Name of Model	Equations †
Spherical Model	$\gamma_s(h;a,b,c)=\begin{cases} a+(b-a)\left[1.5\frac{h}{c}-0.5\left(\frac{h}{c}\right)^3\right] & if\ h\le c \\ b & otherwise \end{cases}$
Exponential Model	$\gamma_e(h;a,b,c)=a+(b-a)\left\{1-\exp\left(\frac{-h}{c}\right)\right\}$
Gaussian Model	$\gamma_g(h;a,b,c)=a+(b-a)\left\{1-\exp\left(\frac{-h^2}{c^2}\right)\right\}$
Linear Plateau Model	$\gamma_l(h;a,b,c)=\begin{cases} a+(b-a)\frac{h}{c} & if\ h\le c \\ b & otherwise \end{cases}$
† In all models; $\gamma = 0$ when $h = 0$	

Table 1. Commonly used models for semivariogram fitting in soil science

2. Theory

Geostatistics can explain the spatial variability and the patterns of a variable in field from its autocorrelation. It describes the relationship between measurements at different locations (or times) separated by certain distance (or time). For example, a soil property $Z(x_i)$ measured at location x_i, where i =1, 2, …, n and n is the number of samples along a transect. The continuity of this relationship can be investigated from h scatterplots, which can be done by plotting the measured values $Z(x_i)$ as horizontal axis and $Z(x_i+h)$ as vertical axis, where i =1, 2, …. The h scatterplot will show a cloud of points, where every point represents one pair of sample locations $Z(x_i)$ and $Z(x_i+h)$ (Pannatier, 1996; Zawadzki and Fabijanczyk, 2007). For each h, half of the mean value of the squared difference $Z(x_i)$ - $Z(x_i+h)$ is defined as semivariance (Matheron, 1962).

$$\gamma(h) = \frac{1}{2N(h)} \sum_{i=1}^{N(h)} \left[Z(x_i) - Z(x_i + h) \right]^2 \tag{1}$$

The variance (σ_i^2) of the squared difference $Z(x_i)$ - $Z(x_i+h)$ can be calculated using Eq. (2).

$$\sigma_i^2 = \frac{1}{N(h)-1} \sum_{i=1}^{N(h)} \left(a_i - \bar{a}_i \right)^2 \tag{2}$$

where, $a_i = [Z(x_i) - Z(x_i + h)]^2$ and $N(h)$ is the number of pairs in a cloud for a particular h.

The experimental semivariogram can be fitted to a mathematical model. Not all the mathematical functions that seem to fit the observed values can be considered as semivariogram model. One important criterion for the semivariogram models is to be "positive-definite" to ensure the nonnegative covariance values restricting the models to be permissible for fitting (Isaaks and Srivastava, 1989; Goovaerts, 1997; Deutsch and Journel, 1998). There are a number of permissible semivariogram models including the most commonly used spherical, exponential, Gaussian, and linear plateau (Table 1).

Among the fitting procedures, the weighted nonlinear least square method is the most robust and reliable method (Cressie, 1985). This procedure minimizes the residual sum of squared errors (RSS) between experimental semivariance data and the models by optimizing the model parameters: nugget, sill and range values. The RSS can be calculated using Eq. 3.

$$RSS = \sum_{i=1}^{m} w_i \left[\tilde{\gamma}_i - \gamma_i \right]^2 \tag{3}$$

where, m is the number of lags, $\tilde{\gamma}_i$ is the semivariance value for lag i, γ_i are the corresponding model predictions, and w_i are weighting factors. The weighing factor used in calculating RSS is related to the variance associated with the semivariance calculation (Jian et al., 1996).

$$w_i = \frac{1}{\sigma_i^2} \tag{4}$$

where, σ_i^2 is the variance calculated in Eq. 2 for lag i. In this weighted least squares approximation the fit can always be improved by diminishing the residual sum of squared errors with addition of parameters to the models (Webster and McBratney, 1989). In case of the same number of parameters in the most commonly used semivariogram models, the small and comparable RSS values indicate the comparable performance of different models. A solution can be achieved by fitting each model using the least square approximation and then comparing the goodness of fit for each model (Webster and McBratney, 1989). The performance of a model can be evaluated and the fitting can be said as the best with the lowest Akaike Information Criterion (AIC) (Akaike, 1973). AIC is an estimate of the expected Kullback-Leiler information (a ruler to measure the similarity between the statistical model and the true distribution) lost by using a model to approximate the process that generated observed pattern (Burnham and Anderson, 2002; Johnson and Omland, 2004). It is also defined as Eq. 5:

$$AIC = -2\ln(\text{maximized likelihood}) + 2 \times (\text{number of parameters}) \tag{5}$$

Or it can be estimated as in Eq. 6.

$$AIC = -2\ln\left[L(\hat{\theta}|y\right] + 2p \tag{6}$$

where, p is the number of parameters and $\ln[L(\hat{\theta} \mid y]$ is the maximized likelihood and can be estimated from Eq. 7.

$$\ln\left[L(\hat{\theta}|y\right] = -\frac{n}{2}\ln\left(\frac{RSS}{n}\right) \tag{7}$$

where, n is the number of data points. With close performance, different models with same number of parameters having the same physical meaning produces different optimized parameter values. A single process in field will be represented by different values based on the selected models. In this case the estimated parameter values will be associated with uncertainty, which can be reduced by making weighted average based on the performance of the selected model. If there are R numbers of models with parameters θ_j (j = 1, 2, 3,...,p), the estimated parameters can be calculated from the Eq. 8 (Burnham and Anderson, 2002; Johnson and Omland, 2004).

$$\hat{\hat{\theta}} = \sum_{j=1}^{R} W_j \hat{\theta}_j \tag{8}$$

where, W_j are the weights assigned to a particular model. The weights can be calculated from the AIC values for each model (Eq. 9).

$$W_j = \frac{\exp\left(-\Delta_j/2\right)}{\sum_{j=1}^{R} \exp\left(-\Delta_j/2\right)} \tag{9}$$

where, $\Delta_j = AIC_j - AIC_{\min}$ and AIC_j and AIC_{min} are the AIC values for a particular model and the lowest AIC value model.

To explain the uncertainty associated with the parameter, we need to calculate the variance associated with each optimized parameter. If a model has p parameters, we can calculate $p \times p$ covariance matrix. From the calculation of Hessian Matrix (the second derivative matrix of χ^2 merit function) we can estimate the standard error in fitted parameters. In the calculation of Hessian Matrix (Press et al., 1992), let us assume we have to fit a model:

$$\gamma = \gamma\left(h;a,b,c\right) \tag{10}$$

The χ^2 merit function, that will be equal to RSS (Eq. 11),

$$\chi^2\left(a,b,c\right) = \sum_{i=1}^{m} \frac{1}{\sigma_i^2} \left[\tilde{\gamma}_i - \gamma\left(h;a,b,c\right)\right]^2 = RSS\left(a,b,c\right) \tag{11}$$

The χ^2 gradient for the parameters, z, will be zero when χ^2 is minimum. The components can be calculated as Eq. (12).

$$\frac{d\chi^2}{dz_k} = -2\sum_{i=1}^{N} \frac{\left[\tilde{\gamma}_i - \gamma\left(h;a,b,c\right)\right]}{\sigma_i^2} \frac{d\gamma\left(h;a,b,c\right)}{dz_k} \text{ where, } k = a,b,c. \tag{12}$$

An additional derivative will give,

$$\frac{d^2\chi^2}{dz_a dz_b} = 2\sum_{i=1}^{N} \frac{1}{\sigma_i^2} \left[\frac{d\gamma\left(h;a,b,c\right)}{dz_a} \frac{d\gamma\left(h;a,b,c\right)}{dz_b} - \left\{\tilde{\gamma}_i - \gamma\left(h;a,b,c\right)\right\} \frac{d^2\gamma\left(h;a,b,c\right)}{dz_a dz_b} \right] \tag{13}$$

It is conventional to remove the factors of 2 by defining

$$\alpha_{ab} \equiv \frac{1}{2}\frac{d^2\chi^2}{dz_a dz_b} \tag{14}$$

For a linear equation, the second derivative term can be dismissed because it is zero or small enough when compared to the term involved in first order derivative. So the α_{ab} formula can be written as (Press et al., 1992);

$$\alpha_{ab} = \sum_{i=1}^{N} \frac{1}{\sigma_i^2}\left[\frac{d\gamma(h;a,b,c)}{dz_a}\frac{d\gamma(h;a,b,c)}{dz_b}\right] \tag{15}$$

So for a three parameter (a, b, c) semivariogram model we will get a 3×3 Hessian Matrix, (Eq. 16).

$$[\alpha] = \begin{bmatrix} \sum_{i=1}^{N}\frac{d\gamma}{da}\frac{d\gamma}{da} & \sum_{i=1}^{N}\frac{d\gamma}{da}\frac{d\gamma}{db} & \sum_{i=1}^{N}\frac{d\gamma}{da}\frac{d\gamma}{dc} \\ \sum_{i=1}^{N}\frac{d\gamma}{db}\frac{d\gamma}{da} & \sum_{i=1}^{N}\frac{d\gamma}{db}\frac{d\gamma}{db} & \sum_{i=1}^{N}\frac{d\gamma}{db}\frac{d\gamma}{dc} \\ \sum_{i=1}^{N}\frac{d\gamma}{dc}\frac{d\gamma}{da} & \sum_{i=1}^{N}\frac{d\gamma}{dc}\frac{d\gamma}{db} & \sum_{i=1}^{N}\frac{d\gamma}{dc}\frac{d\gamma}{dc} \end{bmatrix} \tag{16}$$

The approximation of the covariance matrix can be done by inversing the matrix α (Press et al., 1992) (Eq. 17).

$$[C] = [\alpha]^{-1} \tag{17}$$

This matrix component indicates the variance associated with each parameter for different models. At the same time we need to calculate the variance associated with the averaged parameter. The parameter variance for each model and the weight assigned to the model are used to calculate the variance of the averaged parameter (Eq. 18) (Burnham and Anderson, 2002)

$$\hat{\text{var}}\left(\hat{\bar{\theta}}\right) = \sum_{j=1}^{R} W_j\left[\text{var}\left(\hat{\theta} \mid g_j\right) + \left(\hat{\theta} - \hat{\bar{\theta}}\right)^2\right] \tag{18}$$

where, W is the weights assigned to R number of models, g, $\hat{\theta}$ is the estimated model parameter values from fitting, and $\hat{\bar{\theta}}$ is the average of estimated model parameter values.

Characteristics	Sand (Zeleke and Si, 2005)	Copper (Atteia et al., 1994)
Minimum	52.50	3.55 (1.27)
Maximum	78.75	166.40 (5.11)
Mean	64.23	23.58 (2.87)
Median	64.41	17.20 (2.84)
Mode	58.75	16.40 (2.80)
Skewness	0.28	3.02 (0.42)
Kurtosis	-0.73	11.88 (0.11)
Standard Deviation	6.06	22.24 (0.72)
Variance	36.76	494.48 (0.52)

Table 2. Descriptive statistics of the data sets used for model fitting

3. Materials and methods

Demonstration of this average parameter estimation procedure was done using two soil parameters, which were chosen from two different experimental datasets. One soil property, sand content, was collected from a regularly spaced (3 m) 128 point transect (Zeleke and Si, 2005). The sampling site was located at Smeaton, Saskatchewan, Canada (53°40′N latitude and 104°58′W longitude). The hydrometer method was used to determine the particle sizes (Gee and Bauder, 1986). Another soil property, copper (Cu) content, was selected from a dataset consisting of 359 topsoil samples collected by the Swiss Federal Institute of Technology in the Swiss Jura (Atteia et al., 1994). The data points were selected according to a regular configuration of a mesh of 250 m × 250 m with several clusters from an area of approximately 1450 ha. The cluster samples were collected following nested sampling design with 100, 40, 15, and 6 m sampling interval. Copper content was measured using a direct current plasma spectroscopy (ARL, Spectran V) with other 5 heavy metals (Atteia et al., 1994).

4. Results and discussion

The exploratory information about the dataset is given in Table 2. The skewness of the sand 0.28, which indicates a near normal statistical distribution of the dataset. The copper content is highly skewed (skewness = 3.02) and exhibits a log normal statistical distribution. Therefore, we decided to normalize the copper content using natural logarithm transformation. The exploratory information of logarithm transformed values for copper content is presented in parentheses in Table 2.

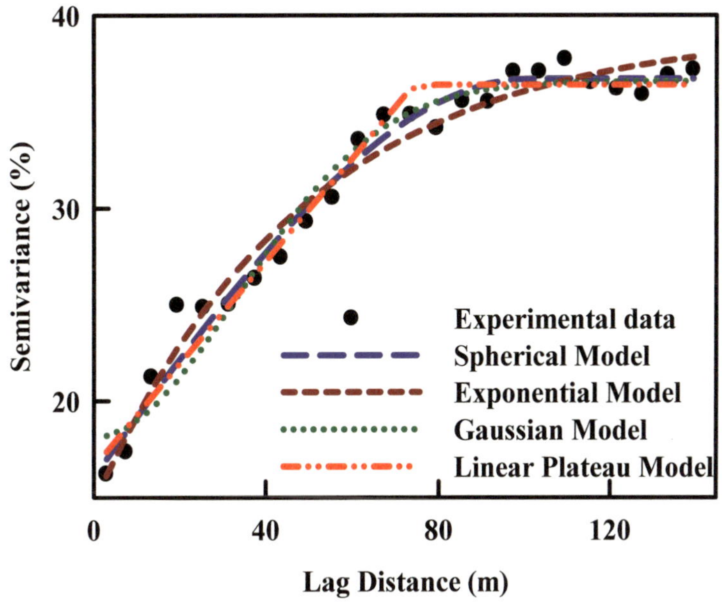

Figure 2. Empirical semivariogram and fitted models for sand content data.

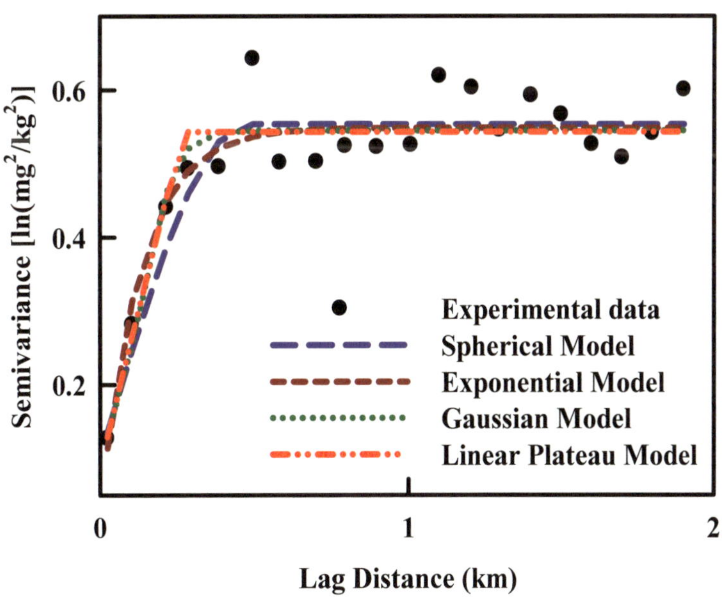

Figure 3. Empirical semivariogram and fitted models of logarithm-transformed copper content data.

Semivariance for sand content (Fig. 2) and logarithm transformed copper content (Fig. 3) are calculated and plotted as a function of lag interval. For the semivariance calculation of regularly spaced sand samples, the lag interval is 6 m, which is twice the minimum sampling interval. The maximum lag distance used for this calculation is 144 m and the minimum number of pairs used for the semivariogram calculation is 24. The semivariance of the copper content is calculated using minimum and maximum lag distance of 0.1 km and 2 km respectively, which leads to at least 20 pairs of data points. The experimental semivariograms are fitted with four most commonly used semivariogram models (Table 1) following weighted least square estimation. The fitted semivariogram models and the experimental semivariograms are shown in Fig. 2 and Fig. 3 for sand and copper, respectively. The optimized parameters (nugget, sill and range) values for each of the models are presented in Table 3. The nugget value optimized by different models for the sand content varies from 14.67 % (exponential model) to 18.13 % (Gaussian model). The value of sill varies from 36.40 % (linear plateau model) to 39.27 % (exponential model). The optimized range value that indicates the distance over which the processes are spatially dependent, for the sand content varies widely from 47.10 m (Gaussian model) to over 100 m (Spherical model). The variance (values in the paresis in Table 3) associated with each parameters are approximated from the calculation of Hessian Matrix (Press et al., 1992). The variance associated with the parameter estimate is very high as the semivariance is calculated from the semivariance cloud, which spreads over a range of values. The nugget, sill, and range have their own physical significance and different models should result in the same set of parameter values. However, due to model uncertainty, four models have different set of optimized parameter values for the sand content. The range of optimized parameter values (Table 3) from different models clearly indicate how uncertain our models and parameters are.

Similarly there are a wide range of optimized parameter values from different models for copper content. The optimized nugget values vary widely in different models (0.027 $\ln(mg^2/kg^2)$ for exponential model to 0.122 $\ln(mg^2/kg^2)$ for Gaussian model) (Table 3). Wide range of sill and range values clearly indicates the uncertainty associated with the parameters. Different parameter values for different models make the interpretation of the parameters and the semivariogram structures difficult and uncertain. The ratio of the nugget semivariance (V_N) to the sill semivariance (V_S) gives a measure of the strength or degree of spatial structure (Cambardella et al., 1994). Therefore, it is sometime difficult to explain the semivariogram structure of a particular property with different nugget and sill values for different models.

In this situation the goodness of fit for each model is calculated. The residual sum of squared error (RSS) between the experimental data and the model data is presented in Table 3. The comparable RSS values indicate that the four models are comparable. From the RSS value, the maximum likelihood is calculated for each model. Akaike Information Criterion (AIC) for different models is calculated from the maximum likelihood of the models and is presented in Table 4. The AIC values for the spherical (-167.269) and exponential models (-167.187) fitted for sand semivariogram are very close indicating a close performance of the models. The performance of the linear plateau and Gaussian models are also acceptable. The

AIC values for all four models are very similar for copper content (vary from -111.739 to -113.739). The similar AIC values for the models indicate that all four models are almost equivalent. Though the performance of the models is very comparable, the optimized parameter values from different models are not similar. In this situation, the selection of parameter value from a particular model can be associated with high uncertainty.

Parameter	Model	Nugget	Sill	Range	V_N/V_S	RSS
		---%---	---%---	---m---		
Sand	Spherical	16.08 (257.45)	36.71 (231.66)	100.94 (29610.00)	0.438	0.021
	Exponential	14.67 (423.95)	39.27 (1081.00)	49.15 (32550.00)	0.374	0.021
	Gaussian	18.13 (198.24)	36.57 (214.91)	47.10 (5233.00)	0.495	0.040
	Linear Plateau	16.57 (217.15)	36.40 (181.53)	74.37 (8379.00)	0.455	0.026
		ln(mg²/kg²)	**ln(mg²/kg²)**	**--km--**		
Copper	Spherical	0.101 (0.074)	0.553 (0.037)	0.480 (0.767)	0.183	0.067
	Exponential	0.027 (0.177)	0.548 (0.039)	0.132 (0.133)	0.049	0.061
	Gaussian	0.122 (0.062)	0.544 (0.034)	0.171 (0.097)	0.224	0.061
	Linear Plateau	0.089 (0.080)	0.543 (0.033)	0.271 (0.235)	0.164	0.063

Table 3. Optimized parameter values and their variance (in paresis) for different semivariogram models for sand content and copper

The model averaging is conducted to reduce parameter uncertainty. Weights are assigned to the models based on their performance and importance. The model with lowest AIC value indicates best model. Based on the smallest AIC, ΔAIC is calculated from the difference between AIC value of a particular model and the smallest AIC value out of four models. The ΔAIC is used to measure the likelihood of each model. The value of the likelihood of models helps in assigning weights for different models. The spherical and the exponential models for the sand semivariogram have the highest weight, which are almost equal. This indicates an equivalent performance of these two models. The Gaussian and linear plateau models have very small weights, indicating less importance of the models than that of the other two. Whereas for copper, the weights assigned to different models are close enough to explain the performance as equivalent. Based on the weights assigned to each model, the average value of the parameter and the variance associated with each parameter is estimated. Sand has an average nugget, sill and range value of 15.42 %, 37.92 % and 75.53 m respectively. The variance associated with each averaged parameters are presented in the paresis in Table 4. The higher weight of the spherical and exponential model indicates that two models have large contribution to the average value of parameters. Similarly, the variance of the averaged parameters will have large contribution from spherical and exponential models. Whereas for copper, the contribution to the average value by each models is comparable. The average value of nugget, sill and range for copper are 0.081 $\ln(mg^2/kg^2)$, 0.546 $\ln(mg^2/kg^2)$, and 0.219 km respectively. The variance of each averaged parameter is also calculated and presented in Table 4 (in paresis). The averaged variance reduced the uncertainty associated with the parameters by taking the average.

Parameter	Model	AIC value	Weight	Average parameter value		
				Nugget	Sill	Range
Sand	Spherical	-167.269	0.4946	15.42 %	37.92 %	75.53 m
	Exponential	-167.187	0.4748	(335.79)	(635.06)	(31007.42)
	Gaussian	-151.279	0.0001			
	Linear Plateau	-161.687	0.0303			
Copper	Spherical	-111.739	0.1230	0.081 $\ln(mg^2/kg^2)$	0.546 $\ln(mg^2/kg^2)$	0.219 km
	Exponential	-113.703	0.3285			
	Gaussian	-113.658	0.3212	(0.106)	(0.036)	(0.235)
	Linear Plateau	-112.964	0.2270			

Table 4. Parameter values obtained through model averaging of commonly used models and their variance (in paresis).

When different models optimize parameters with a wide range of values, the weighted average of the estimated parameter values provide more concise information and better understanding about the parameters and thus the spatial structure. The variance of the optimised parameters clearly indicates the reduced uncertainty of the parameters. This weighted average value of the parameters provides a better prediction about the underlying processes by reducing the uncertainty associated with the parameters and the bias in their selection. Better understanding of the parameters provides guidance for sampling and insights on experimental design.

5. Conclusion

The performance of the commonly used semivariogram models, used for fitting the experimental semivariogram, is comparable. Different models optimize a particular parameter differently, which indicates the uncertainly associated with the parameter. The uncertain parameter value makes the prediction of spatial processes difficult and uncertain. A model averaging procedure is used to obtain the parameter value and to reduce model parameter uncertainty.

Two soil properties (sand content and copper content) are taken as examples for demonstrating the average parameter estimation procedure. The sand content is measured at regular sampling interval along a transect, whereas the copper content is measured at variable sampling intervals in a two-dimensional field. The semivariogram is calculated for both properties and fitted with four most commonly used mathematical models (spherical, exponential, Gaussian and linear plateau). Weighted least square estimation is used for fitting these models to the experimental semivariogram. The goodness of fit for each model is calculated from their residual sum of squares. The parameter for each models are optimized during the fitting procedure. The likelihood of each model is calculated based on the Akaike Information Criterion (AIC) for each model. Different weights were assigned to each model based on their performance and importance from the AIC values. These weights are used for obtaining the weighted average of the optimized parameters. The weighted average of the estimated parameters reduced the uncertainty associated with the parameters and the bias in their selection. The average parameter values and reduced uncertainty provide more concise information about the spatial structure and consequently provide better guidance for sampling, experimental design, and interpolation and mapping.

Acknowledgements

The project was funded by a CSIRO Post Doctoral Fellowship and the University of Saskatchewan.

Author details

Asim Biswas[1,2*] and Bing Cheng Si[2]

*Address all correspondence to: asim.biswas@csiro.au

1 Department of Natural Resource Sciences, McGill University, Canada

2 Department of Soil Science, University of Saskatchewan, Saskatoon, Saskatchewan, Canada

References

[1] Akaike H (1973) Information theory as an extension of the maximum likelihood principle. In: Second International Symposium on Information Theory. Petrov, B.N., Csaki, F., Eds, Akademiai Kiado, Budapest, 267-281 p.

[2] Atteia O, Dubois JP, Webster R (1994) Geostatistical analysis of soil contamination in the Swiss Jura. Environ. poll. 86, 315-327.

[3] Burgess TM, Webster R (1980) Optimal interpolation and isarithmic mapping of soil properties. I. The semivariogram and punctual kriging. J. soil sci. 31, 315-331.

[4] Burnham KP, Anderson DR (2002) Model selection and multimodal inference: A Practical Information-Theoretic approach, Springer.

[5] Burrough PA (1983) Multiscale sources of spatial variation in soil. I. The application of fractal concepts of nested levels of soil variogram. J. soil. sci. 34: 577-597.

[6] Cambardella CA, Moorman TB, Novak JM, Parkin TB, Karlen DL, Turco RF, Konopka AE (1994) Field scale variability of soil properties in central Iowa soils. Soil sci. soc.am. j. 58, 1501-1511.

[7] Carvalho JRP, Silveira PM, Vieira SR (2002) Geostatistics in determining the spatial variability of soil chemical characteristics under different preparations. Pesq. agropec. bras. 37, 1151-1159.

[8] Corwin DL, Hopmans J, de Rooij GH (2006) From Field- to Landscape-Scale Vadose Zone Processes: Scale Issues, Modeling, and monitoring. Vadose zone j. 5, 129-139.

[9] Cressie N (1985) Fitting variogram models by weighted least squares. J. int. assoc. math. geol. 17, 563-586.

[10] Cressie NAC (1991) Statistics for spatial data. John Wiley & Sons, NY, 900 p.

[11] David M (1977) Geostatistical Ore reserve estimation. Elsevier, Amsterdam.

[12] Deutsch CV, Journel AG (1998) GSLIB Geostatistical software library and user's guide. 2nd eds. Oxford Uni. Press, NY.

[13] Fagroud M, Van Meirvenne M (2002) Accounting for soil spatial correlation in the design of experimental trails. Soil sci. soc.am. j. 66, 1134-1142.

[14] Foussereau X, Hornsby AG, Brown RB (1993) Accounting for variability within map units when linking a pesticide fate model to soil survey. Geoderma 60, 257-276.

[15] Gee GW, Bauder JW (1986) Particle size analyses. In: A. Klute (Eds.) Method of soil analyses. Part 1: Physical and Mineralogical Methods. ASA, Madison, Wisconsin, USA.

[16] Goderya FS (1998) Field scale variations in soil properties for spatially variable control: a review. J. soil contam. 7, 243-264.

[17] Goovaerts P (1997) Geostatistics for natural resources evaluation.Oxford Uni. Press., NY.

[18] Goovaerts P (1998) Geostatistical tools for characterizing the spatial variability of microbiological and physio-chemical soil properties. Biol. fertil.soils 27, 315-334.

[19] Hajrasuliha SW, Baniabassi N, Metthey J, Neilsen DR (1980) Spatial variability of soil sampling for salinity studies in South-West Iran. Irrig. sci. 1, 197-208.

[20] Heuvelink GBM, Pebesma EJ (1999) Spatial aggregation and soil process modelling. Geoderma 89, 47-65.

[21] Hoeting JA, Madigan D, Raftery AE, Volinsky CT (1999) Bayesian model averaging: a tutorial. Statist. sci. 14, 382-417.

[22] Isaaks EH, Srivastava RM (1989) An introduction to applied geostatistics, Oxford University Press, Toronto, Canada.

[23] Jian X, Olea RA, Yu Y (1996) Semivariogram modelling by weighted least squares. Comp. Geosci. 22, 387-397.

[24] Johnson JB, Omland KS (2004) Model selection in ecology and evolution. Trends ecol. evol. 19, 101-108.

[25] Lambert DM, Lowenberg-Debeor J, Bongiovanni R (2004) A comparison of four spatial regression models for yield monitor data: a case study from Argentina. Precis.agr. 5, 579-600.

[26] Matheron G (1962) Traité de Géostatistique Appliqué, Tome 1. Memoires du Bureau de RecherchesGéologiquesetMiniéres, Paris.

[27] McBratney AB, Bishop TFA, Teliatnikov IS (2000) Two soil profile reconstruction techniques. Geoderma 97, 209–221.

[28] McBratney AB, Webster R (1986) Choosing functions for semi-variograms of soil properties and fitting them to sampling estimates. J. soil sci. 37, 617-639.

[29] Nielsen DR, Biggar JW, Erh KT (1973) Spatial variation of field measured soil water properties. Hilgardia 42, 214-259

[30] Nielsen DR, Wendroth O (2003) Spatial and temporal statistics: sampling field soil and their vegetation. Catena Verlag, Reiskirchen, Germany.

[31] Pannatier Y (1996) VARIOWIN: Software for spatial data analysis in 2D. Springer-Verlag, NY.

[32] Papritz A, Dubois JP (1999) Mapping heavy metals in soil by (non-) linear kriging: an empirical validation. In: Gómez-Herńandez, J. et al. (Eds.) geoENV II-Geostatistics for environmental applications. Kluwer Academic Publ. Dordrecht, The Netherlands. 429-440 p.

[33] Prakash MR, Singh VS (2000) Network for ground water monitoring-a case study. Environ. geol. 39, 628-632.

[34] Press WH, Teukolsky SA, Vetterling WT, Flannery BP (1992) Numerical recipies in C. The art of scientific computing.Cambridge University press, NY.

[35] Si BC, Kachanoski RG, Reynolds WD (2007) Analysis of soil variability. In:E.G. Gregorich (Eds.) Soil sampling and methods of analysis. 1163-1191 p.

[36] Trangmar BB, Yost RS, Uehara G (1985) Application of geostatistics to spatial studies of soil properties.Adv. agro. 38, 45-94

[37] Van Kuilenburg J, De Gruijter JJ, Marsman BA, Bouma J (1982) Accuracy of spatial interpolation between point data on soil moisture supply capacity, compared with estimates from mapping units. Geoderma 27, 311-325.

[38] Vauclin M, Vieira SR, Vachaud G, Neilsen DR (1983) The use of cokriging with limited field soil observations. Soil sci. soc.am. j. 47, 175-184.

[39] Vieira SR (2000) Geostatistics in studies of spatial variability of soil. In: Novais, R.F. et al. (Eds.) Topics in soil science. Viçosa, Brazilian soc. soil sci. Vol 1, 1-54 p.

[40] Vieira SR, Millete J, Topp GC, Reynolds WD (2002) Handbook for geostatistical analysis of variability in soil and climate data. In: Alvarez, V. et al. (Eds.) Topics in soil science. Viçosa, Brazilian soc. soil sci. V.2.1-45 p.

[41] Webster R, McBratney AB (1989) On the Akaike Information Criterion for choosing models for variograms of soil properties. J. soil sci. 40, 493-496.

[42] Wilding LP, Bouma J, Goss D (1994) Impact of spatial variability on modeling. In: Bryant, R., Arnold, R.W., (Eds.) Quantitative modelling of soil forming processes. SSSA Special Publication #39. Soil Sci. Soc. Am. Inc. Madison, WI. 61-75 p.

[43] Yost RS, Uehara G, Fox RL (1982) Geostatistical analysis of soil chemical properties of large land areas. I. Semivariograms. Soil sci. soc.am. j. 46, 1028-1032.

[44] Zawadzaki J, Fabijanczyk P (2007) Use of variogramsfro field magnetory analysis in upper Silesia industrial area. Stud. geophys. geod. 51, 535-550.

[45] Zeleke TB, Si BC (2005) Scaling relationships between saturated hydraulic conductivity and soil physical properties. Soil sci. soc. am. j. 69, 1691-1702.

Chapter 5

Scale Dependence and Time Stability of Nonstationary Soil Water Storage in a Hummocky Landscape Using Global Wavelet Coherency

Asim Biswas and Bing Cheng Si

Additional information is available at the end of the chapter

http://dx.doi.org/10. 5772/52340

1. Introduction

Soil water is one of the most important limiting factors for semi-arid agricultural production and a key element in environmental health. It controls a large number of surface and subsurface hydrological processes that are critical in understanding a broad variety of natural processes (geomorphological, climatic, ecological) acting over a range of spatio-temporal scales (Entin et al. , 2000). Knowledge on the behavior of soil water storage and its spatio-temporal distribution provides essential information on various hydrologic, climatic, and general circulation models (Beven, 2001; Western et al. , 2002), weather prediction, evapotranspiration and runoff (Famigleitti and Wood, 1995), precipitation (Koster et al. , 2004) and atmospheric variability (Delworth and Manabe, 1993).

The distribution of soil water in the landscape is the response of a number of highly heterogeneous factors and processes acting in different intensities over a variety of scales (Goovaerts, 1998; Entin et al. , 2000). The heterogeneity in factors and processes make the spatial distribution of soil water highly heterogeneous in space and time and create a challenge in hydrology (Quinn, 2004). Therefore, a large number of samples are needed in order to characterize the field averaged soil water with certain level of precision. However, if a field or watershed is repeatedly surveyed for soil water, some sites or points are consistently wetter or consistently drier than the field average. Vachaud et al. (1985) were the first to examine the similarity of the spatial pattern in soil water storage over time and termed this phenomenon time stability. The time stability is defined as a time invariant association between spatial location and classical statistical measures of soil water, most often the mean (Grayson and Western, 1998). Authors used the Spearman's rank correlation to explain the similarity in the overall spatial patterns between two measurement series and the cumulative probability function of relative mean differences to examine the rank similarity of the individual locations over time [Vachaud et al. , 1985]. Various authors have used this concept to examine the similarity between the spatial patterns of soil water storage over a range of investigated area, sampling scheme, sampling depth, investigation period and land use (Kachanoski and de Jong, 1988; Grayson and Western, 1998; Hupet and Vanclooster, 2002; Tallon and Si, 2004; Martínez-Fernández and Ceballos, 2005; Starks et al. , 2006; Cosh et al. , 2008; Hu et al. , 2010b). However, information on the similarity between the spatial patterns of soil water within a season (intra-season), between seasons (inter-season), or within a season of different years (inter-annual) is not very common (Biswas and Si, 2011a).

Kachanoski and de Jong (1988) used the spatial coherency analysis to identify the similarity of the scales of the spatial patterns of soil water distribution over time and named the phenomena temporal persistence. Their study indicated loss of time stability at the scale < 40 m during the recharge period, which was attributed to topography. The spatial coherency analysis is based on the spectral analysis (Jenkins and Watts, 1968; Kachanoski and de Jong, 1988), which approximates the spatial data series by a sum of sine and cosine functions. Each function has an amplitude and a frequency or period. While the squared amplitude represents the variance contribution, the frequency component can be used to represent the spatial scale of ongoing processes (Webster, 1977; Shumway and Stoffer, 2000; Brillinger, 2001). The spectral analysis or frequency domain analysis is based on the assumption of second order stationarity (i. e. the mean and the variance of the series are finite and constant). However, more often than not, the soil spatial variation is nonstationary. Nonstationarity in the spatial distribution of soil water storage was also mentioned by Kachanoski and de Jong (1988). Nonstationarity restricts direct application of spatial coherency analysis to examine the similarity in the spatial patterns of soil water storage at different scales or scale-specific time stability, which calls for a new method.

Wavelet analysis can deal with localized features and thus nonstationarity by partitioning the spatial variations into locations and frequencies (Lark and Webster, 1999; Grinsted et al. , 2004; Si and Farrell, 2004, Yates et al., 2006; Biswas et al. , 2008), therefore providing an opportunity to study the spatial variation in soil water storage at multiple scales. While, the global wavelet analysis can deal with the scale specific variations, the global wavelet coherency analysis elucidates the scale specific correlation between any two spatial series. Therefore, the global wavelet coherency can be used to examine the similarity in the spatial patterns of soil water storage measured at two different times at multiple scales and study the scale-specific time stability. The objective of this study was to examine the scales of time stability of nonstationary soil water storage at different seasons in a hummocky landscape using the global wavelet coherency.

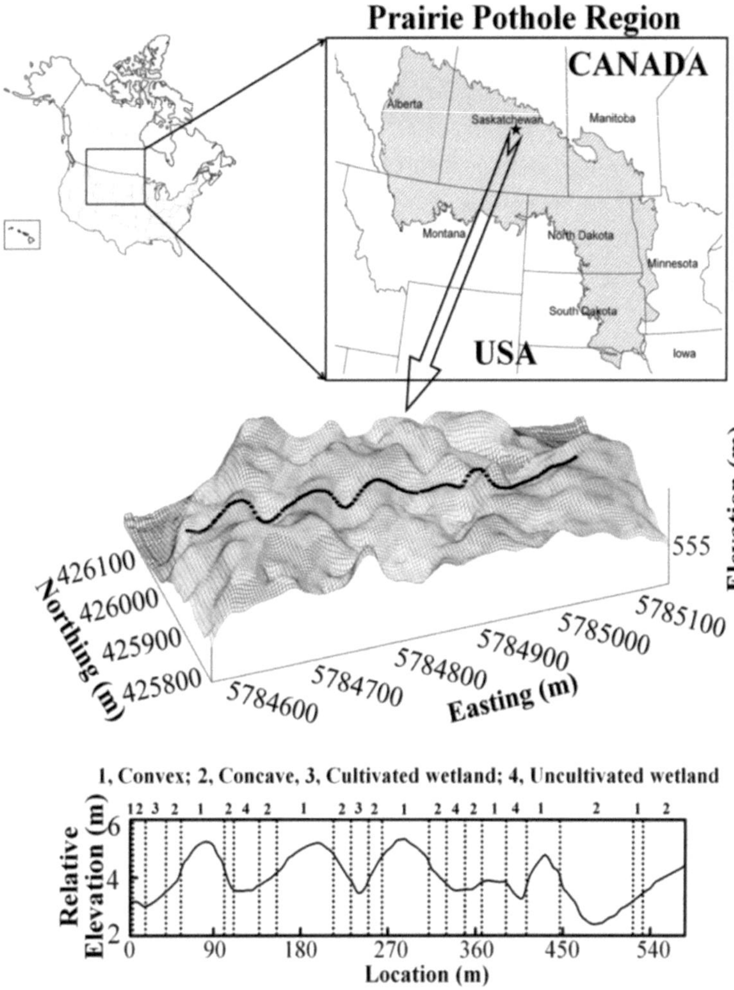

Figure 1. Geographic location of the study site at St Denis National Wildlife Area within Prairie Pothole Region of North America along with the 3-dimensional and cross sectional view of the transect and different landform elements along the transect. CX indicates convex, CV indicates concave, CW indicates cultivated wetlands and UW indicates uncultivated wetlands

2. Theory

Wavelet analysis (Mallat, 1999) is used to divide a spatial series into different frequency components and study each component using a fully scalable window or wavelet. It calculates localized variations by shifting the standard function (mother wavelet) along the spatial series. The detail theory of the wavelet analysis is available elsewhere (Farge, 1992; Kumar and Foufoula-Georgiou, 1993, 1997; Torrence and Compo, 1998) and is beyond the scope of this chapter. There are different types of wavelet transform including discrete wavelet transform (DWT), continuous wavelet transform (CWT), wavelet packet transform (WPT), maximal overlap discrete wavelet transform (MODWT). These are suite of tools and can be used for certain purposes with some advantages and disadvantages. In this study, we use the continuous wavelet transform (CWT), where the wavelet coefficients at consecutive scales and locations can carry common information and provide a redundant representation of the signals information content and thus the detailed scale information (Farge, 1992; Lau and Weng, 1995; Keitt and Fischer, 2006; Furon et al., 2008). The detailed theory of the CWT can be found in various text books including Mallat (1998) and Chui (1992). Briefly, the CWT for a spatial series (Y_i) of length N (i= 1, 2, ...,N) with an equal sampling interval of δx, can be defined as the convolution of Y_i with the scaled (s) and translated (x) wavelet (Torrence and Compo, 1998). Wavelet coefficients, $W_i^Y(s)$ can be calculated as

$$W_i^Y(s) = \sqrt{\frac{\delta x}{s}} \sum_{j=1}^{N} Y_j \psi\left[(j-i)\frac{\delta x}{s}\right] \tag{1}$$

where $\psi[\ ]$ denotes wavelet function. Out of many wavelet functions, the Morlet wavelet was used in this study because of enhanced spatial and frequency resolution. Morlet wavelet can be represented as (Torrence and Compo, 1998)

$$\psi(\eta) = \pi^{-1/4} e^{i\varpi\eta - 0.5\eta^2} \tag{2}$$

where, i is the complex number and equal to $\sqrt{-1}$, ω is the dimensionless frequency and η is the dimensionless space. The imaginary part conserved in the wavelet transform with Morlet wavelet can be used to identify the dominant orientation of variations in a random field. The energy associated with a scale and location can be measured from the magnitude of the wavelet coefficient. The wavelet power spectrum can be defined as $|W_i^Y(s)|^2$, which is the space-frequency-energy representation of a spatial series. The global wavelet spectrum is the average of local wavelet spectra over all locations and is given by,

$$\bar{W}^Y(s) = \frac{1}{N} \sum_{n=0}^{N-1} \left|W_n^Y(s)\right|^2 \tag{3}$$

Similarly, the global wavelet spectra of another spatial series Z will be

$$\bar{W}^Z(s) = \frac{1}{N} \sum_{n=0}^{N-1} \left|W_n^Z(s)\right|^2 \tag{4}$$

The cross wavelet spectra between two spatial series Y and Z can be calculated as

$$W_i^{YZ} = W_i^Y W_i^{Z*} \tag{5}$$

where, $W_i^Z *$is the complex conjugate of W_i^Z. The global cross wavelet spectra can be calculated as

$$\bar{W}^{YZ}(s) = \frac{1}{N}\sum_{n=0}^{N-1}\left|W_n^{YZ}(s)\right| \tag{6}$$

While the global wavelet cross spectra are similar to the covariances in the spatial domain, the global wavelet coherency spectra are similar to the coefficients of determination in the spatial domain for two variables. The global wavelet coherency spectra can be calculated as

$$R^2(s) = \frac{\left|S\left(\bar{W}^{YZ}(s)\right)\right|^2}{S\left(\bar{W}^{Y}(s)\right)S\left(\bar{W}^{Z}(s)\right)} \tag{7}$$

where, S is the smoothing operator, $S(\overline{W}^{YZ}(s))$ is the smoothed global cross wavelet spectra of spatial series Y and Z, $S(\overline{W}^{Y}(s))$ and $S(\overline{W}^{Z}(s))$ are the smoothed global wavelet spectra of the spatial series Y and Z, respectively. In calculating wavelet coherency, it is necessary to smooth global cross wavelet spectra beforehand; otherwise, it will always be equal to 1 (Torrence and Compo, 1998; Maraun and Kurths, 2004). The coherency should be calculated on expected values. However, in most cases, there is only one realization of a spatial series, thus a coherency value has only one degree of freedom. By smoothing the coherency, one can overcome this problem and increase the degrees of freedom. In this study, we have used a boxcar window of size 5 (5 sample point average) to smooth the global wavelet and cross wavelet spectra.

Like the coefficient of determination, the global wavelet coherency spectra (R^2) range from 0 to 1 and measure the correlation between two spatial series at each scale or within a particular frequency band. The closer the coherency values to one, the more similar the spatial patterns at a particular frequency or scale (= sampling interval / frequency).

The significance test for the wavelet coherency spectra can be carried out by calculating the confidence interval from an assumed theoretical distribution (Koopmans, 1974). However, the cutoff points for the test of hypothesis $R^2 = 0$ vs. $R^2 > 0$ can be conducted for $s \neq 0$ from the F distribution (Koopmans, 1974):

$$R^2(s) = \frac{F_{2,N-2}(\alpha)}{n-1+F_{2,N-2}(\alpha)} \tag{8}$$

or

$$R^2(s) = 1-(1-\alpha)^{\frac{1}{2m}} \tag{9}$$

where, α is the significance level and $2m + 1$ is the width of the boxcar window. Therefore, m is the number of terms in each symmetrical half of the boxcar window. If the calculated coherency $\hat{R}^2(s)$ is greater than the theoretical value $R^2(s)$ at a particular scale (s), then the calculated coherency is significantly different from zero at the specified α. In this study, we have used $m = 2$, therefore the cutoff point at $\alpha = 0.99$ is 0.684.

3. Materials and methods

A field experiment was conducted at St Denis National Wildlife Area, (52°12′ N latitude, 106°50′ W longitude), which is approximately 40 km east of Saskatoon, Saskatchewan, Canada (Fig. 1). A detailed information on the study site, soil water measurement and calibration of measurement instruments can be found in Biswas et al. (2012) and Biswas and Si (2011a, b). Briefly, the landscape of the study site is hummocky with a complex sequence of slopes (10 to 15%) extending from different sized rounded depression to complex knolls and knobs (Pennock, 2005) and is typical of the Prairie Pothole Region of North America (Fig. 1). The dominant soil type of the study site is Dark Brown Chernozem (Mollosiol in USDA soil taxonomy), which is developed from moderately fine to fine textured, calcareous, glacio-lacustrine deposits and modified glacial till (Saskatchewan Centre for Soil Research, 1989). The climate of the study area is semi-arid with the mean annual air temperature of 2°C and the mean annual precipitation of 360 mm, of which 84 mm occurs as snow during winter (AES, 1997). The annual precipitation of the site during 2006, 2007, 2008, and 2009 were 489 mm, 366 mm, 331 mm, and 402 mm, respectively (Fig. 2). Year 2010 received 645 mm rainfall only during the spring and summer months (April to September), which is almost double the long-term average annual precipitation (Environment Canada, 2011).

Figure 2. Monthly distribution of total precipitation in the year of 2006-09 along with the long term normal (90 year average)

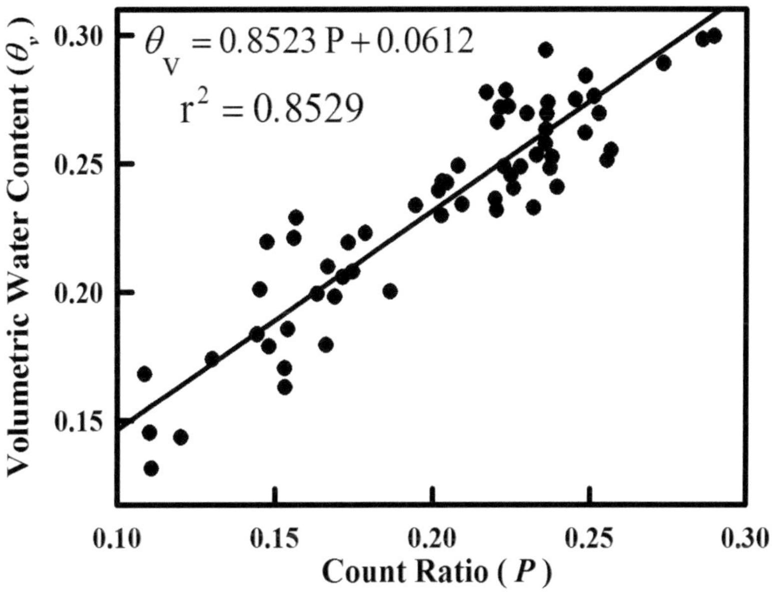

Figure 3. Site specific neutron probe calibration equation completed over three year time (2007-09). P indicates the ratio of the actual neutron count to the standard neutron count.

Soil water storage was measured along a transect of 576 m long with equally spaced 128 points (4. 5 m sampling interval). The transect was established over several knolls and seasonal depressions representing different landform cycles (Fig. 1). Topographic survey of the site was completed using Light Detection and Ranging (LiDAR) survey of the study area at 5 m ground resolution. Different landform elements were also identified as convex (CX), concave (CV), cultivated wetlands (CW) and uncultivated wetland (UW) (Fig. 1). The vegetation of the study site was mixed grass including *Agropyronelongatum, Agropyronintermedium, Bromusbiebersteinii, Elymusdauricus, Festucarubra, Onobrychisviciifolia, Elymuscanadensis, Agropyrontrachycaulum*and *Medicago sativa,* which was seeded in 2004 and allowed to grow every year. Surface 0-20 cm soil water was measured using time domain reflectometry (TDR) probe and a metallic cable tester (Model 1502B, Tektronix, Beaverton, OR, USA). A neutron probe (Model CPN 501 DR Depthprobe, CPN International Inc. , Martinez, CA, USA) was used to measure the soil water down to 140 cm at 20 cm vertical intervals. Soil cores at selected locations within 1 m around the neutron access tube were taken at different moisture conditions and the soil water content of each 10 cm interval were determined by gravimetric methods. The volumetric water content (gravimetric water content × bulk density) and the neutron counts were used to calibrate the neutron probe. The resulting calibration equation (θ_v = 0. 8523 P + 0. 0612 with n = 101 and r^2 = 0. 86, where P is the ratio of neutron count to standard neutron count) was used to convert the neutron probe count ratio to volumetric soil water at different depths and different locations (Fig. 3). Because neutron probe is prone to error for surface soil water measurements, the average soil water content at the surface 20-cm layer was measured using vertically installed time domain reflectometry probe and a metallic cable tester (Model 1502B, Tektronix, Beaverton, OR, USA). A standard calibration equation (θ_v=0.115$\sqrt{k_a}$−0.176, where k_a=$[L_2 / L]^2$ is the dielectric constant, L_2 is the distance between the arrival of signal reflected from the probe-to-soil interface and the signal reflected from the end of the probe curves (measured from waveform) and L is the length of the TDR probe) following Topp and Reynolds (1998) was used to derive the water content from the TDR recordings. Soil water content was measured 25 times at different seasons during a year over a five-year period (2007 – 2011). Based on the season, the measurements were divided into three groups: spring, summer, and fall. Though the analysis was completed for all the measurements in all the years, the space restriction in this chapter and the for demonstration purposes, only the result from 2008 and 2009 were used. The results from these years were very similar to the results from other years and can be generalized over the measurement period.

Wavelet analysis was completed using the MATLAB (The MathWorks Inc.) code written by Torrence and Compo (1998) and is available online at http://paos. colorado. edu/research/ wavelets/. The graphs were prepared in SigmaPlot (Systat Software Inc.).

4. Results

High water storage was found at the locations of 100 to 140 m and 225 to 250 m from the origin of the transect, which were situated within depressions (Fig. 4). On contrary, the knolls stored less water. However, the difference of stored water in depressions and on knolls reduced from spring to fall within a year. For example, the range was 43. 42 cm for 20 April 2009 during spring and was 20. 81 cm for 27 October 2009 during fall season (Fig. 4). Apparently, the mean and variance of the first half of transect was quite different from that of the second half. It is evident that soil water along the transect is non-stationary.

Spearman's rank correlation coefficients were used to examine the similarity of the overall spatial pattern. High rank correlation coefficients between any two-measurement series indicated time stability of overall spatial pattern of soil water storage. There was very strong intra-season time stability. For example, the rank correlation coefficient was 0. 98 between the measurement series on 2 May 2008 and 31 May 2008 (spring) and was 0. 99 between the measurement series on 23 August 2008 and 17 September 2008 (summer). Similarly, there was also strong inter-annual time stability. For example, the rank correlation coefficient was 0. 96 between the measurement series on 2 May 2008 (spring) and 7 May 2009 (spring) and was 0. 97 between the measurement series on 22 October 2008 (fall) and 27 October 2009 (fall). However, a relatively low rank correlation coefficient was observed between the measurement series from two different seasons. For example, the rank correlation coefficient was 0. 89 between the measurement series on 31 May 2008 (spring) and 23 August 2008 (summer), and 0. 85 between the measurement series on 31 May 2008 (spring) and 22 October 2008 (fall). However, the correlation coefficient was 0. 99 between the measurement series on 23 August 2008 (summer) and 22 October 2008 (fall) indicating strong similarity between summer and fall measurements. The correlation coefficients gradually decreased with the increase in time between measurements indicating the decrease in the degree of time stability over time. For example, the rank correlation coefficient was 0. 97 between the measurement series on 20 April 2009 and 7 May 2009, which gradually decreased to 0. 82 between the measurement series on 20 April 2009 and 27 October 2009.

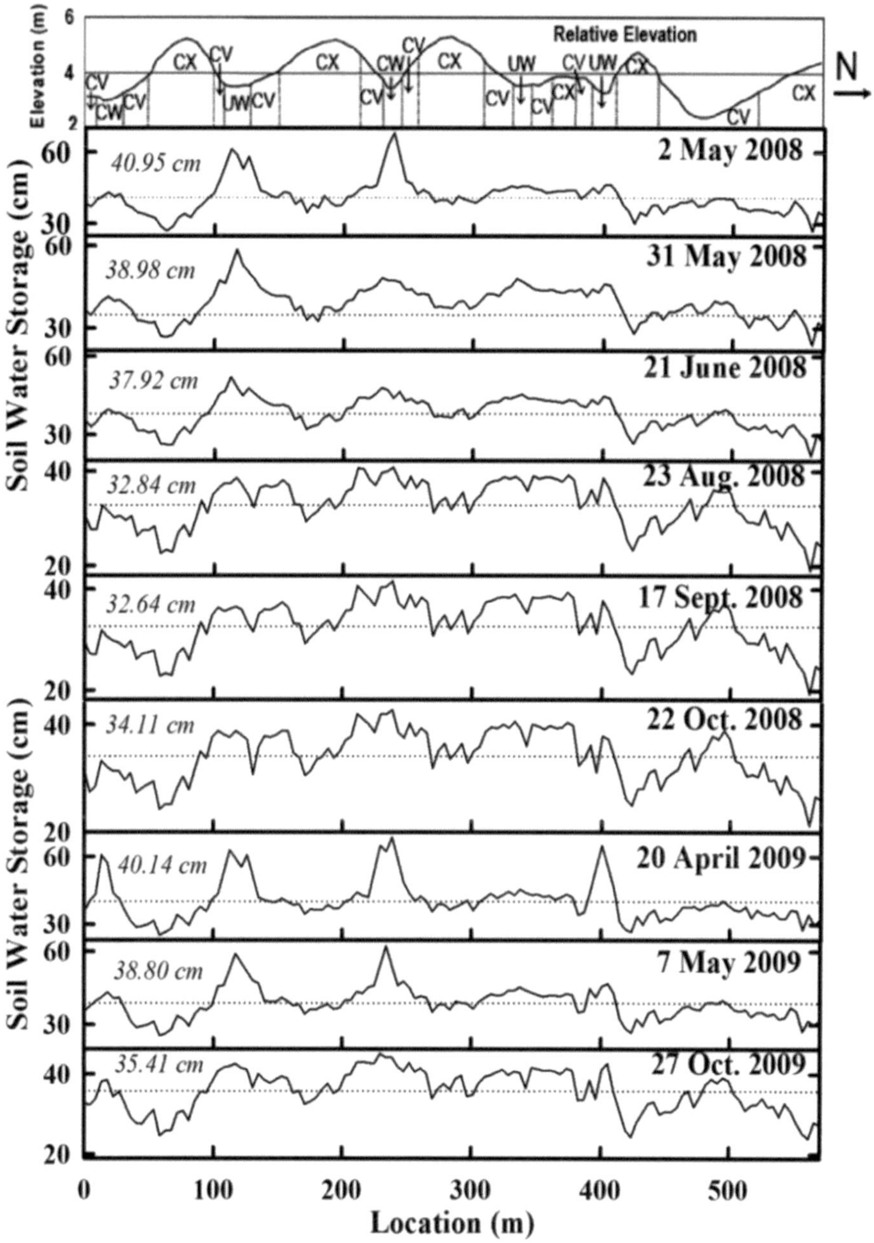

Figure 4. Spatial distribution of selected soil water storage series along the transect. The value in *italics* presents the average soil-water storage.

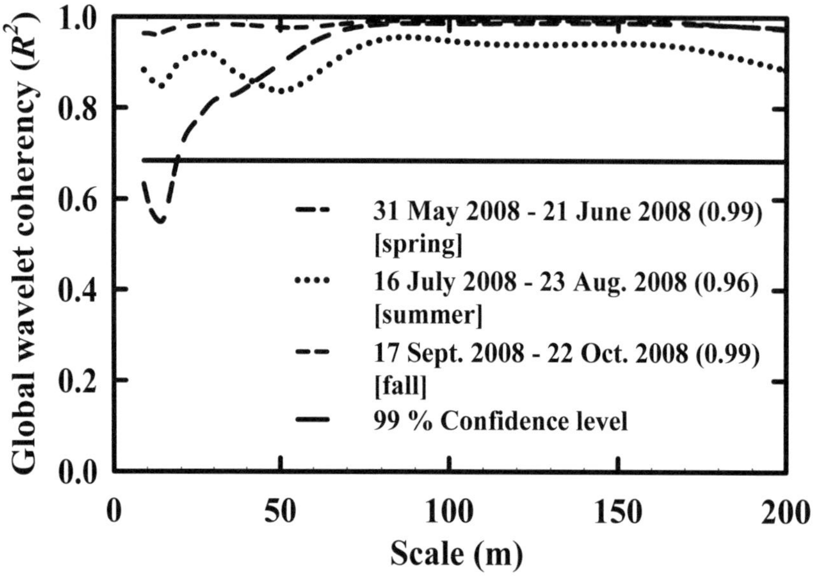

Figure 5. Global wavelet coherency spectra of intra-season (spring, summer and fall of 2008) time stability. Values in the parentheses are the rank correlation coefficients.

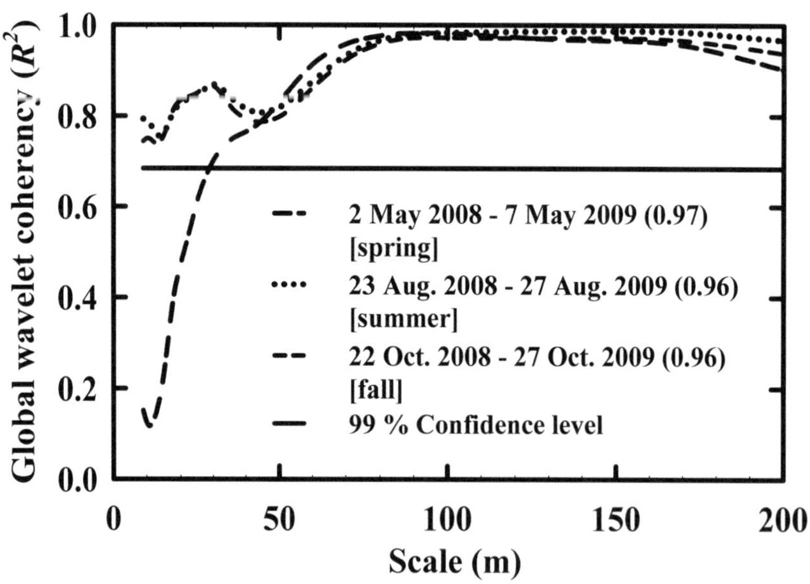

Figure 6. Global wavelet coherency spectra of inter-annual (spring, summer, and fall of 2008 and 2009) time stability. Values in the parentheses are the rank correlation coefficients.

Because of the nonstationarity of soil water along the transect, the scale specific similarity in the spatial pattern of soil water storage was examined using global wavelet coherency. There was strong intra-season time stability during summer or fall compared to spring (Fig. 5). Statistically significant strong coherency at all scales during summer and fall indicated that the spatial patterns present at different scales in summer were also present in fall. Similarly, large significant coherency between the measurement series of spring 2008 and spring 2009 or summer 2008 and summer 2009 indicated strong inter-annual time stability (Fig. 6). However, non-significant coherency at the scales < 20 m in the spring indicated the loss of intra-season time stability (Fig. 5). Similarly, there was loss of inter-annual time stability between springs of different years at the scales < 30 m (Fig. 6). However, the time stability was lost at scales < 65 m between the spring and summer and < 70 m between the spring and fall measurement series (inter-season) (Fig. 7). There was strong time stability between the summer and fall at all scales (Fig. 7). The minimum scale of statistically significant coherency was the lowest within a season and gradually increased with the increase in time between measurements (Fig. 8). For example, the minimum scale of significant coherency was 25 m between the measurement series on 20 April 2009 and 7 May 2009 and was 40 m between the measurement series on 20 April 2009 and 27 May 2009. The correlation between the measurement series on 20 April 2009 and 27 October 2009 was not significant at almost all scales except from 80 to 120 m (Fig. 8).

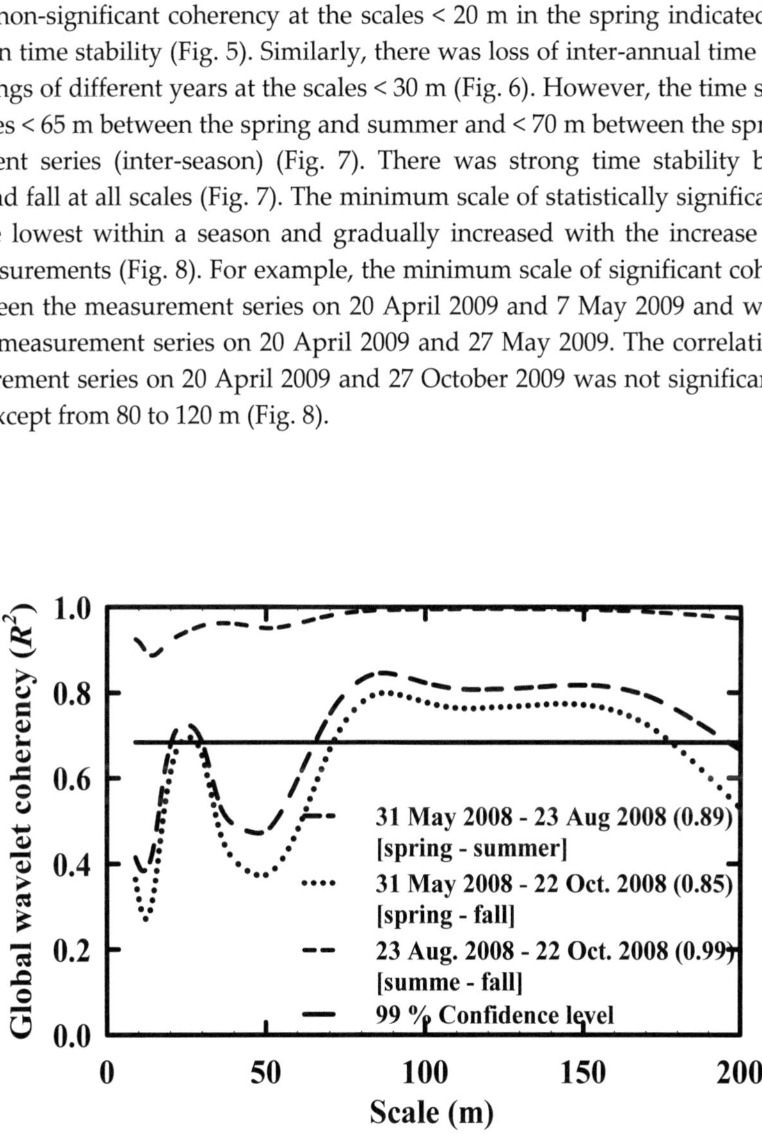

Figure 7. Global wavelet coherency spectra of inter-season (spring-summer, spring-fall, and summer-fall of 2008) time stability. Values in the parentheses are the rank correlation coefficients.

5. Discussions

In our study area, depressions receive snowmelt runoff water from the surrounding uplands and store more water compared to knolls during the spring (Gray et al. , 1985; Winter and Rosenberry, 1995). In addition, uneven distribution of drifting snow in the landscape also contributes to the high water storage in depressions (Woo and Rowsell, 1993; Hayashi et al. , 1998; Lungal, 2009). Therefore, the alternate knolls and depressions along the transect created a spatial pattern in soil water storage inverse to the spatial pattern of elevation (Fig. 4). This spatial pattern with topography persisted through summer until fall (Fig. 4) as the depressions always stored more water than knolls. However, the variable demand of evapotranspiration reduced the difference in the maximum and minimum soil water storage (range) over time within a year. This phenomenon repeated every year.

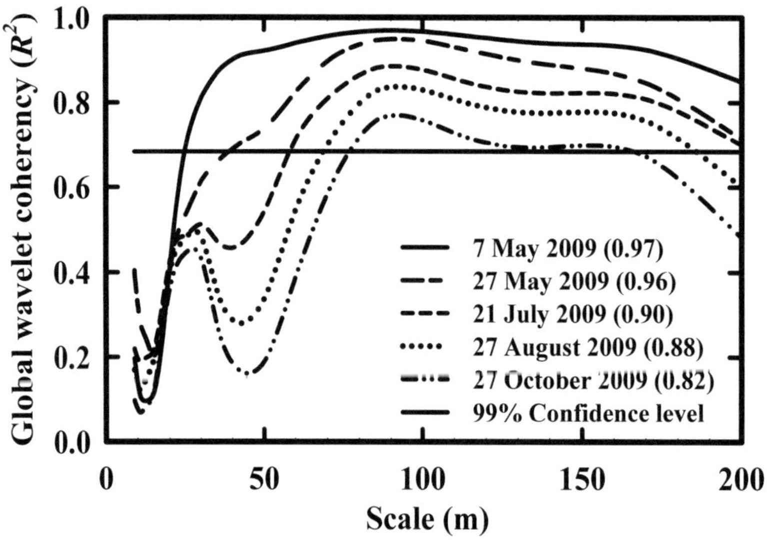

Figure 8. Global wavelet coherency spectra between the soil water measurement on 20 April 2009 and the measurements at different time within the year 2009. Values in the parentheses are the rank correlation coefficients.

At the absence of vegetation during spring, the soil water is lost mainly through surface evaporation, and to a lesser extent, the ground water interaction (Hayashi et al. , 1998; van der Kamp et al. , 2003). Evaporation may be higher in south-facing slope than in north-facing slope, but the difference in evaporation due to aspects may not be able to diminish the spatial patterns of soil water storage due to nonlocal controls (e. g. , macro-topography: knolls and depression) at large scales (Kachanoski and de Jong, 1988; Grayson et al. , 1997). However, the spatial patterns created from the micro-topography were not strong enough to dominate the differential evaporation created from the difference in local controls (e. g. , surface roughness, soil texture). Therefore, there was a loss of intra-season time stability at small scales during the spring. The loss of time stability at small scales was also observed between the spatial series from the spring of different years (Fig. 6). However, the influence of these small-scale processes was not strong enough to change the overall spatial pattern. Therefore, a strong intra-season and inter-annual time stability was observed in the overall spatial patterns (identified from Spearman's rank correlation coefficients) irrespective of moisture conditions in different seasons (Martínez-Fernández and Ceballos, 2003).

However, with the establishment of vegetation, runoff from rainfall became the rare event and therefore the macro-topography was not able to reestablish the spatial patterns weakened by differential water uptake by vegetation (e. g. , more water uptake in depression than on knolls). Though macro-topography is still important, it is the interactions between vegetation and macro-topography that determined the spatial patterns of soil water. Because the interactions were relatively similar in later summer and early fall, the spatial patterns of soil water storage were very similar in summer and fall at all scales (Fig. 5). The intra-season and inter-annual time stability of the overall and the scale specific spatial patterns at all seasons contradicted the results of various authors who mentioned that the wet season showed strong time stability (Gómez-Plaza et al. , 2000; Qiu et al. , 2001; Hupet and Vanclooster, 2002), while others pointed out to the dry season (Robinson and Dean, 1993; Famiglietti et al. , 1998).

In addition, the processes controlling the spatial pattern of soil water storage may not change abruptly. For example, the vegetation growth is gradual and so is the evapotranspiration demand. In our study, large coherency between the measurement series on 20 April 2009 and 7 May 2009 indicated strong intra-season time stability. However, with the increase in time difference between measurements relatively weaker inter-season time stability was resulted (Fig. 8). This result contradicted the findings of Grayson et al. (1997), who indicated that the local controls (e. g. , vegetation, soil texture) are dominant in dry season (evapotranspiration > precipitation), while nonlocal controls (e. g. , topography) dominate redistribution of water during wet period or moisture surplus conditions (evapotranspiration < precipitation). However, the consistent coherency with reduced magnitude in our study indicated the change in the degree of the same control, but not a switch to different controls.

This study was different from Kachanoski and de Jong (1988), who used spatial coherency analysis to identify the scale specific similarity of the spatial patterns of soil water storage. The spatial coherency analysis assumed that the data series was stationary. However, authors identified the nonstationary nature of soil water storage (Kachanoski and de Jong, 1988) and divided the transect to create piecewise stationary series. Conversely, the global wavelet coherency analysis was able to deal with nonstationary soil water series. The wavelet analysis is well established to deal with nonstationarity. In addition, the conclusion of Kachanoski and de Jong (1988) was based on one-year measurement of soil water, which may not be universal as the precipitation variability over years may create a different experimental situation in different years. In this study, we have confirmed our conclusion based on five years of measurement of soil water storage.

Time stability is a result of multiple factors. Due to difference in the intensity of different factors, time stability of soil water and its scale dependence can be different. Instead of over-generalizing time stability, we classified time stability into intra-season, inter-season, and inter-annual time stability, because of the similar intra-season and inter-annual hydrological processes, but different inter-season hydrological processes. Therefore, the conclusion of this study may lead to improved prediction of soil water from reduced number of monitoring sites, thus allowing improved runoff and stream flow prediction in scarcely gauged basins.

6. Summary and conclusions

The similarity in the overall spatial pattern of soil water storage was first examined by Vachaud et al. (1985) and termed as the time stability of the spatial pattern. Kachanoski and de Jong (1988) extended the concept of time stability to the scale dependence of time stability using spatial coherency analysis. However, the stationarity assumption of the spatial coherency analysis restricts the use of this method for nonstationary spatial series. We have used global wavelet coherency analysis to examine the scale dependence of intra-season, inter-season and inter-annual time stability of nonstationary soil water spatial patterns.

There was strong intra-season time stability of the overall and scale specific spatial pattern. The time stability was lost at the scales < 20 m within the spring and < 30 m between the spring measurements from different years. However, strong time stability was present at all scales during the summer and fall, when the high evapotranspiration demand created similar spatial patterns. Similar processes in the summer and fall resulted strong inter-season time stability. However, not so similar processes in spring created weaker inter-season time stability between the spring and summer or the spring and fall. There was loss of time stability at the scales < 65 m and < 70 m between the spring and summer and the spring and fall, respectively. However, the change in the scales of time stability was not abrupt; rather it gradually decreased with the increase of time difference between measurements. The change in the similarity of the spatial patterns of soil water storage over time at different scales is an indicative of the change in the hydrological processes operating at those scales. Therefore, the analysis outcome can be used to identify the change in the sampling domain as controlled by the hydrological processes operating at different scales delivering the maximum information with minimum sampling effort.

Acknowledgments

The project was funded by a CSIRO Post Doctoral Fellowship and the University of Saskatchewan. Authors appreciate help from Amanda, Danny, Jason, Khizir and Henry in field data collection.

Author details

Asim Biswas[1,2*] and Bing Cheng Si[2]

*Address all correspondence to: asim.biswas@mcgill.ca

1 Department of Natural Resource Sciences, McGill University, Canada

2 Department of Soil Science, University of Saskatchewan, Saskatoon, Saskatchewan, Canada

References

[1] AES (1997) Canadian daily climate data for Western Canada. In [CD-ROM]. Atmospheric Environment Service, Environment Canada, Downsview, ON.

[2] Beven K (2001) How far can we go in distributed hydrological modeling? Hydrol. earth sys. sci. 5: 1–12.

[3] Biswas A Si, BC (2011a). Scale and locations of time stability of soil water storage in a hummocky landscape. J hydrol. 408: 100-112.

[4] Biswas A, Chau HW, Bedard-Haughn AK, Si BC (2012) Factors controlling soil water storage in the Hummocky landscape of the prairie pothole region of North America. Can. j. soil sci. (in Press), 92, 649-663.

[5] Biswas A, Si BC (2011b) Identifying scale specific controls of soil water storage in a hummocky landscape using wavelet coherency analysis. Geoderma165: 50-59.

[6] Biswas A, Si BC (2011c) Revealing the controls of soil water storage at different scales in a Hummocky landscape. Soil sci. soc. am. j. 75: 1295–1306.

[7] Biswas A, Si BC, Walley FL (2008) Spatial relationship between δ15N and elevation in agricultural landscapes. Nonlinear proc. geoph. 15: 397–407.

[8] Brillinger, D. R. (2001). Time series: Data analysis and theory. Soc. Ind. Appl. Math. Philadelphia, PA. 540 p.

[9] Chui, C. K. (1992). An introduction to wavelets. Academic Press, New York, NY.

[10] Cosh MH, Jackson TJ, Moran S, Bindlish R (2008) Temporal persistence and stability of surface soil moisture in a semi-arid watershed. Remote sens. environ. 112: 304–313.

[11] Delworth T, Manabe S (1993) Climate variability and land-surface processes. Adv. water resour. 16: 3–20.

[12] Entin JK, Robock A, Vinnikov KY, Hollinger SE, Liu S, Namkhai A (2000) Temporal and spatial scales of observed soil moisture variations in the extra tropics. J. geophys. res. 105: 11865–11877.

[13] Environment Canada (2011) National climate data and information archive. Available at www. climate. weatheroffice. gc. ca. (Verified on 25/02/2012). Environ. Can. , Fredericton, NB.

[14] Famiglietti JS, Rudnicki JW, Rodell M (1998) Variability in surface moisture content along a hill slope transect: Rattlesnake Hill, Texas. J. hydrol. 210: 259–281.

[15] Famiklietti, J. S. , & Wood, E. F. (1995). Effects of spatial variability and scale on aerially-averaged evapotranspiration. Water resour. res. 31: 699–712.

[16] Farge M (1992) Wavelet transform and their applications to turbulence. Annul. rev. fluid. mech. 24: 395–457.

[17] Furon AC, Wagner-Riddle C, Smith CR, Warland JS (2008) Wavelet analysis of wintertime and spring thaw CO2 and N2O fluxes from agricultural fields. Agric. forest meteor. 148: 1305–1317.

[18] Gómez-Plaza A, Alvarez-Rogel J, Alabaladejo J, Castillo VM (2000) Spatial patterns and temporal stability of soil moisture a range of scales in a semi-arid environment. Hydrol. proc. 14: 1261–1277.

[19] Goovaerts P (1998) Geostatistical tools for characterizing the spatial variability of microbiological and physico-chemical soil properties. Biol. fertil. soils 27: 315–334.

[20] Gray, D. M., Landine, P. G., & Granger, R. J. (1985). Simulating infiltration into frozen prairie soils in streamflow models. Can. j. earth sci. 22: 464–472.

[21] Grayson, R. B., & Western, A. W. (1998). Towards areal estimation of soil water content from point measurements: time and space stability of mean response. J. hydrol. 207: 68–82.

[22] Grayson RB, Western AW, Chiew FHS, Blöschl G (1997) Preferred states in spatial soil moisture patterns: Local and nonlocal controls. Water resour. res. 33: 2897–2908.

[23] Grinsted AJ, Moore C, Jevrejeva S (2004) Application of the cross wavelet transform and wavelet coherence to geophysical time series. Nonlin. proc. geoph. 11: 561–566.

[24] Hayashi M, van der Kamp G, Rudolph DL (1998) Water and solute transfer between a prairie wetland and adjacent uplands, 1. Water balance. J. hydrol. 207: 42–55.

[25] Hu W, Shao MA, Han F, Reichardt K, Tan J (2010). Watershed scale temporal stability of soil water content. Geoderma doi:10. 1016/j. geoderma. 2010. 04. 030.

[26] Hupet F, Vanclooster M (2002) Intraseasonal dynamics of soil moisture variability within a small agricultural maize cropped field. J. hydrol. 261: 86–101.

[27] Jenkins, G. M., & Watts, D. G. (1968). Spectral analysis and its application. Holden-Day, San Francisco, CA.

[28] Kachanoski RG, de Jong E (1988) Scale dependence and the temporal persistence of spatial patterns of soil-water storage. Water resour. res. 24: 85–91.

[29] Keitt TH, Fischer J (2006) Detection of scale-specific community dynamics using wavelets. Ecology 87: 2895–2904.

[30] Koopmans, L. H. (1974). The Spectral Analysis of Time Series. Academic Press, New York.

[31] Koster RD, Dirmeyer PA, Guo Z, Bonan G, Chan E, Cox P, Gordon CT, Kanae S, Kowalczyk E, Lawrence D, Liu P, Lu CH, Malyshev S, McAvaney B, Mitchell K, Mocko D, Oki T, Oleson K, Pitman A, Sud YC, Taylor CM, Verseghy D, Vasic R, Xue Y, Yamada T (2004) Regions of strong coupling between soil moisture and precipitation. Science 305: 1138–1140.

[32] Kumar P, Foufoula-Georgiou E (1993) A multicomponent decomposition of spatial rainfall fields: 1. Segregation of large- and small-scale features using wavelet transforms. Water resour. res. 29: 2515–2532.

[33] Kumar P, Foufoula-Georgiou E (1997) Wavelet analysis of geophysical applications. Rev. geoph. 35: 385–412.

[34] Lark RM, Webster R (1999). Analysis and elucidation of soil variation using wavelets. Eur. j. soil sci. 50: 185–206.

[35] Lau KM, Weng H (1995) Climate signal detection using wavelet transform: How to make a time series sing. Bull. am. meteor. soc. 76: 2391–2402.

[36] Lungal, M. A. (2009). Hydrological response to spring snowmelt and extreme rainfall events of different landscape elements within a Prairie wetland basin. Master's Thesis, Univ. of Saskatchewan, Saskatoon, SK, Canada.

[37] Mallat S (1999) A wavelet tour of signal processing. Second edition. Academic Press, New York, NY.

[38] Maraun D, Kurths J (2004) Cross wavelet analysis: Significance testing and pitfalls, Nonlin. proc. geophy. 11: 505-514.

[39] Martínez-Fernández J, Ceballos A (2003) Temporal stability of soil moisture in a large-field experiment in Spain. Soil sci. soc. am. j. 67: 1647–1656.

[40] Martínez-Fernández J, Ceballos A (2005) Mean soil moisture estimation using temporal stability analysis. J. hydrol. 312: 28–38.

[41] Pennock, D. J. (2005). Field Handbook for Saskatchewan soils. Dept. of Soil Sci. , Univ. of Saskatchewan, SK. 4-5 p.

[42] Qui Y, Fu BJ, Wang J, Chen LD (2001) Soil moisture variation in relation to topography and land use in a hill slope catchment of the Loess Plateau, China. J. hydrol. 240: 243–263.

[43] Quinn P (2004) Scale appropriate modeling: representing cause-and-effect relationships in nitrate pollution at the catchment scale for the purpose of catchment scale planning. J. hydrol. 291: 197–217.

[44] Robinson M, Dean TJ (1993) Measurement of near surface soil water content using a capacitance probe. Hydrol. proc. 7: 77–86.

[45] Saskatchewan Centre for Soil Research (1989) Rural municipality of Grant, Number 372: Preliminary soil map and report. Publication# SK372. Saskatchewan Centre Soil Res. , Univ. of Saskatchewan, Saskatoon, SK.

[46] Shumway, R. H., & Stoffer, D. S. (2000). Time series analysis and its applications. Springer, New York, NY.

[47] Si, B.C., & Farrell, R.E. (2004). Scale dependent relationships between wheat yield and topographic indices: A wavelet approach. Soil sci. soc. am. j. 68: 577–588.

[48] Starks, P. J., Heathman, G. C., Jackson, T. J., & Cosh, M. H. (2006). Temporal stability of soil moisture profile. J. hydrol. 324: 400–411.

[49] Tallon, L. K., & Si, B. C. (2004). Representative soil water benchmarking for environmental monitoring. J. environ. inform. 4: 31–39.

[50] Topp, G. C., & Reynolds, W. D. (1998). Time domain reflectometry: a seminal technique for measuring mass and energy in soil. Soil till. res. 47: 125–132.

[51] Torrence C, Compo GP (1998) A practical guide to wavelet analysis. Bull. am. meteor. soc. 79: 61–78.

[52] Vachaud G, Desilans AP, Balabanis P, Vauclin M (1985) Temporal stability of spatially measured soil-water probability density-function. Soil sci. soc. am. j. 49: 822–828.

[53] van der Kamp G, Hayashi M, Gallen D (2003) Comparing the hydrology of grassed and cultivated catchments in the semi-arid Canadian prairies. Hydrol. proc. 17: 559–575.

[54] Webster R (1977) Spectral analysis of gilgai soil. Aust. j. soil res. 15: 191–204.

[55] Western AW, Grayson RB, Blöschl G (2002) Scaling of soil moisture: A hydrologic perspective. Ann. rev. earth planet. sci. 30: 149–180.

[56] Winter, T. C., & Rosenberry, D. O. (1995). The interaction of ground water with prairie pothole wetlands in the cottonwood lake area, east-central North Dakota, 1979–1990. Wetlands 15: 193–211.

[57] Woo, M. K., & Rowsell, R. D. (1993). Hydrology of a prairie slough. J. hydrol. 146: 175–207.

[58] Yates, T. T., Si, B. C., Farrell, R. E., & Pennock, D. J. (2006). Wavelet spectra of nitrous oxide emission from hummocky terrain during spring snowmelt. Soil sci. soc. am. j. 70: 1110–1120.

Chapter 6

Improving Soil Primary Productivity Conditions with Minimum Energy Input in the Mediterranean

Elsa Sampaio and Júlio C. Lima

Additional information is available at the end of the chapter

http://dx.doi.org/10.5772/52466

1. Introduction

As stated by the editors Brandt, C. and Thornes, J., 1996, "Both the vegetation and land use in the Mediterranean areas of Europe strongly reflect human activity since at least the Bronze Age". Putting it briefly for the recent past *ca.* ninety-year period, the natural landscape covered by the original "silvalusitana" has been deforested for agricultural purposes and substituted by monocultures of cereal crops, mainly wheat and barley, aiming at transforming the overall Alentejo region, southern continental Portugal, into the "granary of Portugal" (Pedroso, M. et al., 2009).

The so transformed natural landscape resulted in what has been called cereal steppe or pseudo-steppe, since the 1930′s, when the "wheat campaign" started (Pedroso, M. et al., 2009). However, such a "new" land-use system soon revealed to be aggressive to the environment and of discussable sustainability for the soil physics, chemistry and ecology. Negative impacts to the landscape physiognomy and dynamics and also soil quality have been declared when it has been realized that the initial and abundant soil organic matter (SOM) content (say then, the original (accumulated) soil's capital") was inexorably exhausted due to its induced unbalance. In fact, the regional, Mediterranean climate is prone to accelerate the SOM mineralization process in combinations with the intensification of the crop production system through, for example, the action of mechanical soil ploughing practices.

The persistence in space and time of such un intensive rain-fed, land use system for cereals production has let to the lessening of (1) crops dry matter (DM) production per hectare (*ha*), or yield; (2) rain-water use efficiency (WUE); (3) the light/radiation use efficiency (LUE) and (4) the overall efficiency of the energy-input into the agro-ecosystem per unit crop DM produced. The promotional motivation for national self-sufficiency of food, mainly wheat (and other correlated cereals) has generated the dominant, monoculture production system already described above. The way the production system has evolved ultimately originated the present, semi-anthropogenic rural landscape with sparse tree-cover and whose general physiognomy still reflects a history of severe soil erosion by run-off and increasing land use extensification, continuous biophysical degradation of soils, regional desertification risk and environmental hazards.

The sustainable exploitation of the natural resources in agro-ecosystems for improving the primary and secondary production of food, fibbers and bio-energy must be explicitly assumed as a common framework for policies and decision-makers worldwide. A fundamental pillar that should support such a collective goal (e.g., of food safety) is the ecological principal of sustainability of the agro-ecosystems by putting in practice a set of rational principles for agriculture, soil and on farming water management.

The present-time context of "global changing" that affects the "Earth [as an auto-regulated] system" (Journal of Earth System Science, 2012: http://www.springer.com/earth+sciences +and+geography/journal/12040) claims a great effort to improve the rational exploitation of the natural resources that constitute the primary agro-ecosystems input-factors, namely, the atmospheric radiant energy (solar radiation), carbon dioxide (CO_2) gas and liquid water, as rain (precipitation) (Brandt, C. and Thornes, B., 1996; Bolle, H. et al., 2006; IPCC, 2001; Le Houérou, H., 1996;Monteith, J., 1972). These are the three basic "raw materials" supporting plant photosynthesis, the biophysical process which is responsible for the energetic autonomy of any plant and terrestrial ecosystem (Monteith, J., 1972).

In the particular case of the rain-fed Mediterranean agro-ecosystems, the typical seasonality of the annual distribution and inter-annual variability patterns of precipitation, often generate situations of soil-water (soil-moisture) shortage and thus intense plant water-stress during their developmental cycles that spans in spring-summer, rainless season (e.g., Brandt, C. and Thornes J.,1996; Tenhunen, J. et al., 1987).

By the way, based on the ratio "mean annual precipitation (MAP)/MA-Potencial Evapotranspiration (PET)" – the aridity index, where PET = ETo, as calculated using the Penman-Monteith formulation – the Mediterranean-type climates are semi-arid climates (Le Houérou, H.,1996; World Climate Classification System, 2009), which means these areas are under desertification risk, according to updated biophysical indicators (Brandt, C. and Thornes J.,1996; Bolle, H. et al., 2006).

One realizes that the general diagnosis of desertification is the continuous biophysical degradation of the affected ecosystem; also the actual ecosystem's equilibrium state is reflected, for instance, on the destruction of soil aggregates, soil structure disorganization, lessening of the soil organic matter, soil-water contents and increasing soil erosion (e.g. Brandt, C. and Thornes J., 1996). Altogether these "active agents" trigger a series of desertification-promoting processes, interconnected through positive feed-back loops, leading the ecosystem to a state of significant lesser overall biomass production. All these environmental threats ultimately induce a decrease in the ecosystem maximum carrying capacity, relative to that a healthy ecosystem could bear with the same genotypes, under the same edaphic and climatic conditions. Several case-studies in the Mediterranean are described in the text-book edited on this issue (Brandt, C. and Thornes J., 1996).

In the context of this chapter, the potential maximum net primary production (NPPp) of the most sustainable agro-ecosystem is analysed via the soil-water balance (SWB), the actual available soil-water (ASW) for the soils having different effective soil-depths (and water residence time) for plant rooting, in connection to the precipitation regime and infiltration depth (Rodriguez-Iturbe and Porporato, A., 2004). Thus, this chapter focuses on the analysis of the technical viability of alternative agricultural practices to be adopted as intended to reverse soil degradation processes in terms of their impacts on the soil physical-hydraulic properties and SWB terms as compared to the control-system's prior state of evolution.

The perspective outlined, as above, supported as the main objective of this study the adoption of some intervening measures towards the restoration of degraded soils to ultimately re-establish a good level of soil fertility indexes for plant health and integrated land management. The integrated approach should be able to augment the overall ecosystem WUE, basically in part due to a higher soil-water reservoir capacity than it was in the former soil-degraded phase. Note that, for a given crop, the WUE is a quasi-inverse function of the atmospheric air vapour pressure deficit (VPD); WUE is higher when plants are experiencing some level of water stress (Tenhunen, J. et al., 1987). The soil-water reservoir capacity (a proxy of effective soil depth for plant rooting), along with the structural stability, is the main soil feature to be promoted in the first phase of a soil restoration programme, if any. This is followed by installing crops as part of a rotational scheme.

Relative to agricultural activity under present Mediterranean, water-controlled agro-ecosystems, the parameter "plant water-use efficiency" of the cultivated varieties (cultivars) subjected to the traditional, intensive land-cultivation system (*i.e.*, TS), may be a key criterion to deal with, if one's objective is to improve agriculture practicing based on adequate water and soil management. This goal is more important in ecosystems whose biophysical diagnostic reveals a degraded soil/substrate, such that it requires either conservation or remediation practice to mitigate or reverse the positive feed-back (FB^+) loops of the complex, natural dynamics of the desertification process (Lima, J. and Sequeira, E., 2004).

2. Problem statement

2.1. From the traditional to conservation agriculture

After farmers and agronomists, and policymakers have realized the negative impacts of the cereal monoculture system through the decreasing productivity per *ha* cultivated land-area and of soil buffering capacity for water conservation, they decided to implement mitigation measures against the negative pressure the production system was exerting on the environment. The mitigation measures involved the adoption of a crop-rotational scheme that is now designated the "traditional rotational system" (TS) of cereal crops production in the Alentejo. In effect, the cultural rotation has been based on cultivated varieties (*cultivars*) of wheat (*Triticum* sp.) and barley (*Scale* sp.) as the autumn/winter crops, followed by spring/summer ploughed-field, often "covered" with the water-stress resistant, oil-seed crop sunflower (*Helianthus annuus*L.). Eventually this last "(crop-)covered", ploughed-field technique was just substituted by fallow (consisting of, and according to the Webster's Dictionary, as "*n*": *a*) "usually cultivated land that is allowed to lie idle – i.e. not occupied, useless - during the growing season, for one or two consecutive years"; or *b*) the tilling of land without sowing it for a season; as "*vb*": "to plough, harrow, and break up land without seeding to destroy weeds and conserve soil moisture"; as "*adj*": "left untilled or unsown after ploughing").

Yet, by the middle eighties, the paradigm of land-use management has changed again following the integration of Portugal into the (former) EC, since the land-area of cereals (wheat and other species) has retreated over 35% under the "set-aside" agro-environmental political measure. Then, farmers with lands under site-aside should receive compensatory pecuniary payments aiming at protecting and stabilizing farmers′ incomes (Agri-environmental indicators, 2010) according to the CAP (Common Agriculture Policy) framework and legal mechanisms for election. Pedagogically, the set-aside principle of rural landscape management has opened an intense, intellectual debate based on the tension that arose around the "intensification/extensification" antagonism, dealing with rain-fed agro-silvo-pastoral production systems in the Alentejo (Pinto-Correia, T. and Mascarenhas, J., 1999). The fact is that technically the "set-aside" meant the conversion of acreages previously under the traditional system (TS) into abandonment, which is an extreme, useless extensification situation from the farmer's standpoint.

Lands under set-aside are presently claimed for an alternative, ecologically sustainable model for primary and secondary production systems. Ideally, in such a system, plants should be integrated in a food, fibre and energy-crop production system of minimum energy-input by specifically adopting no-tillage technique and null conventional fertilizer application (Jørgensen, U. and Kirsten, S., 2001; Dumanski, R. et al., 2006).

In extending the idea of energy-conversion efficiency improvement, the more adequate agro-technical itinerary, that is intended to be adopted, should minimize the fuel-oil-derived energy input into the agro-ecosystem through the application of "the modern agriculture technologies to improve production while concurrently protecting and enhancing the land resource on which production depends" (Dumanski, R. et al., 2006). This quotation introduces and greatly defines the concept of "conservation agriculture" (CA) whose main concepts and proposed techniques are thoroughly revised by Dumanski, R. et al., 2006 and summarized at the thematic Food Agriculture Organization's (FAO) and European Conservation Agriculture Federation (ECAF) websites on this issue.

http://dx.doi.org/10.5772/52466

Apart present time legislation framework on items related to conservation agriculture, the acreage under CA is increasing worldwide, but slowly in Europe (Dumanski, R. et al., 2006; ECAF, 2012). As examples upon local CA programs, a cost-benefit analysis of a case-study in Portugal is available for Castro Verde, in the inner-Alentejo (Pedroso, M. et al., 2009). For a broader scale, Kassam, A. et al.,(2012), offer a comparative overview of the importance of trade-off between CA practice (that is recommended) and the return value attributed to some selected features linked to the agro-environmental services, in countries having Mediterranean climate, that leads to a soil organic matter content that is chronically low to very low (Zdruli, P. et al., 2004; Bach, G. et. al., 2008).

The "cornerstone of conservation agriculture is the zero-tillage (seeding) technique and other related soil conservation practices" (Dumanski, R. et al., 2006). A recent inventory analysis shows that CA is increasing worldwide and particularly in countries having dry Mediterranean climate, including Portugal (Kassam, A. et al., 2012). The, now revised, alternative perspective offered under CA practicing allows *the* rational management of the natural, local resources for crop production, as well as their use and conservation, like soil fertility, available soil-water and solar radiation energy inputs. Philosophically, one should think of "any fraction of the available water and of solar radiation that enter an agro-forestry ecosystem could not be lost". This philosophy implies a land surface area unit fully covered by active vegetation all along the year and, in practice, it is equivalent to say that the ASW and net solar radiation (R_n) waste has to be minimized and their efficient use maximized under ecosystems management, while guaranteeing an adequate plant nutritional state too.

From the territory management viewpoint, the reclamation of the "unproductive" lands under set-aside into an ecologically, and operationally sustainable agricultural use can be done through the establishment of an adequate "technical itinerary" aiming at stabilizing in time the subsequent crop production. In turn, this "re-built" production system should promote cultural biodiversity (under rotational agriculture) to "compensate" (or substitute) biomass production for fluctuating opposite behaviours of different species, or cultivars, while responding to local climate and features of soil fertility during a given hydrologic year. This is an expectable situation if the crops′ respective life cycles can partially overlap, as the case study involving oilseed rape (*Brassica napus*L.) and sunflower in a fertile, clayey soil, has shown at Beja, Portugal (Lourenço, M. et al., 2000).

Under CA this plant production scheme is thought to optimize both WUE and LUE. These are two complementary, key eco-physiological parameters which can be used to describe the growth of plants. To accomplish this strategy in Mediterranean, rain-fed areas, fertile soils that can be selected are deep clayey soils with high water-holding capacity (WC) and cation-exchange capacity (CEC), for indicating only two indicators of soil fertility. Besides, the selection of plants species (the commonly used, or alternative, crops) of actual genomes should consider their propensity for (1) high values of both WUE and LUE; (2) deep root systems; (3) duration of the growing season and associated lower-temperature threshold and high relative growth rate (RGR) interactions and short lost-time for emergence (Goudriaan, L. and Monteith, J.,1990; Monteith, J., 1993), also (4) multi-purpose specimens and (5) cultural biodiversity, for instances, for biological control of plagues.

Finally, as water is certainly the major factor limiting crop production in the Mediterranean areas, attention is essentially drawn on the efficient water use in crop production (Monteith, J. 1993).

2.2. Water-Use Efficiency (WUE) of crop production

The WUE may be determined at the leaf level, for instantaneous gases exchange measurements (stomatal conductance, transpiration and photosynthetic rates), or long-term amount (gram, g), of plant-assimilated dry mater (biomass), divided by the concurrently amount of transpired water (kg) (Tanner, C, and Sinclair T., 1983). The magnitude and units of the WUE-parameter depends on the chosen computational base (or reference) selected to express it numerically, examples being per unit molar-mass of glucose ($C_6H_{12}O_6$) or CO_2 molecules or an atom of carbon (C). The choice determines a mass-based scale of WUE, its "absolute scale" being a function of the atmospheric air vapour-pressure deficit (VPD, say, in kPa), which in turn is a function of the air temperature at the reference height of 2-m above plants/vegetation canopy. By the way, Lindroth, A. and Cienciala, E. 1996 found a unique relationship between the WUE and the VPD (the argument-variable) valid for leaf, tree and stand levels for willow (*Salix viminalis*L.) forests, in Sweden.

On the other hand, the product VPD × WUE defines the "normalized scale of WUE" (or NWUE), which is quite invariant over a specified range of VPD, plant species and both C_3 and C_4 photosynthetic pathways (Jørgensen, U. and Kirsten S., 2001; Monteith, J., 1993; Tanner, C, and Sinclair T., 1983). Else, these three bibliographical references offer typical values of the NWUE for different geographical locations, in units of g [dry matter (DM)] per kg [water] times VPD (in kPa).

3. Material and methods

3.1. Location and characterization of the experimental site

The field trials were carried out on two regional, representative soil types under two different land-cultivation systems for crop production, in the rain-fed areas of the inner-Alentejo-Province, southern continental Portugal. The study site (*Herdade da Almocreva*) is located nearby Beja town and reference geographical coordinates (relative to the International Meridian) are 37º 59′ 48′′ N; 37º58′50′′ S; 7º 58′ 00′′ W; – 7º 55′ 32′′ E.

3.1.1. Climate

The Köppen′s climatic classification scheme characterizes the overall Portuguese continental territory as having a warm, temperate subtropical Mediterranean climate of the type Cs; this type of climate identifies a hot, dry summer season in which the sum of the monthly precipitation of the driest consecutive three months (June – August) is less than 30 mm (Table 1).

Months	Mean air temperature (ºC)	Precipitation (mm)	Relative Humidity (%) at12:00 h	Insolation (h)	Wind speed at screen height (Km/h)	Frost frequency (Nr of days)
Jan	9,5	83,2	82	145,8	15,6	3,6
Feb	10,2	83,0	78	152,9	16,4	2,5
Mar	11,8	80,2	70	183,3	15,9	1,1
Apr	13,8	48,9	61	235,5	15,3	0,2
May	17,1	35,0	54	291,2	15,4	0,0
Jun	20,7	26,2	49	310,0	15,1	0,0
Jul	23,6	1,2	41	367,9	15,7	0,0
Aug	23,8	2,5	41	345,1	15,8	0,0
Sep	21,8	18,8	49	252,6	14,2	0,0
Out	17,6	67,0	62	202,6	14,7	0,0
Nov	12,8	73,7	72	160,9	14,6	0,0
Dec	9,9	85,7	79	147,7	15,2	0,6
Year	16,1	605,6	61	2795,5	15,3	3,7

Table 1. Mean monthly values of climatic variables recorded at Beja (37º 59´ N; -7º 57´W) (1951-1980):

Soil Vc / TS		**Depth limits of soil layers (cm)**			
Classes of soil particles	Units	0.0 – 10	11 – 20	21 – 30	31 – 40
Coarse sand	%	5.7	4.4	5.5	3.1
Fine sand	%	11.1	11.4	11.6	10.2
Silt	%	20.1	20.4	19.9	25.5
Clay	%	63.1	63.9	63.0	61.2
Texture class		*Clayey*	*Clayey*	*Clayey*	*Clayey*
Coarser elements	%	*9.1*	*10.8*	*17.6*	*45.6*
Soil Bvc / TS		**Depth limits of soil layers (cm)**			
Classes of soil particles	Units	0.0 – 10	11 – 20	21 – 30	31 – 40
Coarse sand	%	7.5	7.2	6.4	5.3
Fine sand	%	13.4	18.0	12.5	10.2
Silt	%	23.4	19.2	22.3	24.6
Clay	%	55.7	55.6	58.8	59.9
Texture class		*Clayey*	*Clayey*	*Clayey*	*Clayey*
Coarser elements	%	3.9	6.2	7.4	9.2

Soil Vc / DS (10)		**Depth limits of soil layers (cm)**					
Classes of soil particles	Units	0.0 – 10	11 – 20	21 – 30	31 – 40		
Coarse sand	%	5.7	3.8	3.5	4.2		
Fine sand	%	11.9	12.1	10.7	11.8		
Silt	%	18.8	19.3	18.0	19.1		
Clay	%	63.6	64.8	67.8	64.9		
Texture class		*Clayey*	*Clayey*	*Clayey*	*Clayey*		
Coarser elements	%	10.7	10.1	8.2	33.5		
Soil Bvc / DS (10)		**Depth limits of soil layers (cm)**					
Classes of soil particles	Units	0.0 – 10	11 – 20	21 – 30	31 – 40	41 – 50	51 – 60
Coarse sand	%	7.1	6.3	5.8	4.9	4.5	3.2
Fine sand	%	17.4	17.5	16.3	10.8	12.7	15.3
Silt	%	23.5	24.6	22.5	19.9	20.0	20.2
Clay	%	52.0	51.6	55.4	64.4	62.8	61.3
Texture class		*Silty clay*	*Silty clay*	*Silty clay*	*Clayey*	*Clayey*	*Clayey*
Coarser elements	%	7.1	6.7	6.6	6.2	5.4	5.1

Table 2. Mass fractions of the mineral components and coarser elements

Soil Vc / TS		**Depth limits of soil layers (cm)**					
Parameters	Units	0.0 – 10	11 – 20	21 – 30	31 – 40		
Bulk density	g cm^{-3} (= Mg m^{-3})	1.53	1.55	1.57	1.64		
Resistance	kgf cm^{-2} (=0.098 MPa)	32	35	39	37		
Soil Bvc /TS		**Depth limits of soil layers (cm)**					
	Units	0.0 – 10	11 – 20	21 – 30	31 – 40		
Bulk density	g cm^{-3} (= Mg m^{-3})	1.45	1.59	1.63	1.66		
Resistance	kgf cm^{-2} (=0.098 MPa)	34	39	42	31		
Soil Vc / DS (10)		**Depth limits of soil layers (cm)**					
	Units	0.0 – 10	11 – 20	21 – 30	31 – 40		
Bulk density	g cm^{-3} (= Mg m^{-3})	1.53	1.46	1.45	1.51		
Resistance	kgf cm^{-2} (=0.098 MPa)	49	46	40	36		
Soil Bvc / DS (10)		**Depth limits of soil layers (cm)**					
	Units	0.0 – 10	11 – 20	21 – 30	31 – 40	41 – 50	51 – 60
Bulk density	g cm^{-3} (= Mg m^{-3})	1.36	1.53	1.51	1.52	1.45	1.39
Resistance	kgf cm^{-2} (=0.098 MPa)	50	49	46	45	41	39

Table 3. Bulk density and resistance to the penetrometer

http://dx.doi.org/10.5772/52466

Soil Vc /TS		Depth limits of soil layers (cm)					
Consistency limits	**Units**	**0.0 – 10**	**11 – 20**	**21 – 30**	**31 – 40**		
Plastic limit (soil moisture)	%	24.29	24.48	26.80	21.79		
Liquid limit (soil moisture)	%	44.95	43.39	39.75	39.04		
Soil Bvc / TS		**Depth limits of soil layers (cm)**					
Consistency limits	**Units**	**0.0 – 10**	**11 – 20**	**21 – 30**	**31 – 40**		
Plastic limit (soil moisture)	%	30.09	24.21	26.14	28.07		
Liquid limit (soil moisture)	%	52.41	53.35	54.28	60.42		
Soil Vc / DS (10)		**Depth limits of soil layers (cm)**					
Consistency limits	**Units**	**0.0 – 10**	**11 – 20**	**21 – 30**	**31 – 40**		
Plastic limit (soil moisture)	%	25.03	27.16	34.87	24.64		
Liquid limit (soil moisture)	%	50.72	48.54	51.55	46.06		
Soil Bvc / DS (10)		**Depth limits of soil layers (cm)**					
Consistency limits	**Unit**	**0 – 10**	**11 – 20**	**21 – 30**	**31 – 40**	**41 – 50**	**51 – 60**
Plastic limit (soil moisture)	%	28.63	25.37	25.74	28.01	27.09	27.23
Liquid limit (soil moisture)	%	53.84	53.97	51.35	52.41	62.59	61.82

Table 4. Soil consistency limits

The virtually rainless early-spring and (the subsequent) summer season, followed by the typical winter (markedly from November to March), rainy season and low temperature regime does not promote as an intense as desirable vegetation activity (growth/development) for primary dry mater (biomass) production. In fact, these limiting factors are among the main atmospheric determinants of Mediterranean environments for the selection of both (crop) cultivars and the production systems to implement at the local and regional scales. These local limitation relies on two major aspects: (1) the optimum combination of required heat units (thermal-time) and the available soil water (ASW) for optimal plant growth and development is generally absent and (2) during the rainy season, the combination of short-daylight length and low temperatures are harmful to vegetation activity which is emphasized under soils characterized by deficient drainage, a common negative feature of the soils in the Alentejo.

3.1.2. Relief

The terrains in the study area have generally gentle or very gentle to quasi plane slopes. A set of twenty soil profiles aiming at the study of soil properties had been dogged on very gentle to plane (0 to 2%) mid land-hill positioned in the landscape, facing south-east, the dominant local landscape aspect.

3.1.3. Soil units

In the Portuguese, national main soil survey ever conducted for the Southern continental territory (Cardoso J., 1965), both experimental fields have been installed on the same soil types, namely, a *VerticCambisol* (*Vc*; Portuguese Nomenclature (PN)) and a *Chromic Vertisol* (*Bvc; PN*), as they are classified on the available soil map Nº 43C of the "Carta dos Solos de Portugal", the Portuguese Soil Map (hereafter PSM) on the cartographic scale of 1:50 000. The soil labels "Bvc" and "Vc" stand for the designation used in the Portuguese soil classification system.

Pedo-genetically, these *Vc* and *Bvc* soil types encompass relatively incipient soils with typical pedologic profile that consists of a sequence of horizons A/C or A/B/C of calcareous materials(with sub-layers in some cases) and the mean depth of the C horizon goes down to 40-60 cm below ground. The difference between the two soil units is established through the B horizon, that is, Bc in Vc soils, and Btx in Bvc soils. A prominent feature of the Bvc soils is their higher content in expansive clay minerals of the type 2:1 bi-layer structure, as the montmorillonite in this case, that are self-mulching, a process associated to the alternating seasonal swelling (watering) and shrinking (drying) processes. The high concentration of montmorillonite clay of these soils determine some of their relevant physical properties, such as a high plasticity, hardiness, anisoform structure at the more surfacial layers that changes to the prismatic in the B horizon, slick-in-side planes, surfacial and deep cracking under desiccation; in the B (if any) and C horizon it is common noting the occurrence of manganese concretions. Other properties directly determined by the presence of montmorillonite clay are, for instance, the high "cation-exchange capacity" (*CEC*), soil pH > 7, and the ability to accumulate soil-organic matter.

Both soil types involved in the study were present in the two experimental fields where they have been subjected to the same land-production (land-cultivation) systems next described.

3.2. Characterization of the land-production systems

The traditionally intensive crop-production systems (TS) imposed on both Bvc and *Vc* soils are based on rotational schemes of rain-fed agriculture of cereal crops (*i.e.*, winter-sowing wheat, late-winter-sowing barley and sunflower in mid-spring). However, an alternative agricultural rationale, the low-tillage, or low-ploughing system is now an increasing agricultural practice in dry Mediterranean countries, including Portugal (Kassam, A. et al. 2012), as is the case of direct (say no-tillage) seeding (SD) system. By this reason, the TS and the DS (or zero-tillage system) under a Mediterranean environment will be analysed in this chapter by taking the former as the ecosystem's initial state and, the later, as the alternative system for soil conservation. Both TS and DS are known to differ in the amount of energy input required to produce a unit-mass (or weight) of plant/crop biomass and, also, in their diverse environmental impacts and soil water conservation.

Because the biomass, or plant dry matter production (DM) has not been quantified during the experiments described here, the expected grain yields (Y) and correlated harvest index

(HI) factor of the following crops were collected in the literature (*e.g.*, Ahmad R. and Jabran K., 2007):

Crop species	Mean grain yield (kg/ha)	Harvest index (HI)
Wheat (*Triticumaestivum* L.)	3 000	0.5
Barley (*Hordeumvulgare* L.)	2 000	0.5
Sunflower (*Helianthus annuus* L.)	1 600	0.4

Table 5. Crop species used in the traditional system (TS) and their mean grain yield (Y) and mean maximum harvest index (HI)

In turn, total dry mater (TDM) production will be estimated based on known plant dry mater partitioning coefficients (e.g. Stockle, C. and Nelson, R., 1996). Notice that both TS and DS crop-production systems integrated those same crops and rotational schemes (see below) whilst their main differences relies on the types of soil tillage, weed control and fertilization criteria (Sampaio, E., 2002).

It follows the characterisation of both crop-production systems already referred to.

3.2.1. Traditional Rain-Fed System (TS) = Ecosystem initial state

Crop rotation - (soft wheat→ barley → sunflower)associated to the following schemes dealing with land tillage, cropdisease control and nutrient applications.

- Soft-wheat – The seed-bed was prepared with two cross-deep ploughings down to 30-cm of the soil depth. Seeding proceeded with an in-line seeder and soil-surface roller. A phyto-sanitary prevention measure was taken. As demanded by the crop, it was used a chemical fertilizer containing the three primary (main) macro-nutrients (N, P, and K), namely 250 kg/ha of the "10-14-36" composed fertilizer for a prior fertilization at seeding and 70 kg/ha of N in a covering fertilization at the inner-spring.

- Barley – As for wheat, the seed-bed was prepared with two cross-deep ploughings down to 30-cm of the soil depth and seeding proceeded using an in-line seeder and rolling the ground. A phytosanitary, chemical product was used for preventing plant diseases; mineral fertilization was accomplished using250 kg/ha of "10-14-36" fertilizer at seeding and 46 kg/ha of N to covering in the inner-spring.

- Sunflower– in preparing the seed-bed the soil is initially submitted to deep-ploughing down to 30-cm of the soil depth, followed by shallower steam-ploughing (< 10-cm below the ground surface) and some times passes with cultivator shanks, depending on the actual soil hardiness. Subsequent seeding, croskiling and usually two mechanical weeding during the vegetative period follow. Neither phytosanitary nor nutrient application was made..

3.2.2. Direct-Seeding System (DS) = Ecosystem final state

Crop rotation - (soft wheat→ barley → sunflower)adopting the direct-seeding agro-technique for a continuous period that spans 10 consecutive years. Referring to the mechanical soil preparation, phyto-sanitary treatments and mineral fertilizations, for all crops, the weed-control technique were applied in the pre-emergence phase by using a mixture of an integral, systemic and no-residual herbicide, and an hormonal herbicide against the broad-leaved weedy species. Only 10 days later the direct seeding treatment was able to be executed using a proper seeding-machine.

- Soft-wheat – it was applied an initial, deep fertilization (N and P) at the rate of 230 kg/ha of the "18-46-0" composed fertilizer, at seeding. Early in spring nitrogen was top dressed with 70 kg N per hectare (ha) and also performed a new weed plants control.
- Barley – A deep fertilization (N and P) was made simultaneously at seeding with 200 kg/ha of the "18-46-0" composed fertilizer; it fallowed top dressed (fertilization) with 80 kg/ha of N. In the beginning of spring new chemical control on the weeds took place.
- Sunflower – nor fertilization neither weed control was done in spring.

3.3. Field work

Tasks in the open had been carried out in the two experimental fields on both two types of soil that have been subjected to both (DS and TS) land-production systems. For each of the four ($n = 4$) combinations resulted from the two soil types under the two cultivation systems, it were dogged five ($n = 5$) representative soil profiles to totalize 20 of them. The soil-type × production-system output was as follows:

- Profile type 1 – DS10/Bvc: Soil Bvc subjected to direct seeding (DS) for a consecutive ten (10) year time-period;
- Profile type 2 – TS/Bvc:- Soil Bvc subjected to the traditional system (TS) for more than a 10 year time-period;
- Profile type 3 – DS10/Vc: - Soil Vc and direct seeding (DS) for a consecutive ten (10) year time-period;
- Profile type 4 – TS/Vc:- Soil Vc and traditional system (TS) for more than a 10 year time-period.

Among other properties, also of interest in this study, are the identification of the soil texture classes for soil particles diameter and water properties computations and also the soil-water retention (pedotransfer) function fitting.

3.4. Soil sampling criteria

These soils have been characterized after excavating five representative soil profiles per soil type, thus an overall of 20 soil profiles have been surveyed by sampling specific properties on soil physics, chemistry and hydraulics. In the sixth and tenth (last) year of the beginning of the experiment, the soils survey have been performed in May, when the dried, finest (clayey) soil experienced its seasonal swelling process.

The ensemble of soil properties elected to characterize the ecosystems′ initial and final states (then, of the experiment time span) have been surveyed at depth, with a constant 10-cm vertical increment, from the ground surface down to the upper-boundary of the C-horizon, of variable depth, between 40 cm and 60 cm below ground. The survey involved all the 20 soil profiles sampled for: soil texture and volumetric mass (soil density); soil strength; soil-matric water-retention-curve (WRC); porosity of the porous media,; mechanical consistence (related to the "physical resilience of soils, their ability to resist deformation under stress" (Kimmins, J.,1997), for plasticity and liquid-limits of the soil fine-earth fraction; actual size of the aggregates and aggregation index; the soil organic-matter (SOM) content; typology and size classes for the continuity of such vertical soil pores distribution, as well.

In the field it was evaluated the soil strength and, for statistics, five (n = 5) replications per soil layer (of 10 cm) were taken for averaging results as arithmetic means. Five (n = 5) standard soil-sampling cylinders (internal volume = 90 cm^3) per soil layer were collected, hermetically conditioned in the field and transported to the laboratory for subsequent analysis. Five soil samples, weighting 1 kg, were collected each 10-cm depth increment in all soil profiles.

3.5. Physical analysis

Texture and coarse materials – the mass fractions of the fine-earth mineral components (sand, silt and clay) for soil textural classification and the coarser materials present in the same soil samples were based on sieving separation and mechanical analysis.

Bulk density (BD), the mass per unit volume of soil – despiteits numeric value, the soil BD (10^6 g m^{-3}) is an integrating parameter that results from the combined effect of texture, aggregates structure and dimensions, intrinsic porosity and organic matter content. This parameter has been determined on intact soil samples.

Soil Strength and Penetrometry - a portable cone-shaped penetrometer ("Soil Test, ref. TL 700A") have been used for measuring the strength of the soil material. The lower and upper instrument's scale-limits were, respectively, 0.25 and 4.5 kgf/cm^2, which are equivalent to 24.5 kPa and 441 kPa. In-field readings were taken by pressing the walls of all the 20 soil profiles at right angles, at constant depth-increments of 10 cm.

Soil-Water Tension via pF scale - the ability of the soils to retain water has been evaluated by determining the mathematical relationship between the soil-moisture that remains in the soil samples subjected to prescribed water-extracting suctions, expressed on the pF scale. The usual operational (agronomic) limits of pF are 1.8 (for saturation) and the 4.2, while the soil-moisture under pF 2.54 is currently related to the upper limit for plant water comfort at the "field capacity" (FC). The lower limit is for pF 4.2 at the "permanent plant wilting point" (WP). From lower to higher values, pF has been determined through the suction-method with silt under a "sand-bath", the method of pressure-membranes and the method of pressure-plates. Results are referred to as the arithmetic means of n = 5 statistical elements.

Distribution of Soil Porosity – this determination followed the work of (Blume H., 1984), where it was established a quantitative relationship between the actual hydrated mean pore-diameter and the corresponding pF (retention force) value; the relationship is based on the Hagen equation that assumes the soil pores as having circular shape of variable dimension. For example, (Blume H., 1984), indicated ordered-pairs of the type (pF, ≤ pore-diameter) of (1.8, 50 μm), (2.5, 10 μm) and (4.2, 0.2 μm), which estimates the volume occupied by all pores giving a mean diameter less than the values indicated inside the parenthesis, as determined in the laboratory, and enabling one to estimate its percentage relative to the total soil-volumetric porosity ($\theta_s = 1 - (BD/2.65)$).

Aggregates Distribution and Soil Stability – For this task it was used the sieving method of Yoder (1936), at time, the one supported by the apparatus available in the laboratory. Apart from the existence of several modern methods that could be used for, this early one is reliable enough to generate confidence on the results obtained. The distribution of aggregates and their stability in a soil-ped are related to soil consistence.

3.6. Chemical analysis

Soil Organic Matter (SOM) - the (mass-based) SOM content had been determined by destroying each organic samples via the "humid-oxidation" method. The soil nitrogen (N) content had been determined on air-dried samples using the Kjeldahl method. Based on this, the carbon/nitrogen (C/N) ratio had been calculated, since the relationship C = 0.574 SOM stands.

Cation-Exchange Capacity (CEC)– a solution of barium-chloride and trielanolamine, at pH of 8.2, was used that was combined with the spectro-photometry technique. Here CEC (meq[H+]/100g of oven-dried soil) represents the soil fertility and nutrient buffering effect of the soil. *CEC* is dependent on soil pH.

Phosphorus and Potassium (for plant nutrition) – both nutrients had been determined on air-dried soil samples via the *Van der Paaw* method.

Exchangeable Cations and Base Saturation – it was also analysed via spectro-photometry on lixiviates from soil samples previously saturated with barium. The contents of the four chemical cations, Ca^{2+}, Mg^{2+}, Na^{+} and K^{+}, were determined as well; their total gives the sum of bases (*S*). When expressed as a fraction of CEC, S is converted into the base-saturation degree (*V*).The CE is effectively determined by the colloidal (clay and humus) components of the soils.

Soil reaction or pH– The soil pH is a measure of the hydrogen ion (H^{+}) concentration and it has been determined in a 1:1 aqueous suspension of soil material. The apparatus used was a potentiometer. An aqueous solution that has a pH of 4.5 has $10^{-4.5}$ moles of hydrogen ions [H^{+}] per liter of solution at the (standard) temperature of 25 ºC. Soil pH affects the availability of nutrients for plant nutrition and is, of course, an important chemical descriptor of soil fertility.

3.7. Vertical continuity of the soil porosity (the distribution of space in soils)

Aiming at identifying and evaluating the continuity of the soil porous phase (soil voids) network, it was used the relative frequency distribution of pores (Blume H., 1984) with the morpho-metric classification of bio-pores, soil cracks and irregular void space suggested in Sampaio E. and Sampaio J., 2010.

The method used consists of preparing a blue coloured plaster suspension in water (in one part of plaster per three parts of water in volume) which is brought to infiltrate into the soil profile from the ground surface. After the infiltration cessation and a given time-lag, enough to the strengthening of intruded material, horizontal slices of the soil material are removed from the profile in order to visualize the "filled voids" at depth that are counted and classified by dimension classes and morphometric typology (Sampaio E. and Sampaio J., 2010).

There are a lot of methods that use shaping substances to fill the soil voids through their infiltration into the soil. The plaster have been elected because *i*) it is a lesser degrading (more innocuous) material to the study-soil environment and *ii*) it gives results as good enough to be used in such evaluations (Bouma J., et al., 1982).

Following Schneider J. and Stunke A., 1991, we introduced the innovation of blue-colouring the prepared suspension in order to avoid confusing some soil constituents and an uncoloured suspension.

Soil voids had been distinguished through: *i*) the presence of plaster or *ii*) plaster was absent (this was subdivided into "continuous" and "non-continuous"). Only the continuous voids had been counted because the difference between the total porosity and themselves can be used to evaluate the "non-continuous" voids without the need to perform individuals counting of the last type.

Three types of soil voids (bio-pores; cracks and irregular voids) were distinguished for class dimensions, as follows:

1. **Bio-pores** are all circular-shaped soil-voids that are mainly originated by the soil fauna or plant roots activities that, in turn, generate cylindrical-shaped channels in the soil matrix; generally these channels have a tortuous path –length oriented toward any direction in a soil volume unit, attain variable depths and are of paramount importance in establishing preferential soil-water drainage paths under saturation and may facilitate the soil colonization by plant roots. As a rule, the bio-pores are big enough to exert an important role on the soil-water retention function.

2. **Cracks** (often deep) are all elongated voids (lengths were at least twofold their own widths) that naturally result from the shrinking (under drying), alternating with the swelling (under soil hydration) processes where there was a high expansive, montmorillonite clay content. Ultimately the cracks promote the infiltration of water into the soil only in the beginning of watering after a dry season or year.

3. **Irregular voids,** of a not well defined origin, are thought to result from soil compaction that may lead to the destruction of the last crack types previously describe. As their are

relatively ephemerals, the irregular voids cannot be considered as having relevant importance neither on the typical soil-water movement nor for roots colonization.

All these three void types, say bio-pores (*P*), cracks (*F*, from "fissure") and irregular voids (*Esp*) are distinguished and quantified according to their dimensions into the light of their influence on water deep-drainage, soil-water balance and production of crops. The bio-pores dimension classes so defined were base on their diameters (Ø): large bio-pores (PG): Ø >5 mm; medium bio-pores (PM): 1 < Ø <5 mm; small bio-pores (PP): 0.4 < Ø <1 mm; very small bio-pores (Pmp): 0.15 < Ø <0.4 mm and "minimal" bio-pores (Pm): 0,03 < Ø < 0,15 (criterion of[32]).

In the beginning, the classification of soil voids had been established based on their lengths by(Scneider J. and Stunke A., 1991) and the classes are: large cracks (FL), *i.e.*, greater than 200 mm; medium cracks (FM), which are between 5 and 200 mm; short cracks (FC), that are lesser or equal to 5 mm. Subsequently, the dimension of each void class became classified based on width, as "large" (≥5 mm), "medium" (1 to 5 mm) and "narrow" (≤1 mm). The irregular voids had been not classified into dimension classes; alternatively it has been recorded the individual area that was summed out.

Despite it seems somewhat antiquate the (already described) method used to study the connectivity of voids through the soil porous media profiles, there are evidences that support modern alternative methods using the 3D approach do not describe well the reality, which opens the possibility to improve them (Weerts A., et al, 2001). The 3D modern methods already referred to include, e.g., gaseous diffusion, computer-assisted axial tomography (CAT), Lattice-Boltzmann simulation models, fractal analysis.

From previously prepared horizontal surfaces (or planes) for the observation of the soil profile at depth, digital pictures from horizontal planes of the soil were taken in vertical sequence at a constant 10 cm-depth increment down to the C-horizon. The digital pictures captured by the operator have a square plane representing a 50-cm width (the area is 2 500 cm^2) quadrate; it can be distinguished on each quadrate the blue-coloured plaster in the impregnated voids. The first, uppermost picture of the (first) observation surface was taken from a plane at 1 to 2 cm of depth; the operator proceeds by digging the 10-cm depth increments for more picture takings on different horizontal planes of the soil profile. In this method pictures are only taken after homogenizing the actual soil-plane surface using a spatula, levelling and cleaning. This task ends at the depth where it is no more visible any blue material trait.

Each of these planes was photographed under the maximum magnificence of the camera used (a Nikon F90, with an objective lens of the type AF NIKKOR 35 to 135 mm). The sequence of tasks just described were replicated three times for each of the four combinations (soil type x crop production system) used and under study to obtain a total of twelve photo-sequences.

The soil-surface pictures then collected were subjected to an image analyser "*Sigma ScanPro*" and processed to allow the voids classification according to the outlined typology followed by counting the void classes. This software allows the development of works that comprises

Morphometric measurements;	**Image processing**; contrast enhancements
distance (strait line and curvilinear)	lookup table and pseudo-colour grey and colour
X, Y co-ordinates	transformations
area and perimeter	true and pseudo Clearfield background equalisations
major and minor axis	**Data worksheet**;
pixel counts	import and export ASCII files
shape factor	output data to a printer
Feret diameter	store the results of mathematical transformations
Intensity measurements;	store calibration and lookup table values
displaying a histogram of pixel intensities	**Graphing results**;
pixel intensity measurement	apply and plot basic statistical functions
measurement of average intensity over an area	plot X, Y graphs with linear regression
	Other features;

Table 6. Tasks allowed by the Sigma Scan Pro for classifying the soil voids

It's noteworthy that this software automatically distinguishes and classifies the pores and cracks by their geometric properties. In order to do that it constructs a table through small built-in programs in its own computing sheet (Sampaio, E. and Sampaio, J., 2010).

3.8. Soil-water retention curves

As it is presented, the RC derives from converting the original pF – volumetric soil moisture (water), θ (m^3 m^{-3}), relationship, in which the pF scale is converted into the matric potential, ψ_m in kPa (or J/kg). The mass fraction of soil coarser elements is useful in correcting the available soil-water (ASW) calculated from texture only because of the scarcity of such kind of information, whose lacking overestimates the ASW.

3.9. Estimation of crop biomass production

The potential rate of the maximum net primary production (NPPp, in g(DM) m^{-2} day^{-1}) is estimated for local agro-ecosystems, as thought to be virtually independent on soil type, while plant types are distinguished through their biochemical C_3 or C_4 photosynthetic pathways. Contrary to the solar radiation load, the soil water availability is considered here as the biophysical factor that effectively reduces NPPp to its (positive) actual rate to the condition NPP ≤NPPp. Plant, or ecosystem, total dry matter (TDM) biomass production over time can be readily estimated from available information on NPP, say, the potential maximum TDM (or DMp), the normalized (transpired) water-use efficiency (NWUE), and the local atmospheric air vapour pressure deficit (VPD).

Formally, this is the Tanner-Sinclair approach (Tanner, C. and Sinclair T. 1983) for which the linear relationship NWUE = WUE × VPD holds. Existing database (Jørgensen, U. and Kirsten S., 2001) shows that NWUE is quite conservative for a lot of plant groups having the same photosynthetic mechanisms, C_3 or C_4. Finally, the above-ground component of the total dry mater for wheat is related to the corresponding grain-yield (Y) through the harvest index (HI) factor (section 3.2; White, E. and Wilson F., 2006), while 30% of TDM is allocated to the root system, as done in the CropSyst model (Stockle, C., 1996). Else, because plant dry mater production is directly proportional to the transpired water (TDM = WUE × WT), for healthy plants (Lima, J. 1996; Abreu, J. 1994; Monteith, J. 1993; Tanner, C. and Sinclair, T. 1983), the use of these data in a reverse-parameter modelling approach facilitates the crop water requirements (only WT, here) to be computed, without the imperative need to tackle the soil-water balance problem.

The transpired water during the (daily) time-step expresses the soil water uptake by plant roots integrated at depth, which can be used to truncate the effective rooting depth (Lima, J. et al. 2011). There also exists a linear relationship between the maximum effective rooting depth (Zrx in meter) and the maximum leaf-area index (LAIx: 3.0 to 6.3 $m^2\ m^{-2}$): Zrx (m) = 0.2487LAIx + 0.2734 ($n = 9$; $r^2 = 0.842$) (Stockle, C. and Nelson, R 1996). For an irrigated wheat crop grown in a clayey, *chromic vertisol* soil in Lisbon, observed values of LAIx and Zrx reached 3.0 $m^2\ m^{-2}$ and 0.80 m, respectively (Abreu, J. 1994).

The spatial and time resolution of the model is the square meter and the astronomic day (24 h), while the time-period of interest covers the entire growing season. The criterion for numerical convergence analysis is minimizing the difference between the calculated WT and the soil water holding capacity (WC) for the effective rooting depth.

The DMp is estimated for a winter-wheat (C_3) crop in a clayey soil and local climatic conditions similar to that of Alvalade-do-Sado in the Alentejo, where a maize (*Zea maysL.*) crop has been intensively managed in a modern Fluvisol and is considered here as a reference high crop production system, neither short of water nor for mineral nutrition (Lima, J. 1996). For the maize as a C_4 species, NWUEp = 10.3 kPa g kg^{-1} (Jørgensen, U. and Kirsten S., 2001); this figure is, in turn, converted to the NWUEp for the wheat crop by dividing it by the empirical coefficient 2.0 to scaling from the C_4 to the C_3 photosynthetic mechanism efficiency. This sets NWUEp = 5.1 kPa g kg^{-1} for wheat which eventually can be adjusted for local VPD and water shortage (e.g. lowering soil water holding capacity) under rain-fed production regime.

In this exercise, a six-year (1996-2002) time period of climatic variables data for Évora (80 km distance northern Beja) is used as inputs. For this period, the mean range VPD = 0.88 ± 0.52 kPa were observed for air humidity, while the total evapotranspiration ET (WT + direct evaporation from the soil) for maize varied between 500 and of 600 mm per growing season, typically of 130 days (from 05 May to 13 September, 1995) (Lima, J., 1996). So, 550 mm is limiting the maximum water use of a cereal crop in the Mediterranean

4. Results

4.1. Profiles description

Depht (cm)	Layer	Description
00 - 18	Ap	2,5YR3/3 (Munsell Chart color) Clayey with gravel of compact limestone; structure composed of anisoform angular thin and strong anisoform angular and subangular,medium to coarse, strong, slightly porous, very fine, medium compactness, friable, slightly sticky and slightly plastic; crack when dry, cool, effervescencewithHCl Flat and smooth transition to:
19 - 28	BwC	2,5YR3/5 (Munsell Chart color) Clayey with gravel of compact limestone; structure composed of prismatic and anisoforme angular to subangular, medium and coarse, strong, some very fine pores, firm, slightly sticky to sticky, slightly plastic to plastic; compactness great effervescencewithHCl; Gradual and wavy transition to:
29 -	C	Mixture of limestone friable material yellowish red with fragments of limestone compact.

Table 7. VerticCambisol / Traditional System - Vc / TS

Depth (cm)	Layer	Description
00 - 14	A	2,5YR3/4 (Munsell Chart color) Clayey-loam with some quartz gravel and limestone, structure anisoform angular very fine to fine, strong, medium to fine anastomosing cracks, slightly porous, very fine, medium compactness, consistency very hard, sticky and plastic; many roots thin crack when dry, cool, zero effervescent with HCl. Flat and smooth transition to:
15 - 21	BA	2,5YR3/4 (Munsell Chart color) Similar as above, clayey withsome gravel ofquartz, schist and feldspar materials, structure anisoform angular to subangular, coarse to very coarse, strong. Flat and smooth transition to:
22 – 44	Bt	2,5YR4/5 (Munsell Chart color) with small gray dark reddish spots. Clayey with gravel and some rubble of quartz, schist and feldspar materials; structure composed of prismatic and anisoformsubangular, medium and coarse, moderate to strong, some thin and medium vertical cracks, some very fine pores; consistency very firm, sticky to very sticky, plastic very plastic; great compactness, aggregate faces with film of clay and polished surfaces; additions of organic matter giving a clear stained. Gradual and wavy transition to:
45 – 55	BCca	Lithological materials in an advanced state of decomposition with some clay (YR4 2.5/6 color) that makes strong effervescence with HCl.

Depth (cm)	Layer	Description
		Gradual and wavy transition to:
56 -	Cca	Limestone materials, hard quartzite, schist and sandstone materials.

Table 8. Chromic Vertisol / Traditional System – Bvc / TS

Depth (cm)	Layer	Description
00 - 12	A	2,5YR3/3 (Munsell Chart color) Clayey with fragments of limestone compact, structure composed of granular fine medium to strong and anisoformangularto subangularmedium, strong, medium to fine anastomosing cracks, slightly porous, very fine, medium tobig compactness, consistency very hard, slightly sticky and slightly plastic, with many fine roots, crack when dry, cool, great effervescence withHCl. Flat and smooth transition to:
13 - 19	AB	2,5YR3/4 (Munsell Chart color) Similar to above but clayey; structure composed of prismatic and anisoform angular to subangular, medium to coarse, strong; Flat and smooth transition to:
20 - 28	Bw	2,5YR3/5 (Munsell Chart color) Clayey with gravel of compact limestone; structure composed of prismatic and anisoform angular, medium and coarse, strong, some thin and medium vertical cracks, some very fine pores; consistency very firm, slightly sticky to sticky, slightly plastic to plastic; great compactness, great effervescence withHCl; Gradual and wavy transition to:
29 - 44	CB	Materials with friable limestone nodules in an advanced state of decomposition with enough clay (YR3 2.5/5 color), structure anisoformangular, medium to strong, which makegreat effervescence withHCl Gradual and wavy transition to:
45 -	C	Mixture of friable material of limestone yellowish red with limestone compact fragments.

Table 9. VerticCambisol / Direct Seedalong 10 years – Vc /DS (10)

Depth (cm)	Layer	Descrição
00 - 19	A	2,5YR3/4 (Munsell Chart color)

Depth (cm)	Layer	Descrição
		Clayey-loam with some gravel of quartz and limestone; structure composed of granular medium to fine and anisoform angular to subangular medium to coarse, strong, enough big; consistency very hard, sticky and plastic; many fine roots; cool. Flat and smooth transition to:
20 - 43	BA	2,5YR3/4 (Munsell Chart color) Similar as above clay with some gravel of quartz, schist and feldspar materials, structure composed of granular medium and anisoformsubangular medium to coarse, moderate to strong. Flat and smooth transition to:
44 - 66	Bt	2,5YR4/5 (Munsell Chart color)with small gray dark reddish spots. Clayey with gravel and some rubble of quartz, schist and feldspar materials; structure composed of prismatic and anisoformsubangular, medium and coarse, moderate to strong, medium and fine vertical cracks; very fine pores; consistency very firm, sticky to very sticky, plastic to very plastic; great compactness, aggregate faces with clay film and polished surfaces; additions of organic matter giving a mottled evident. On the basearestock of red clay materials (YR4 2.5/6 color); wet. Gradual and wavy transition to:
67 - 78	BCca	Lithological materials in an advanced decomposition stage with some clay (colour: 2,5YR4/6) present; highly effervescent in contact with HCl. Gradual and wavy transition to:
79 -	Cca	Limestone materials, hard quartzite, schist and sandstone materials.

Table 10. Chromic Vertisol / Direct Seed along 10 years

4.2. Physical and chemical properties of soils (their analytic integration)

Results are sequentially presented as the soils profiles description; mass fractions of fine and coarse components; volumetric mass (bulk density) and the resistance to the penetrometer; hydric limits for consistency; aggregates´ stability; coefficient of aggregation; aggregates´ size; organic matter (SOM) content; elemental N (nitrogen) total; C/N ratio; cation-exchange capacity (CEC); exchangeable cations (EC); base saturation degree (V) pores connectivity and soil-water retention curve (RC).

The measured and estimated physical and chemical parameters of the soils are summarized in Tables 2 to 7.

CEC is an effective chemical descriptor of the soil fertility and, in general, the higher the CEC the highest the fertility. It represents the maximum amount of cations a soil is capable of holding, at a given pH, available for exchanging with the soil solutions. It is expressed in units of milli-equivalent (meq [H^+]) of hydrogen ion per 100 g of dry soil, which is equivalent, and numerically coincident to its S.I. units, the cmol per kg. A clayey soil has higher CEE than a sandy soil and CEC increases if the formation of humus is promoted.

Titration: is a method or the process of determining the strength of a solution or the concentration of a substance in solution in terms of the smallest amount of reagent of known concentration required to bring about a given effect in reaction with a known volume of the test solution.

4.3. Pores connectivity: Vertical distribution of the soil pores type

In order to observe the vertical connectivity of the porous space (biopores, cracks and irregular spaces) along the entire profile data were organized in each soil type and production system by depth layer. Like Graphics 1 to 4 which shows the results obtained for the biopores, others have been made for the cracks and irregular spaces but there results are only referred because they are published in (Sampaio E., 2009).

4.4. Soil-water retention curves

The soil-water retention curves (RC) are presented in Table 8. These RC are presented in order to highlight the interactions between soil types, on one hand, and the land-use systems, on the other hand. For each of the three pF values, the volumetric soil-water content θ (m^3 m^{-3}) has been determined at four our six soil-layers depth. Once the variations of θ was low, only mean values are presented per each ψ_m value of the RC curves.

4.5. Estimation of crop biomass production

At this point the fertility of both (Bvc and Vc) soil types has been characterized through quantitative physical and chemical analysis. However the potential maximum total crop dry matter (DMp) production under rain-fed condition is generally limited in this sort of soils by frequent intra- and inter-annual soil-water (ASW) shortage and chronic low organic mater (SOM) content (Kassam et al., 2012; Zdruli, P. et al., 2004), thus low available elemental nitrogen (N) for plant nutrition and low C/N ratio for soil microbes´ activity.

Now, consider a winter-wheat cultivar that yields 3 000 kg per ha and year in average, under the TS, intensive rain-fed production system (Barros, et al., 2004). According to the procedure and parameters described in section 3.9, the corresponding TDM would be 7 800 kg ha^{-1} per year. But for the available soil-water holding capacity (WC) the effective rooting depth is Zr = 60 cm, for the "Soil Bvc/DS(10)" in Table 5, with increased soil depth after ten years under DS treatment. Dividing TDM by the adjusted WUE = 6.44 g kg^{-1} and setting ASW = 0.2 m^3 m^{-3}, the amount of total transpired water, so returned, is WT = 121 mm to justify that dry mater production, imposed under VPD = 0.8 kPa. This result is just for balancing the calculated rooting depth (Zr = 60.6 cm) to the observed effective mean soil depth (as above), setting the numerical convergence relative error to −0.97%, essentially zero. For the same soil, but under TS treatment foe which Zr = 40 cm, the calculated WT is only 80 mm.

The expected Zrx that conforms the wheat's LAIx is 1.01 m, thus a water deficit of 45% is estimated for the growing season. The practical effect of this is a potential biomass deficit of wheat crop under rain-fed Mediterranean. Note that total demand for water by crops are compared to the net water input into the agro-ecosystem, as calculated through a water balance model (Thornthewaite), in which rooting depth is a key parameter. For the historical climatic period of 1951-1980, for example, the estimated annual total of actual evapotranspiration is ETR = 450 mm. This means that WT just calculated for wheat production under DS(10) and Zr = 60 cm represents the fraction of 0.27 ETR versus 0.18ETR for TS and Zr = 40 cm.

5. Discussion

The soil profiles in both soils subjected to the DS system have evolved pedogenetically, i.e., horizons differentiation, and also have augmented their maximum depth and, thus, the potential rooting depth and water-holding capacity.

Soil texture is analysed in comparing both soil types and the effects of both (DS an TS) production systems on each one. The profiles were very similar in texture but when subjected to the DS system for 10 years the more surfacial horizons in the Bvc soil became coarser than previously, meaning a relative loss of finer elements; the Vc soil maintained its texture unaltered. In this soil, the coarse fraction is more important than in Bvc, ranging between 9% at the upper soil-layer and 46% at 40-cm depth in the former.

In the Vc soil, the vertical distribution of bulk density (BD) showed a relative decrease during the 10-year period under DS system which is attributed here to a less compaction situation than in the ploughed field under TS. Although the increase of BD with depth is natural, under DS it revealed an almost constant distribution. This can be attributed, among other factors, to a more intensive biological activity such as plant root colonization and/or soil fauna activity that promotes water infiltration. In the Bvc the vertical distribution of BD maintained its natural behaviour but at lower values under DS than under TS. Once again, this effect reflects the augmentation of the porosity, or a corresponding decrease in soil compaction under DS. The same applies to the soil resistance (to the penetrometer) but these effects are amplified.

The hydric limits for the workability of both soil types increased and stabilized at depth during the 10-year period under DS.

This parameter is indirectly accessed via the combination of the aggregation coefficient (AC) and size distribution of aggregates. Thus, both soils′ AC showed an effective increase under DS ranging between 23% and 48% in the upper soil-layer and from 36% to 44% at 40-cm depth, for the Vc soil. In turn, the corresponding ranges for the Bvc soil are 27%-45% and 29%-36%, respectively. These figures sustain the experimental evidence that the DS production system is a more conservative agricultural technique in respect to this physical soil fertility descriptor than the ploughed TS is.

The size distribution of the aggregates is embodied in the evaluation of the stability of the aggregates′ parameters. For both Vc an Bvc soils, total percentage of aggregates increased a lot under DS while under TS the soil matric is signifficatively less aggregated, what degenerates a loose soil, so more susceptibility to erosion process for the same land cover. The size classes of aggregates that have been increased under DS are the media (1 – 0.25 mm in diameter) (Table 5); the two soils did not show signifficative difference in this parameter.

Soil Vc / TS		Depth limits of soil layers (cm)			
	Units	**0.0 – 10**	**11 – 20**	**21 – 30**	**31 – 40**
Aggregation coefficient	%	23.40	23.54	38.78	36.50
Diameter (Ø) class (mm)					
"/> 5.00	%	0.07	0.56	0.00	3.00
5.00 – 2.00	%	0.67	1.38	2.03	3.35
2.00 – 1.00	%	1.24	0.71	7.18	3.39
1.00 – 0.50	%	5.29	3.50	14.53	4.93
0.50 – 0.25	%	13.00	12.32	14.85	15.23
0.25 – 0.10	%	12.54	12.57	12.44	9.19
Soil Bvc / TS		**Depth limits of soil layers (cm)**			
	Units	**0.0 – 10**	**11 – 20**	**21 – 30**	**31 – 40**
Aggregation coefficient	%	27.12	27.35	28.16	28.80
Diameter (Ø) class (mm)					
"/> 5.00	%	0.20	0.00	0.57	0.67
5.00 – 2.00	%	0.60	0.94	2.27	1.50
2.00 – 1.00	%	1.29	1.83	1.40	2.57
1.00 – 0.50	%	8.90	9.24	8.65	8.75
0.50 – 0.25	%	9.51	9.85	10.67	10.80
0.25 – 0.10	%	13.10	14.15	13.71	13.81
Soil Vc / DS (10)		**Depth limits of soil layers (cm)**			
	Units	**0.0 – 10**	**11 – 20**	**21 – 30**	**31 – 40**
Aggregation coefficient	%	47.76	60.80	61.08	43.54
Diameter (Ø) class (mm)					
"/> 5.00	%	0.11	0.39	0.07	0.04
5.00 – 2.00	%	0.29	7.29	6.93	1.85
2.00 – 1.00	%	2.64	14.79	15.55	3.52
1.00 – 0.50	%	15.79	18.96	16.72	17.58
0.50 – 0.25	%	18.47	9.57	11.28	15.84
0.25 – 0.10	%	12.92	5.75	6.90	12.47

Soil Bvc / DS (10)		Depth limits of soil layers (cm)					
	Units	0.0 – 10	11 – 20	21 – 30	31 – 40	41 – 50	51 – 60
Aggregation coefficient	%	45.26	32.30	34.70	35.56	45.74	47.44
Diameter (Ø) class(mm)							
"/> 5.00	%	0.00	0.04	0.00	0.00	0.04	0.00
5.00 – 2.00	%	1.08	0.07	0.07	0.22	1.94	0.76
2.00 – 1.00	%	1.58	0.65	1.60	3.33	4.97	4.94
1.00 – 0.50	%	11.26	9.49	13.00	13.16	14.84	16.98
0.50 – 0.25	%	16.16	16.06	16.40	18.08	17.35	21.23
0.25 – 0.10	%	9.93	14.16	12.30	11.72	9.68	13.52

Table 11. Aggregate stability, coefficient of aggregation and aggregates size

It is seen from Table 6 that the SOM content in both soils has increased along the entire profiles, after having been submitted to the conservative (DS) production system, as an adequate alternative system to the TS one; this augmentation is more evident in the upper 30-40-cm soil-layer. This result is expected since the stubble in the field determines less soil-surface heating (relative to bare soil), stabilizes the soil temperature regime over time and conserves water as direct evaporation is diminished (Gill, S. and Jalota, S., 1996). These are conditions that lead the plant residues left on the soil surface to have more opportunity under humification rather than under mineralization. Ultimately the SOM residence time in the soil augments (Zdruli, P. et al. 2004).

Referring to N and C/N ratio, it is interesting to note that, surprisingly, the N content in the Vc soil diminished at the upper soil-layer under DS which is in contrast with the observed increase in C/N and SOM. This apparent duality may essentially reflects a more relative increase in C than in N since under the same environmental conditions the mineralization rate of N is greater than that of C of organic origin. From the crop production perspective, the Vc soil should be supplied with a supplemental fertilization in N in order to prevent competition for N demand between the plants and soil microfauna that are present. Why this only happened in the Vc soil, the results obtained with this study do not permits an explanation. Finally, in Vc soil, the mean measured maximum C/N ratio values ranged between 7.20 ± 1.02 (for TS) and 10.00 ± 0.85 (for DS); in the Bvc soil the differences of that intervals were not significantly different at 5% level (± C.I. 95% probability; $n = 4$ for all cases).

The general pattern of the data on CEC (or T) after a 10-year period shows an increase in both Vc and Bvc soils under DS system, but in Vc this increase has been greater at the soil surface while in the Bvc it happened deeper in the soil profile. As CEC did, the exchangeable bases, CE (Table 7), revealed spatially in a similar fashion in which Ca^{2+} has been the more influent cation. Finally the pattern of the bases saturation degree (V) is a consequence of CE and T behaviours.

Soil Vc / TS		Depth limits of soil layers (cm)			
Parameters	**Units**	**0.0 – 10**	**11 – 20**	**21 – 30**	**31 – 40**
N total	%	0.18	0.11	0.10	0.09
Organic matter (SOM)	%	1.90	1.60	1.32	1.02
Organic C	%	1.10	0.93	0.77	0.59
C/N	%	6.12	8.44	7.66	6.57
Soil Bvc / TS		**Depth limits of soil layers (cm)**			
	Units	**0.0 – 10**	**11 – 20**	**21 – 30**	**31 – 40**
N total	%	0.08	0.07	0.07	0.06
Organic matter (SOM)	%	1.32	1.24	1.18	1.05
Organic C	%	0.77	0.72	0.68	0.61
C/N	%	9.57	10.27	9.78	10.15
Soil Vc / DS (10)		**Depth limits of soil layers (cm)**			
	Units	**0.0 – 10**	**11 – 20**	**21 – 30**	**31 – 40**
N total	%	0.13	0.12	0.10	0.10
Organic matter (SOM)	%	2.18	1.97	1.95	1.64
Organic C	%	1.26	1.14	1.13	0.95
C/N	%	9.69	9.50	11.30	9.50

Soil Bvc / DS (10)		Depth limits of soil layers (cm)					
	Units	**0.0 – 10**	**11 – 20**	**21 – 30**	**31 – 40**	**41 – 50**	**51 – 60**
N total	%	0.09	0.09	0.08	0.08	0.08	0.07
Organic matter (SOM)	%	1.86	1.64	1.34	1.32	0.79	0.65
Organic C	%	1.08	0.95	0.78	0.77	0.46	0.38
C/N	%	11.99	10.57	9.72	9.57	5.73	5.39

Table 12. Soil organic matter, N total and C/N ratio

Soil Vc / TS		Depth limits of soil layers (cm)			
Parameters	**Units**	**0.0 – 10**	**11 – 20**	**21 – 30**	**31 – 40**
CEC (or T)	meq/100g	14.55	13.63	26.84	14.80
S	meq/100g	13.05	11.73	26.84	14.10
V	%	89.69	86.06	100	95.27
Ca	meq/100g	10.2	9.17	23.55	11.24
Mg	Idem	2.13	12.00	2.71	2.39

K	meq/100g	0.28	0.17	0.14	0.09
Na	meq/100g	0.44	0.39	0.44	0.44
Titratable H	meq/100g	1.50	1.90	0.00	0.70

Soil Vc / TS		**Depth limits of soil layers (cm)**			
	Units	**0.0 – 10**	**11 – 20**	**21 – 30**	**31 – 40**
CEC (or T)	meq/100g	26.86	15.80	32.20	17.40
S	meq/100g	25.16	13.80	32.10	14.70
V	%	93.70	87.34	99.67	84.48
Ca	meq/100g	23.23	12.22	30.18	12.39
Mg	meq/100g	1.02	0.92	1.13	0.98
K	meq/100g	0.39	0.27	0.35	0.85
Na	meq/100g	0.52	0.39	0.44	0.48
Titratable H	meq/100g	1.70	2.00	0.10	2.70

Soil Vc / DS (10)		**Depth limits of soil layers (cm)**			
	Units	**0.0 – 10**	**11 – 20**	**21 – 30**	**31 – 40**
CEC (or T)	meq/100g	26.86	15.80	32.20	17.40
S	meq/100g	25.16	13.80	32.10	14.70
V	%	96.67	87.34	99.67	84.48
Ca	meq/100g	23.23	12.22	30.18	12.39
Mg	meq/100g	1.02	0.92	1.13	0.98
K	meq/100g	0.39	0.27	0.35	0.85
Na	meq/100g	0.52	0.39	0.44	0.48
Titratable H	meq/100g	1.70	2.00	0.10	2.70

Soil Bvc/ DS (10)	**Depth limits of soil layers (cm)**						
Parameters	**Units**	**0.0 – 10**	**11 – 20**	**21 – 30**	**31 – 40**	**41 – 50**	**51 – 60**
CEC (or T)	**meq/100g**	14.73	17.87	16.88	27.24	28.34	31.55
S	**meq/100g**	11.73	15.37	13.68	25.34	26.64	28.95
V	**%**	79.70	86.10	81.10	93.10	94.00	91.80
Ca	**meq/100g**	9.72	12.33	10.84	22.88	24.00	25.05
Mg	**meq/100g**	1.29	2.33	2.21	1.45	1.92	2.54
K	**meq/100g**	0.35	0.32	0.17	0.60	0.31	0.17
Na	**meq/100g**	0.37	0.39	0.46	0.41	0.41	0.59
Titratable H	**meq/100g**	3.00	2.50	3.20	1.90	1.70	2.60

Table 13. Cation-exchange capacity (CEC), exchangeable cations (EC) and bases saturation degree (V)

		Vc / TS (a)	Vc / DS.10 (b)	Bvc / TS (c)	Bvc / DS.10 (d)	Bvc / DS.6 (e)	Difference for Vc (b) – (a)	Difference for Bvc (l.u) (d) – (c)
pF	$-\psi_m$ kPa				θ $m^3 m^{-3}$			
1.8	6	0.27 (0.01)	0.36 (0.02)	0.32 (0.01)	0.44 (0.01)	0.38 (0.01)	0.09	0.12
2.54	35	0.19 (0.02)	0.26 (0.02)	0.21 (0.02)	0.35 (0.02)	0.27 (0.02)	0.07	0.13
4.2	1584	0.12 (0.01)	0.16 (0.02)	0.11 (0.01)	0.20 (0.01)	0.15 (0.01)	0.04	0.09

Table 14. Soil-water retention against matric-potential (ψ_m) for both (Vc an Bvc) soils under the traditional (TS) or direct-seeding (DS) system and differences in soil-water retention induced by each system and by time (only Bvc soil).

The titratable H^+ ion increased under DS in both soil types and along the depth of all soil profiles. This means that the alternative, conservation agriculture system tends to increase the hazard of acidifying the soils but in this particular situation it does not represent an effective risk since the soils $pH(H_2O)$ is in the range 7.5-8 (data no shown).

In both soil type it is evident the increase of the vertical pore connectivity at depth but these results are much more representative in soil Bvc. The dimension classes which gain more area with the conservative system are the medium ones (diameter between 1 and 5 mm and between 1 and 0.4 mm) and the biggest class (diameter greater than 5 mm) disappeared in this system. These results should be due to the increasing accumulation of plant residues and soil fauna (i.e. SOM) in this system, in opposition to the contrary effect of less SOM as a consequence of the fragmentation of the residues in the ploughed soil that characterizes the traditional system.

While the ψ_m-derived values correspond to the conventional values at field capacity (FC), *i.e.ca.* 1/3 atmosphere, and permanent wilting point (PWP), at 15 atmosphere, their associated θ and the resulting available soil-water (ASW) fractions (0.07 – 0.15 $m^3 m^{-3}$) are relatively low (the measured mean ASW is 0.11 ±0.03 $m^3 m^{-3}$) for a clayey or a silt-clay soil, as is the case. It can be seen from the results summarized for 21 soils textural classes by Federer, C., et al., 2003 that the ASW values are typically >0.2 $m^3 m^{-3}$ for clay and silt-clay soils, which is at least two-fold greater than the one under consideration. Although in absolute terms, the ASW measured values are relatively low, for comparison purposes, they are safe. In this sense, the differences in available soil water induced by cropping system in both soils are quantitatively important and conservative for the three soil-matric potentials already referred to here. The ASW mean values are 0.07 ± 0.03 and 0.11 ±0.02 $m^3 m^{-3}$ for Vc and Bvc soils, respectively.

In the previous section(s) soil's fertility has been characterized through quantitative physical and chemical analysis. However the potential and the grain crop dry matter production under rain-fed condition is limited in these soils under the local climate by frequent intra- and inter-annual soil-water (ASW) shortage and chronic low organic mater (SOM) content, thus, low available elemental nitrogen (N) for plant nutrition and low C/N ratio for soil microbes' activity. These agro-ecological threats are imposed by climate and anthropic actions and such circumstances may introduce a systematic deviation between the observed mean values of DM and DMp.

The biophysical sustainability of this dry mater estimation must be envisaged in the scope of the soil water-holding capacity for the typical rooting depth of 600 mm of the Bvc soil after a 10-year period under DS system (Zr = 60 cm; see Table 5 for "Soil Bvc/DS(10)"), for instance, whose ASW was set equal to 0.20 m^3 m^{-3} as a clayey soil (Federer, C., et al., 2003). For these conditions the integration at depth gives 121 mm of water for plant uptake (the soil contribution to seasonal ETR) and the need to bring the soil moisture status at field capacity 3.7 times during the growing season to match the annual ETR. According to the results, a direct consequence of the DS system on the Bvc soil hydrologic state was an increased of 0.11 m^3 m^{-3} of the available soil water relative to the TS system (Table 5 for "Soil Bvc/TS"), which means for this last system a 0.09 m^3 m^{-3} of less water then the previous figure and only 80 mm of water (Zr = 40 cm) for plant uptake. Shortly, it would be necessary to refill the soil water reserves ca. 10 times in an annual cycle to achieve ETR for continuous primary production. Based on these calculations it is clear that crops under the TS system are more subjected to water stress risks than under the alternative, DS system

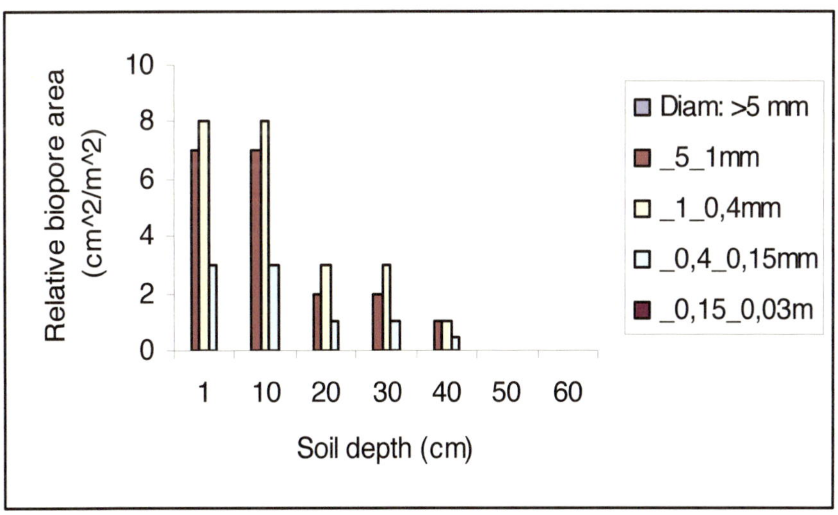

Graphic 1. Relative area of biopores dimension classes in Soil Vc / TS

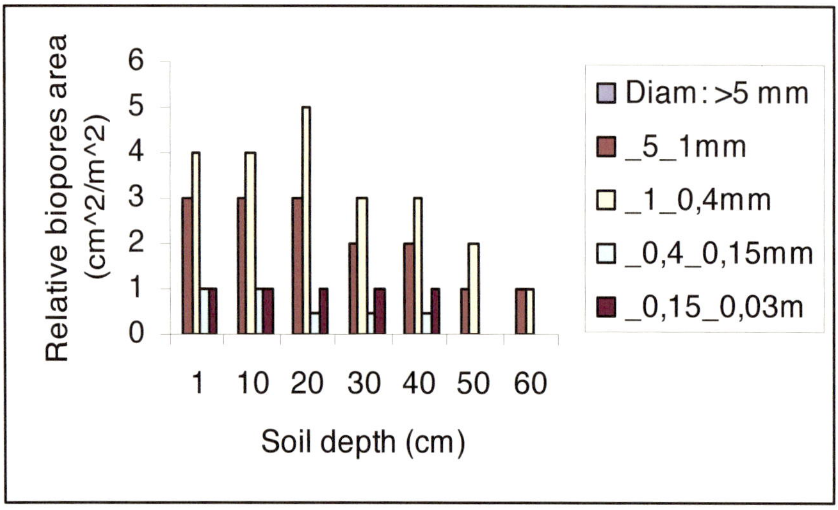

Graphic 2. Relative area of biopores dimension classes in Soil Bvc / TS

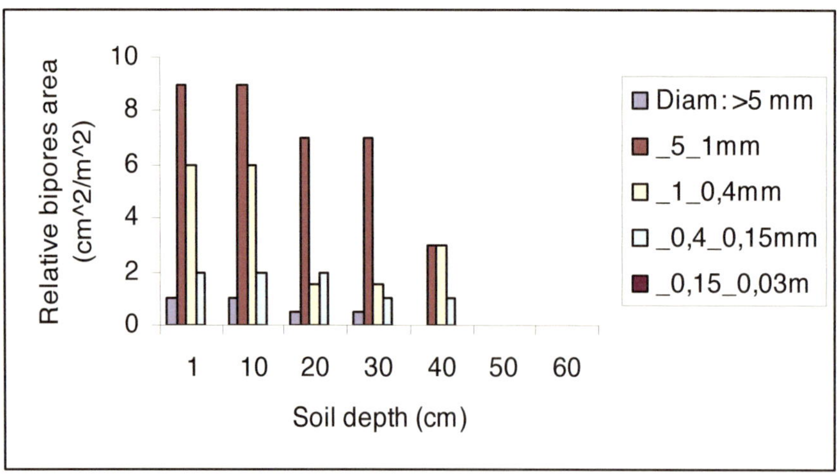

Graphic 3. Relative area of biopores dimension classes in Soil Vc / DS (10)

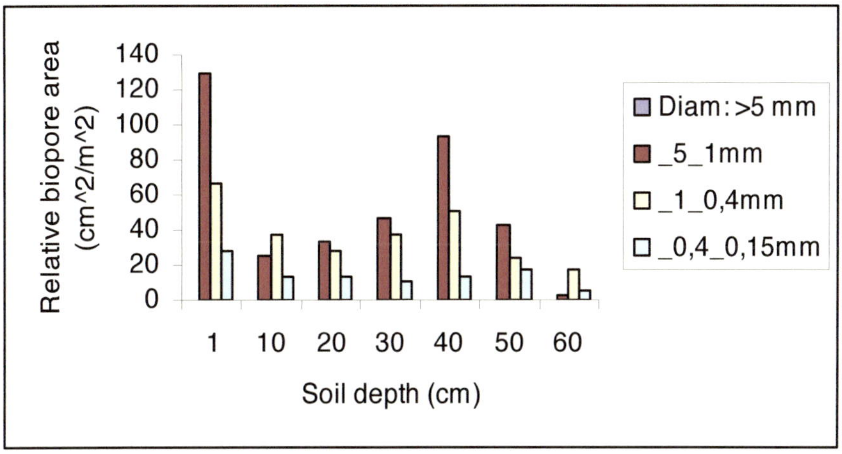

Graphic 4. Relative area of biopores dimension classes in Soil Bvc / DS (10)

6. Conclusions

In this study it has been physically and chemically characterized two of the considered more fertile soils (Vc and Bvc) of the Portuguese soil classification nomenclature, which were the base of two alternative, cereal crop production systems. Results support the conclusion that the traditional (mechanized) system (TS) negatively impacts pedogenesis and also leads the soil a degraded condition in terms of SOM and N contents, C/N ratio; compaction, workability, structural stability, CEC and associates parameters of the soil profile; connectivity of soil pores and water holding capacity. By the contrary, all these soil properties have been ameliorated, independently of the soil type, when subjected to the alternative DS system over the same time (a 10-year period).

Under DS system and contrary to that seen for the Bvc soil, in the Vc soil N decreased in the upper soil-layer, despite the increasing in SOM content and C/N ratio. This finding suggests that to promote an increase in the potential crop production in the Vc soil, a supplemental N-fertilization is needed to compensate that behaviour.

In a general perspective, and in terms of water availability in the ecosystem, the alternative DS system led to an increase in: i) soil pores connectivity at depth; ii) effective soil depth and; iii) water-holding capacity. It is highlighted that the promotion of the soil-water holding capacity under DS is advantageous in water-limited environments because of its characteristic seasonal precipitation regime, as it happens in the Mediterranean. These effects contribute to a longer residence time (WT/WC) of the native soil water and also to a higher buffering effect (also CEC increased) of the porous medium. Consequently, potential groundwater contamination risk is diminished and the water quality for groundwater-dependent ecosystems may be prevented.

Relative to TS, the alternative DS system proved to be effective in ameliorating all the soil properties analysed, with special relief for SOM content, this one having positive effects on other soil fertility parameters; also under DS it was verified an increase in soil depth and so water saving, and this is translated into a marginal gain in crop dry matter. For continuous cropping this marginal gain in DM under DS also represents a potential increase in carbon sequestration capacity of these agro-ecosystems, recalling that C accounts for 40% DM.

Although for the Bvc and Vc clayey soils the crop biomass production under rain-fed regime is the greatest reported for the Alentejo (Barros, J. et al. 2004), it is concluded that the dry mater production is usually lower than the local agro-ecological potential, even the soil is not limiting plant rooting. This deficit in biomass production of a given species (e.g. wheat) can be evaluated by comparing a rain-fed and an irrigated production system. Comparing both DS and TS for dry mater production, the last system is more limiting plant rooting depth and so water availability. Following this, the difference between the actual and the potential dry matter production may stimulate the agriculture enterprise. This activity is recommended to be hold in the scope of the conservation agriculture (CA) which makes use of all modern technologies available to produce (Dumanski, R. et al. 2006), while simultaneously promoting low fuel-based energy inputs, high efficiency of input factors, optimized crop productivity and better soil protection against water erosion. This philosophy is thought to guarantee the ecological sustainability of the whole agro-ecosystem under management.

In the light of the results obtained here, a general improvement of the new crop production system that results from intensive monocultures to conservation agriculture (CA) practices is promising, particularly in what concerns the increase in soil-water holding capacity under a changing environment associated to increased dryness risks (IPCC, 2001). Taking into account that CA is greatly increasing worldwide, and that there exists a real deficit of dry matter production in semi-arid areas of ca. 20-30% (Kassam, A. et al. 2012), the adoption of the CA paradigm (Dumanski, R. et al. 2006) is an ecological, technical and economical secure way for facing the world balance on food demand/supply problem.

Acknowledgments

"This work is funded by FEDER Funds through the Operational Programme for Competitiveness Factors - COMPETE and National Funds through FCT - Foundation for Science and Technology under the Strategic Project PEst-C/AGR/UI0115/2011".

Author details

Elsa Sampaio[1,2*] and Júlio C. Lima[1]

*Address all correspondence to: ems@uevora.pt

1 Geosciences Department, University of Évora, ICAAM - Instituto das Ciências Agronómicas e Ambientais Mediterrânicas, Évora, Portugal

2 University of Évora, Institute of Agrarian and Environment Sciences of the Mediterranean (ICAAM), Portugal

References

[1] Abreu, J. P. M. (1994): Modelo de crescimento e produção da cultura do trigo em condições mediterrânicas. Ph. D. Thesis; Instituto Superior de Agronomia (ISA), Lisboa, Portugal.

[2] Agri-environmental indicators (IRENA indicators), (2010): http://epp.eurostat.ec.europa.eu/portal/page/portal/agri_environmental_indicators/indicators_overview/farm_management_practices

[3] Ahmad R. Hassan B. and Jabran K. (2007): Improving Crop Harvest Index: http://archives.dawn.com/2007/10/01/ebr6.htm

[4] Bach, G. Barros, J.F.C., Calado, J M G.; Brandão, M L C. (2008): The potential of notillage and residue management to sequester carbon under rain-fed Mediterranean conditions. Proceedings of the 5Th International Scientific Conference on Sustainable Farming Systems; Nov-5-7, 2008, inPiest´anySlovakia.

[5] Barros, J.F.C. Carvalho, M. Basch, G. (2004): Response of sunflower (Helianthus annuus L.) to sowing date and plant density under Mediterranean conditions. European Journal of Agronomy, 21 (3): 347-356

[6] Bolle, H.-J.; Eckardt, M.; Koslowsky, D.; Maselli, F.; MeliaMiralles, J.; Menenti, M.; Olesen, F.-S.; Petkov, L.; Rasool, S.I.; Griend, A. (2006): Mediterranean Land-Surfaces Processes Assessed from Space. ISBN 10 3-540-40151-2, Springer BerlinHeidelberg Ney York; © Springer-VerlagBerlin Heidelberg 2006.

[7] Bouma J., Belmans C. F. M. and Dekker l. W., (1982): Water infiltration and redistribution in a silt loam subsoil with vertical worm channels. Soil Science Society of America Journal, 46: 917-921.

[8] Blume H. P. (1984): BodenKundlicheslaborpraktikum. Institut fur Rflanzenesnahvunj und BodenKunde der CAU, Kiel.

[9] Brandt, C. Jane and Thornes, Jonh B. (Ed.) (1996): Mediterranean Desertification and Land Use. ISBN 0-471-942250-2; © 1996 John Willey and Sons Ltd. England (UK).

[10] Cardoso J. C., (1965): Os solos de Portugal, sua classificação, caracterização e génese. 1 – A Sul do Rio Tejo. Secretaria de Estado da Agricultura, Lisboa.

[11] Dumanski, R.; Peiretti, J. Beniets R.; McGarry D. and Pieri C. (2006): The paradigm of conservation agriculture. Proc. World Assoc. Soil and Water Conserv., P1:58-64.

Available at the web-site: http://www.unapcaem.org/publication/ConservationAgri/ParaOfCA.pdf

[12] ECAF (European Conservation Agriculture Federation), (2012): http://www.ecaf.org/index.php?option=com_contentandtask=viewandid=69andItemid=64

[13] Federer, C. A.; Vorosmarty, C. J. and Fekete B. (2003): Sensitivity of annual evaporation to soil and root properties in two models of contrasting complexity, Journal of Hydrometeorology, 4: 1276-1290.

[14] Gill, B.S. and Jalota, S.K. (1996): Evaporation from soil in relation to residue rate, mixing depth, soil texture and evaporativity. Soil Technology 8: 293-301.

[15] Goudriaan, L. and J. L. Monteith (1990): A mathematical function for crop growth based on light interception and leaf area expansion. Ann. Bot. 66: 695-701

[16] INMET (InstitutoNacional de Meteorologia): NormaisClimatológicas, (2012): http://pt.allmetsat.com/clima/index.html

[17] IPCC (2001), Climate Change – Synthesis Report. WMO, UNEP.

[18] Jørgensen, Uffe and Kirsten Schelde (2001): Energy crop water and nutrient use. Prepared for the "The International Energy Agency, IEA Bioenergy Task 17, Short Rotation Crops. Available at the web-site http://infohouse.p2ric.org/ref/17/16275.pdf.

[19] Journal of Earth System Science (2012): http://www.springer.com/earth+sciences+and+geography/journal/12040

[20] Kassam, A. et al., (2012), Conservation Agriculture in dry Mediterranean climate. Field Crops Res. doi:101016/j.fcr.2012.02.023.

[21] Kimmins, J.P. (1997): Ecological Principles: http://www.cfr.washington.edu/classes.esrm.201su/Reading%20assignments/Readings/Ecological%20principles.pdf

[22] Le Houérou, Henry N. (1996): Climate Change, Drought and Desertification. Journal of Arid Environments,34:133-185.Accessed-at: http://www4.nau.edu/direnet/publications/publications_l/files/LeHouerou_1996.pdf

[23] Lima, Júlio C. (1996): Ecofisiologia da oportunidade da rega no milho. M.Sc. Dissertation. Universidade de Évora, Portugal. 112 pp.

[24] Lima, Júlio C. and Sequeira, Eugénio M. (2004): A conservação do solo e da água como base da diversidade biológica dos ecossistemas agrícolas/ "Water and soil conservation as the base for biological diversity of rural ecosystems", p. 63-66, Proceedings of the Intern.Congr. "Rural ecosystems and biological richness" (LPN Ed.: www.lpn.pt), hold at Castro Verde, Portugal; Oct. 31st –Nov. 1st, 2004.

[25] Lima, J.; Corte-Real, J. and Sampaio, E. (2011): Delimiting tree territory from soil-water balance equation: a case study. Book of Abstracts, p. 58. Intern.Conf. Ecohydrology and Climate Change; Tomar-Portugal, 15-17.09.2011. www.ecohccc2011.ipt.pt

[26] Lindroth, A. and Cienciala, E. (1996): Water-use efficiency on short rotation Salix viminalis at leaf tree and stand scales. Tree Physiology 16: 257-262

[27] Lourenço, Maria Ermelinda V.; José E. D. Regato; Suzana F. Dias, Maria José S. Dubraz C. Vivas (2000). Avaliação de Culturas Alternativas não-alimentares. Relatório final do Projecto PAMAF Nº. 1016, 56 pp. Universidade de Évora.

[28] Marta-Pedroso, Cristina; Helena Freitas and Tiago Domingos (2009): A estepe cerealifera de Castro Verde (Cap. 16). In Ecossistemas e bem-estar humano – avaliação para Portugal do Millenium Ecosystem assessment. Pereira, H. M. et al. (ed.), ISBN 978-972-592-274-3, Lisboa, Portugal.

[29] Monteith, J. L. (1972): Solar radiation and productivity in tropical ecosystems. J. App. Ecol. 9: 747-766 (December 1972).

[30] Monteith, J.L. (1993): The exchange of water and carbon by crops in a Mediterranean climate Irrig. Sci. 14: 85-91.

[31] Pinto-Correia, T. and Mascarenhas, J. (1999).Contribution to the extensification/intensification debate: new trends in the Portuguese Montado. Landsc Urban Plan 45: 125-131.

[32] Rodriguez-Iturbe and AmilcarePorporato (2004): Ecohydrology of Water-controlled Ecosystems: soil moisture and plant dynamics; Cambridge University Press, ISBN-13 978-0-521-03674-0; Cambridge, New York.

[33] Sampaio, Elsa Paula (2002): A evolução do perfil cultural dos solos sujeitos a diversos tratamentos. Ph. D. Thesis, Universidade de Évora, Évora, Portugal.

[34] Sampaio E. (2009): Estudio de las Prácticas Culturales, Porosidad del Suelo y Gestión Hídrica, en el Combate de la Desertificación. InformaciónTecnológica. 20 (3), 101-112.

[35] Sampaio E. and Sampaio J. (2010): Advances in morphometry of soil macro-porosity through simple techniques of mathematics, International Agrophysics. 24 (3), 303-311

[36] Schneider J. and Stunke A. (1991): Vergleich von Vertisolstandortenunterdireksaat und traditionellerbearbeitung in SdlichenAlentejo/Portugal. UniversitatHohenheim, Stuttgart.

[37] Stockle, Claudio O. (1996): CropSyst – Cropping System Simulation Model Manual. WashingtonStateUniversity; www.bsyse.edu/

[38] Tanner, C, B. and T. R. Sinclair (1983): Efficient water use in crop production: research or re-search? In: Taylor H; Jordan WR, Sinclair TR (eds) Limitations to water use in crop production. ASA, CSSA, SSSA, Madison, Wis, USA, 1-27.

[39] Tenhunen, J. D.; F.M. Catarino; OLL Lange and W. C. Oechel (1987): Plant Response to Stress – Functional Analysis in Mediterranean Ecosystems. NATO ASI Series; Series G: Ecological Sciences, Vol. 15; ISBN 3-540-16082-5. Springer BerlinHeidelberg Ney York; © Springer-VerlagBerlinHeidelberg, 1987.

[40] Weerts, A. H.; Kandhai, D.; Bouten, W. and Sloot, P. M. A. (2001): Tortuousity of an unsaturated sandy soil estimated using gas diffusion and bulk soil electrical conductivity: comparing analogy-based models and Lattice-Boltzmann simulations. Soil Science Society of America Journal. 65: 1577-1584.

[41] White, E.M. and F.E.A. Wilson (2006): Responses of grain yield, biomass and harvest index and their rates of genetic progresses to nitrogen availability in ten winter wheat varieties. Iris Jo. Agric. Food Res. 45: 8-101.

[42] Wösten, J.H.M.; Pacheps, Ya.A.and Rawls, W.J. (2001): Pedotransfer functions: bridging the gap between available basic soil data and missing soil hydraulic characteristics. Journal of Hydrology 251 (2001): 123-150.

[43] WCCS/World Climate Classification System (2009): http://www.physicalgeography.net/fundamentals/7v.html

[44] Zdruli, Pandi; Robert J.A. Jones and Luca Montanarella (2004): Organic Matter in Soils of Southern Europe. European Soil Bureau Research Report No. 15, Office for Official Publications of the European Communities. Luxembourg. EUR 21083 EN, 16 pp.

Chapter 7

Soil Behaviour Characteristics Under Applied Forces in Confined and Unconfined Spaces

Seth I. Manuwa

Additional information is available at the end of the chapter

http://dx.doi.org/10.5772/52774

1. Introduction

Soil strength has been regarded as important characteristics that affect many aspects of agricultural soils, such as the performance of cultivation implements, root growth, least-limiting water range and trafficabilty [1]. They further reported that characterization of soil strength is usually made by measuring the response of a soil to a range of applied forces.

Soil compaction may be defined as the densification of unsaturated soil due to reduction in air volume without change in mass wetness [2].Soil compaction occurs in unsaturated soils when subjected to mechanical forces [3]. While soil compaction is essential in many engineering works (especially civil engineering) it is undesirable in agricultural production to a large extent. Compaction reduces the soil permeability to water, so that run off and erosion may occur and adequate recharge of ground water is prevented. Compaction reduces regeneration of the soil, so that metabolic activities of roots are impaired. Compaction increases the mechanical strength of the soil, so root growth is impeded. It is known that in agricultural system, the risk of soil compaction increases with the growth of farm size, increased mechanization and equipment weight, and the drive for greater productivity. Soil compaction also has negative effects on the environment by increasing runoff and erosion thereby accelerating potential pollution of surface water by organic wastes and applied agrochemicals [4]. All of these effects may reduce the quality and quantity of food and fibre grown on the soil. Therefore, the knowledge of soil compaction is increasingly important and desirable within agriculture and environmental protection.

The state of soil compactness is expressed in several ways: bulk density (expressed on a wet or dry basis), porosity and apparent specific gravity [5]. Accurate compaction behaviour equations will provide a means to predict compaction. The ability to predict compaction is the first requirement for attaining control of compaction. Considerable research has been performed in attempts to develop soil compaction behaviour equations [6, 7, 8, 9, 10, 11, and 12]. Others have also reported on effects of organic matter and tractor passes on compaction and yield of crops [13, 14].

The aim of this chapter therefore was to observe the behaviour of textured soils under uni-axial compression in confined spaces and also in unconfined spaces under the influence of agricultural machines as it is affected by applied pressure and water content and also to model the behaviour using regression analysis for the purpose of prediction.

2. Soil under pressure

When forces are applied to soil it changes its stateof compactness to have greater density. The pressure can be applied pressure or natural forces. Applies pressure can be due to animals like horses and cattle (160 kPa), man (95 kPa), sheep (60 kPa). The pneumatic wheels of vehicles can be inflated to 100 kPa for tractors and 200 kPa for for trailers and may produce local stresses twice these values [3]. The problem therefore is that over the years arable soils have suffered degradation as a result of excessive compaction of surface and sub surface soil horizon through overuse of agricultural machinery. This trend has global phenomenon as man labours to meet the ever increasing demand for food, feed and fibre for the increasing population of the world.

2.1. Application area

Soil compaction has many areas of application in human endeavour apart from agriculture and forestry. When a seed is sowed the soil has to be sufficiently firmed around it (compacted) to to enable it germinate. Arable terrains require appropriate level of compaction for it to be effectively trafficked by running gears like wheels and tracks. In sports the field must have the required strength or bearing capacity (a measure of some level of compaction) for it to be good for sporting activities.

2.2. Research course

In this chapter soil compaction is considered in two regimes: in confined and unconfined spaces. The former is typical of studies under laboratory conditions while the latter signifies field conditions. The three experimental soils studied under uni-axial compression have textures of sandy clay loam, loamy sand and silt loam according to the USDA Soil Textural Classification, while the soil studied under field condition was sandy clay. In Nigeria, published data on compressibility and compaction of agricultural and forestry soils are scarce. The development and implementation of practical guidelines in order to manage soil compaction for a wide range of machinery types and forestry soils depend upon an understanding of the relative importance of applied pressure and water content during the compaction process.

http://dx.doi.org/10.5772/52774

3. Materials and method

3.1. Sites of samples

i. Confined compression test: soil samples were taken from three sites for the confined compression tests.

  a. Akure sandy clay loam soil: the site was a portion of agricultural land under fallow at the Federal University of Technology, Akure, Nigeria (7^0 15^1N, 5^0 15^1E), and elevation 210 m. The soil is Oxic paleustalf (Alfisol) or ferric Luvisol (FAO). A mini soil pit was dug to expose the profile. The site was designated Experimental Bin Soil (EBS). Three horizons EBS1, EBS2, and EBS3, from top to bottom respectively, were identified and samples taken from each. The thickness of the three horizons from top to bottom was 8, 15 and 15 cm, respectively.

  b. Igbokoda loamy sand soil: the soil used in this study was the major soil predominant in Igbokoda (headquarters of Ilaje Local Government area), Ondo State, South Western Nigeria. The soils were collected from the main agricultural production areas. This region is predominantly rainforest zone. Rainfall is in the order of 150 cm to 300 cm per annum, and mean annual temperature ranges from 25 to 28^0C. Soil samples were taken from freshly cut faces in open shallow pit to dept of 40 cm.

  c. Ilorin silt loam soil: the soil sample was taken from the arable soils of National Centre for Agricultural Mechanization (NCAM) Ilorin, Kwara State, Nigeria (8.30 N 4.32 E). The soil was Regosols (FAO). The soil samples were collected from the first 35 cm of soil profile; each sample was dug to a radius of 15 cm and then mixed thoroughly to get a homogeneous mixture, and then taken to the laboratory for further processing and analysis.

ii. Unconfined Compression Test

The study was carried out in August 2010 in an experimental plot located at The Federal University of Technology, Akure (FUTA) Step-B (Science and Technology Education Post – Basic) project site Akure, with geographical coordinate of 7° 10 North and 5° 05 East. The site is a two-hectare arable land and was manually cleared to avoid compaction due to machinery impacts on the land. The experimental plot of length and width 48.00 m by 12.00 m, respectively, was divided into three transects each of 16 m by 4 m. There was also a control plot of the same dimensions as transect. Soil samples were carefully collected from the test plots for textural and soil analysis.

3.2. Analytical methods

Particle size analyses of all the experimental soils for both confined and unconfined samples were performed using hydrometer method [16]. Organic matter content of the soils was determined using the [16] method. Other physical and chemical properties of the soils were also determined using standard methods.

3.3. Confined compression tests

i. Akure sandy clay loam soil: sample were collected, air-dried and ground to pass through a 2- mm sieve. The moisture content level for compaction for each soil sample was chosen according to the consistency limits of the soils determined by the procedure described [15]. Soil sample was placed in aluminium sleeve 75 mm in diameter and 50 mm long similar to that described 11].The cylinder was then placed on a 5 mm perforated metal base before the soil was added and tapped gently tapped to allow settling of the soil particles. Such prepared soil samples in the cylinders were then subjected to applied pressures 17.5, 100, 200, 300, 400, 500 and 618 kPa in turn applied by an Hydraulic Universal Testing machine (SM100, model No CPI-60) of 0.1 kN sensitivity, manufactured by TEC QUIPMENT Ltd, Nottingham, England). A hand pump of the machine was used to generate the required pressure on the soil. The soil samples were allowed to rebound before final heights were measured. The depression in the soil surface from the rim of the cylinder was measured at three points around the core by using a vernier caliper. These measurements were used to calculate bulk density in each test.

ii. Igbokoda Loamy Sand Soil: similarly, samples were air-dried and pulverized to pass through a 2 mm sieve. Each soil was wetted to a range of water contents between saturation and wilting point. For each sample, approximately 2 kg or air-dry soil was poured into a plastic tray. The soil was wetted with an atomizer and thoroughly mixed to bring the soil to the desired water content. The tray was then placed in a plastic bag and the sample was allowed to equilibrate for 48 hr. After equilibration, a soil sample at particular water content was placed in steel sleeve, 90 mm in diameter and 100 mm long. The cylinders were then placed on a 5 mm perforated metal base with cheesecloth before the soil was added. The cylinders were gently tapped to allow settling of the soil particles. Soil samples in the cylinders were then subjected to applied pressures of 0, 17.5, 100, 175, 289.5, 404, 511 and 618 kPa respectively, applied by a Universal Testing machine (SM100, model No CPI-60) of 0.1 kN sensitivity, manufactured by TEC QUIPMENT Ltd, Nottingham, England), consisting of an hydraulic ram connected to a piston. The compression force was read through a digital load meter. A hand pump was used to generate the required pressure on the soil. The pressure was maintained for a few second and then released. The samples were allowed to 'rebound' before final heights were measured. Water and air could escape around the piston and through fine perforations in the base plate. The depression in the soil surface from the rim of the cylinder was measured at three points around the core by using a vernier scale. This procedure was similar to that reported by research [10]. These measurements were used later to calculate bulk density.

iii. Ilorin Silt Loam Soil: samples were collected were each air-dried and ground to pass through a 2-mm sieve. The moisture levels for compaction tests were chosen according to the consistency limits of the soils determined by the procedure described by [15]. Compaction test was performed by filling the proctor mould with a known mass of soil and placed under a uni-axial compression apparatus (Universal Testing Machine (UTM), manufactured by the Testometric Co. Ltd., U. K.). Compression was carried out at a steady speed of 30 mm/min. Soil samples in the mould were subjected in turn to 75, 100, 150, 200, 300, 400, 500, 600 kPa. The soil displacement and mass were recorded for each compaction treatment. The mass was used to calculate bulk density of the compacted soil sample. The proctor mould was 16.8 cm height and 10 cm diameter. A circular thick metal plate was placed on the compression end of the UTM to effect uniform compaction in the proctor mould. After each compaction test, the change in depth of compressed soil was measured with the aid of a digimatic vernier caliper.

3.4. Penetration resistance measurement

Penetration resistance or Cone index (CI) was determined using a Rimick CP20 recording penetrometer (model CP 20 ultrasonic, Agridry Rimik Pty Ltd, Toowoomba), with a standard 30^0 cone of 322-mm^2 base area and a penetration rate was less than 10 mm/s. Measurements were taken at two depths 5 and 10 cm of the proctor mould and the average of the readings was taken as the representative value of cone index at that test.

3.5. Shear strength measurement

Shear strength values of experimental soils were observed using a shear vane tester of 19 mm vane. Measurements were taken at two depths of 5 and 10 cm in the Proctor's mould and the average recorded to represent the shear strength of the specific sample treatment.

3.6. Unconfined compression test

i. **Experimental treatments and layout:** the machinery used for this specific study was a medium Bobcat 430 Excavator of 31.9 kW power used to induce the necessary compaction on the experimental plots. The characteristics of the excavator are presented in Table 1. Seven treatments were imposed on plots 48 m long x 4 m wide, while the experimental variable was traffic frequency of 0, 1, 3, 5, 7, 9 and 11 excavator passes on the same tracks, with 3 m wide buffer zones between plots to avoid interactions. Plots were completely randomized having three replications. Statistical analysis was performed utilizing Excel 2007. An analysis of variance was carried out on the data and means were analyzed by Duncan's multiple range test.

ii. **Soil response variables:** experimental variables related to soil compaction include dry bulk density measured by the cylinder core soil sampler method, soil moisture content measured with moisture meter model PMS- 714, Taiwan with 0.1% resolution over the depth ranges of 0-10, 10-20, 20-30, 30-40 mm. Bulk density was the average of five measurements. Moisture content was verified by gravimetric method. Soil penetration resistance PR, or cone index CI was determined using a Rimick CP20 recording penetrometer (model CP 20 ultrasonic, Agridry Rimik Pty Ltd, Toowoomba), with a standard 30^0 cone of 322-mm^2 base area. The penetration rate was less than 10 mm/s. Data were recorded at depth increments of 50 mm. Cone index was the average of 15 measurements per plot. The resistance of the soil was measured in the centre line of the foot print of the excavator tract. Rut depth of the excavator foot print was measured using a profile meter similar to that reported [18]. The bar was placed across the wheel tracks, perpendicular to the direction of travel and rods position to conform to the shape of the depression. Rut depth was calculated as the average depth of 40 readings on the 1 metre bar. The estimated rut depth at any traffic frequency was calculated as the mean of the total number of sections' ruts at that frequency.

4. Results and discussion

4.1. Confined compression

4.1.1. Sandy clay loam soil

The physical and some other properties of the experimental sandy clay loam soil are shown in Table 2.Soil samples EBS1 and EBS2 have similar characteristics yet not the same as they are different physically especially in colour which may be due to presence of more dead organic material in the EBS1 than in the EBS2. Bulk density increased with increase in applied pressure, initially at a decreasing rate and further reaching an asymptotic value especially at applied pressure of about 600 kPa and beyond. Most relationships observed between soil bulk density and applied pressures were nonlinear. At moisture contents less than 12.9 %, it was observed that bulk density change little with applied pressure in the soil as reflected from the third horizon. There was little change in bulk density between 0 and 600 kPa pressure. This can be adduced to relatively high clay content of the third horizon (EBS3), and its least organic carbon content. A clay soil is naturally more compact than a sandier or silty soil; hence its bulk density changes little under applied pressure. Generally, a very strong correlation was observed for relationship between applied pressure and bulk density, R^2 values ranged from 0.92 to 0.99. The relationships between applied pressures for horizon 1 are represented by equations (1) to (5) for moisture contents of 10.4, 8.4, 4.7, 3.5 and 1.6 %, respectively.

Soil Type	Texture Sand silt clay			Bulk density Mg/m^3	Sat hyd cond mm/min	Clay ratio	Clay + silt	O.C %
	%	%	%					
Sandy clay loam (EBS1)	54	21	25	1.43	1.21	33.3	46	1.41
Sandy clay loam (EBS2)	54	21	25	1.55	1.22	33.3	46	1.22
Sandy clay loam (EBS3)	52	17	31	1.39	0.85	44.9	48	0.87

Table 2. Physical Properties of experimental sandy clay loam Soils

$$BD = 0.8654X^{0.114} \left(R^2 = 0.9967 \right) \tag{1}$$

$$BD = 0.09763X^{0.083} \left(R^2 = 0.9805 \right) \tag{2}$$

$$BD = 0.079Ln\left(X\right) + 1.089\left(R^2 = 0.9938\right) \tag{3}$$

$$BD = 0.0005X + 1.189\left(R^2 = 0.9312\right) \tag{4}$$

$$BD = 1.342e0.0002X\left(R^2 = 0.9899\right) \tag{5}$$

where BD = bulk density, Mg/m^3 and

X = applied pressure, kPa

Similarly, the relationships between applied pressure and bulk density for horizon 2 (EBS2) are represented by equations (6) to (11) below representing those for moisture contents of 17.0, 10.2, 7.4, 5.1, 3.6, and 2.2%, respectively.

$$BD = 0.187Ln\left(X\right) + 0.907 \left(R^2 = 0.9562\right) \tag{6}$$

$$BD = 0.163Ln\left(X\right) + 0.736 \left(R^2 = 0.9968\right) \tag{7}$$

$$BD = 1.115X^{0.066} \left(R^2 = 0.9826\right) \tag{8}$$

$$BD = 0.091Ln\left(X\right) + 1.065\left(R^2 = 0.9842\right) \tag{9}$$

$$BD = 1.267X^{0.042} \left(R^2 = 0.9455\right) \tag{10}$$

$$BD = 1.403e^{0.0001X} \left(R^2 = 0.9539\right) \tag{11}$$

X = applied pressure, kPa

Furthermore, the relationships between applied pressure and bulk density for horizon 3 (EBS3) are represented by equations (12) to (16) below representing those for moisture contents of 12.95, 7.7, 6.4, 5.5 and 2.3%, respectively.

$$BD = 0.225Ln(X) + 0.604 \ (R^2 = 0.9787) \tag{12}$$

$$BD = 1.006X^{0.079} \ (R^2 = 0.9859) \tag{13}$$

$$BD = 1.222X^{0.051} \ (R^2 = 0.9287) \tag{14}$$

$$BD = 1.416X^{0.019} \ (R^2 = 0.9028) \tag{15}$$

$$BD = 1.425X^{0.015} \ (R^2 = 0.9954) \tag{16}$$

where BD = bulk density, Mg/m^3 and

X = applied pressure, kPa

The above models (equations 1 to 16) were best fitted equations to the experimental values of the bulk density-applied pressure relationships. They enable prediction of bulk density of the soil to be possible in the range of moisture content and applied pressure considered in this study. The study showed that four types of models (power, logarithmic, exponential and linear functions) could be used to describe the relationship of bulk density and applied pressure at a specific moisture level, judging from their high coefficients of determination that ranged from 0.9028 to 0.9968. However, the best model at each moisture content level was reported in this paper. Power function seemed to be the best and most popular model, followed by logarithmic, exponential and linear function in that order. Substituting extra values of applied pressure and moisture content, which were not used in establishing the models, validated the models. There was high correlation ($r > 0.9936$) between predicted and measured values of bulk density and the errors less than 5.6 %. Other researchers had reported similar result. Researchers [20] fitted exponential curves for loamy soils while [21] reported logarithmic curves. Researchers [3] fitted logarithmic and power functions for sandy loam and clay soils at different moisture content levels and applied pressure. Figures 1, 2 and 3 indicate that irrespective of soil horizon of the soil sample and applied pressure, soil compactability increased as the moisture content increased to a limit. Generally, bulk density increased with moisture content irrespective of the value of applied pressure. Compaction caused squeezing out of water from the pores, especially the macrospores. The implication is that tillage whether manual or mechanical should be done when the soil is moist rather than wet condition. From this work, moisture content of 10% and above is unfavourable since soil compaction at these levels of moisture content would enhance soil compaction as indicated by excessively high soil bulk density. Considering the bulk density value of 1.5 Mg/m^3 which is indicated to be critical for crop production [22], the moisture contents of 10.0 (Figure 1), 17.0 (Figure 2) and 12.9% (Figure 3) are liable to produce high bulk density if pressure is applied at 200 kPa and above. The recommended moisture content for minimizing soil compaction (for sandy clay loam) or soil bulk density is less than 10% and the applied pressure should not exceed 200kPa, as indicated by this study.

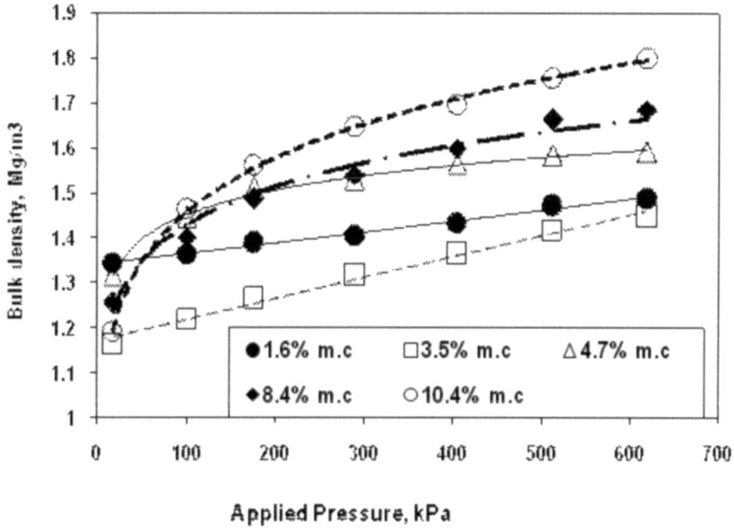

Figure 1. Effect of applied pressure, moisture content on bulk density of sandy clay loam soil (EBS1), Layer1 (54, 21, 25)

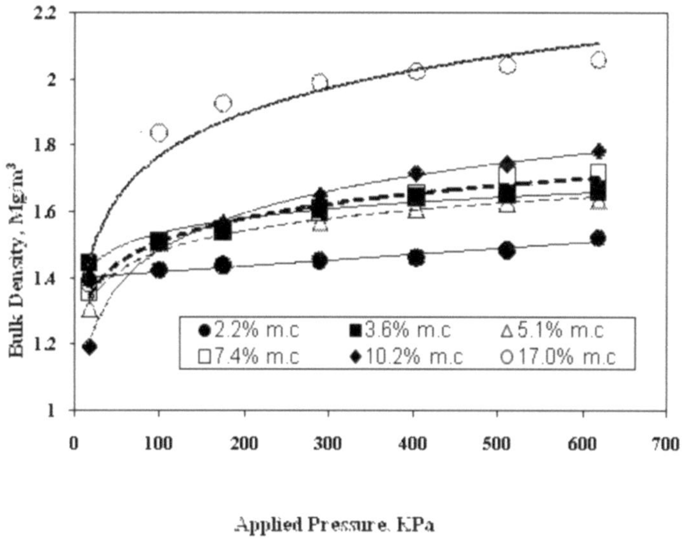

Figure 2. Effect of applied pressure, moisture content on bulk density of sandy clay loam soil (EBS2), Layer2 (54, 21, 25)

The compatibility of the soils was also considered in terms of the minimum increase in compaction (when the soil was dry) and maximum increase in compaction (when the soil was wet). For the soil in the uppermost horizon (EBS1), the minimum and maximum increases in compaction were 5.5 and 61.5% respectively. For the soil in the middle horizon (EBS2), the values were 3.9 and 53.6% while for the bottom horizon; the values are 4.5 and 76.9% respectively. From the foregoing, it was obvious that sandy clay loam soils are sensitive to water content at the time of compaction. This is in agreement with the findings reported by [11]. The results also show that the percentage increase in soil compaction for the sandy clay loan is related more to changes in water content than to applied pressure. Scheduling tillage operations on this type of soil in drier condition will have greatest benefit in lowering soil compaction not withstanding changes in ground pressure.

4.1.2. Loamy sand soil

Physical and some other properties of the loamy sand soil are shown in Table3. Values of bulk density (ordinate) were plotted against applied pressure (abscissa). The best-fit models at each moisture level were established using regression analysis with the regression models and their coefficients shown (Table 4). The models vary from linear to intrinsically linear models: Linear fit (LINFIT); Exponential fit (EXPFIT); Logarithmic fit (LOGFIT) and power fit (PWRFIT). It was observed that experimental soil showed similar behaviour under compression when it is very dry and when it is almost saturation. This report is similar to that reported elsewhere [10]. The linear model describing the compression at those moisture levels could explain this. At intermediate moisture contents, the models are only intrinsically linear. The high coefficients of correlation of the models showed that the models are adequate for the prediction of the soil behaviour. It was reported [11] that the values for applied pressure should be regarded as relative and that the pressure required in a uni-axial test to produce the same bulk density in the field was considerably higher [23].

Soil type	Texture			Bulk density Mg/m^3	Sat. hydraulic conductivity mm/min	Clay ratio %	Clay + silt %	LOI %
	Sand %	Silt %	Clay %					
Loamy Sand	85	3	12	1.56	5.3	13.6	15	1.12

Table 3. Physical properties of the experim ental loamy sand soil

Effect of moisture content on bulk density: The variation of bulk density with water content at the different levels of applied pressure is governed by a quadratic relationship of the type shown in Equation (17). It is observed that bulk density decreased with increase in water content at any particular applied pressure, to a minimum density and then increased again. The range of applied pressure considered (17.5 to 618 kPa) covered the range common in agricultural productions. For the purpose of predictions, best models were established to fit the data of bulk density variation with moisture content. By simply "eyeballing" the data in the scatter diagram, it was apparent that a parabolic (or quadratic model) was the best model. This was confirmed by comparing the coefficients of determination of possible models:

$$y = A + Bx + Cx^2, \tag{17}$$

where y = bulk density (Mg/m^3), x = moisture content, % (db), A,B,C = regression coefficients, R^2 = coefficient of determination. The coefficients of the model are presented in Table 5. The values of the R^2 were high enough to make the models suitable for predictions.

MC (%) db	Model type	Correlation coeff. (r)	Model Equation
0.01	(LINFIT)	0.9956	$\rho_b = 1.5913+7.898E\text{-}5\sigma$
0.1	(EXPFIT)	0.9908	$\rho_b = 1.4405EXP\ (1.043E\text{-}4\sigma)$
1.6	(LINFIT)	0.9830	$\rho_b = 1.3697 + 1.411E - 4\sigma$
1.9	(LOGFIT)	0.9943	$\rho_b = 1.09158+ 5.413E\text{-}21n\sigma$
4.5	(PWRFIT)	0.9977	$\rho_b = 1.127\sigma^{0.0397}$
4.7	(PWRFIT)	0.9900	$\rho_b = 1.216\sigma^{0.0327}$
8.0	(LOGFIT)	0.9938	$\rho_b = 1.0416 + 0.071n\ \sigma$
10.4	(PWRFIT)	0.9828	$\rho_b = 1.3086\sigma^{0.0252}$
13.0	(LINFIT)	0.9919	$\rho_b = 1.4728 + 0.000213\sigma$

LINFIT = Linear fit ; EXPFIT = Exponential fit LOGFIT = Logarithmic fit; PWRFIT = Power fit

Table 4. Bulk density and applied pressure relationships at different moisture content for Igbokoda loamy sand soil.

Pressure (kPa)	Model Coefficients			
	A	B	C	R^2
17.5	1.4827	-0.0717	0.00567	0.6312
100	1.499	-0.0415	0.00337	0.6421
175	1.4991	-0.034	0.00291	0.7344
289.5	1.5121	-0.0315	0.00291	0.214
404	1.5377	-0.0296	0.00265	0.6643
511	1.5377	-0.0281	0.00254	0.7516
618	1.5525	-0.0312	0.00281	0.6452

Table 5. Regression coefficients of quadratic models ($y = A + Bx + Cx^2$) of the effect of moisture content on bulk density at different applied pressures

Multiple regression of moisture content and applied pressure on bulk density: multiple regression analysis was carried out to assess the effects of water content, applied pressure on bulk density. The range of pressure was 17.5 to 511 kPa and 0 to 13% (db) water content. The multiple regression equation was:

$$y = 1.4359 + 0.00017\,X_1 - 0.0039\,X_2,\quad R^2 = 0.6114 \tag{18}$$

where y = bulk density (Mg/m^3), X_1 = applied pressure (kPa), X_2 = water contents (%) db, R^2 = coefficient of determination. The multiple regression analysis (Equation 18) showed that the effect of moisture content was greater than that of applied pressure on the bulk density, judging from the higher coefficient of the independent variable, water content.

4.1.3. Silt loam soil

i. **Physical properties:** The soil studied was a silt loam soil according to the USDA textural classification of soils. Table 6 shows some physical and chemical properties of the soil. The consistency limits of the soils are presented in Table 6. Plasticity index is an index of workability of the soil and a large range of plasticity index implies a need for large amounts of energy to work the soil to a desired tilth.

Property	Values
Sand (%)	22.6
Silt (%)	62.8
Clay(%)	14.5
Silt + clay (%)	77.3
Texture (%)	Silt loam
Organic carbon (g/kg)	2.03
Organic matter (%)	3.51
Total nitrogen g/kg)	0.18
pH in H_2O (1:2)	7.93
Ca^{2+} (cmol/kg)	0.17
Mg^{2+}(cmol/kg)	1.70
Na^{+}(cmol/kg)	0.17
K^{+}(cmol/kg)	0.29
P (mg/kg)	4.00
Plastic limit (%)	9,2
Liquid limit (%)	40
Plasticity index	30.8

Table 6. Some physical and chemical properties of Ilorin silt loam soil

ii. **Strength properties:** shear strength and cone index are indicators of soil strength. Shear strength is the resistance of soil to shearing or structural failure. The shear strength of an individual clod decreases with wetting, but more importantly, the strength of the bulk soil increases with increasing moisture up to the lower plastic limit at which each particle is surrounded by a film of water which acts as lubricant. Soil strength drops sharply from that point to the upper plastic limit, where the soil becomes viscous. The effects of moisture content and applied pressure on shear strength of the experimental soils are presented in Figure 4. Shear strength increased with increase in applied pressure and moisture content until a peak value was reached after which the value of the parameter declined with further increase in moisture content. This is a typical soil behaviour which has been reported by other researchers. The peak value occurred at higher moisture content as the applied pressure increased. The maximum shear strength of the soils at applied pressure of 600 kPa was 1025 kPa at moisture content of 9.1 % (db). Similarly, the effect of moisture content and applied pressure on cone index of the soil is presented in Figure 5. The relationship is similar to that exhibited by shear strength. The maximum cone index at applied pressure of 600 kPa was 1325 kPa at moisture content of 5.0 % (db). However, the effects of moisture content and applied pressure on bulk density showed different behaviour from those of shear strength and cone index (Figure 6). Bulk density increased with increase in moisture content for all the applied pressure in the range of moisture content considered, however the increase was with a peak at about the plastic limit. Bulk density increased with increase in moisture contents up to a maximum because addition of water increased cohesion. Bulk density decreased at higher moisture content after the peak value because further addition of water created greater water pressure which reduced soil compressibility. The maximum bulk density at applied pressure of 600 kPa was 2.1 Mg/m^3 at moisture content of 10.0 % (db). This moisture content was significant because it was the moisture content at which the soil reached maximum bulk density. This is in agreement with other researchers' report that soils with high amount of fine particles (clay plus silt) are more susceptible to compactability [17]. The regression models that describe the behaviour of soil parameters shown in Figures 4 to 6 are presented in Table 7. The models are largely nonlinear and they agree well with those reported by other researchers [6, 7, 8, 9, 11, and 12]. Regression models (Table 8) were also established to show relationships between compaction indices such as shear strength, cone index and bulk density at applied pressures of 75, 300 and 600 kPa representing a range of low to medium and high pressures. The relationships varied from linear to exponential and to polynomial functions. The results also found a linear correlation between cone index and shear strength at a low applied pressure of 75 kPa. This agrees well with the findings of Vanags et al. [1] who reported linear relationship between cone index and surface shear resistance of soil.

Dependent variables	Independent variables	Predictive models	R^2	Applied pressure, kPa	Model type
SS	MC	$y = -0.797x^2 + 21.77x + 21.63$	0.9733	75	polynomial
SS	MC	$y = -1.384x^2 + 35.46x + 52.5$	0.9394	150	polynomial
SS	MC	$y = -2.54x^2 + 55.09x + 191.1$	0.7899	300	polynomial
SS	MC	$y = -3.875x^2 + 97.17x + 39.08$	0.468	600	polynomial
CI	MC	$y = -1.056x^2 + 25.1x + 83.75$	0.8061	75	polynomial
CI	MC	$y = -1.53x^2 + 34.2x + 177.3$	0.8384	150	polynomial
CI	MC	$y = -2.63x^2 + 52.79x + 342.1$	0.7914	300	polynomial
CI	MC	$y = -4.543x^2 + 82.54x + 698.9$	0.729	600	polynomial
BD	MC	$y = 1.133e^{0.022x}$	0.9661	75	exponential
BD	MC	$y = 31.4x + 1228.7$	0.9624	150	linear
BD	MC	$y = 32.02x + 1304.5$	0.8976	300	linear
BD	MC	$y = -0.001x^2 + 0.059x + 1.395$	0.9442	600	Polynomial

CI = cone index; BD = bulk densityMC = moisture content; SS = shear strength

Table 7. Relationships between dependent and independent variables Ilorin silt loam soil

Dependent variables	Independent variables	Predictive models	R^2	Applied pressure, kPa	Model type
CI	SS	$y = 1.184x + 24.52$	0.6651	75	linear
CI	SS	$y = 0.0006x^2 + 0.97x + 60.5$	0.7914	150	polynomial
CI	SS	$y = 177.59e^{0.0025x}$	0.9687	300	exponential
CI	SS	$y = 0.001x^2 - 1.406x + 866.8$	0.2600	600	polynomial
CI	BD	$y = -627.3x^2 + 1890x - 122$	0.4711	75	polynomial
CI	BD	$y = -0.0014x^2 + 4.42x - 3108.8$	0.8094	150	polynomial
CI	BD	$y = -0.0017x^2 + 5.30x - 3108.8$	0.4531	300	polynomial
CI	BD	$y = -10128x^2 + 36518x - 31362$	0.7790	600	polynomial

CI = cone index; BD = bulk densityMC = moisture content; SS = shear strength

Table 8. Relationships between cone index, shear strength and bulk density of Ilorin silt loam soil

http://dx.doi.org/10.5772/52774

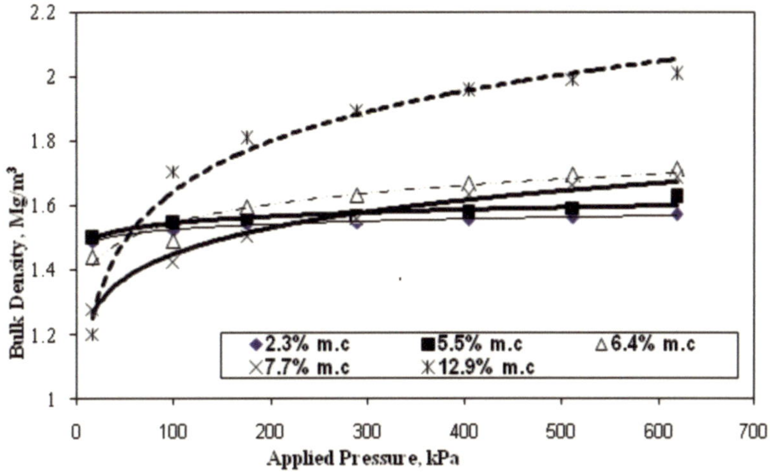

Figure 3. Effect of applied pressure, moisture content on bulk density of sandy clay loam soil (EBS3), Layer3 (52, 17, 31)

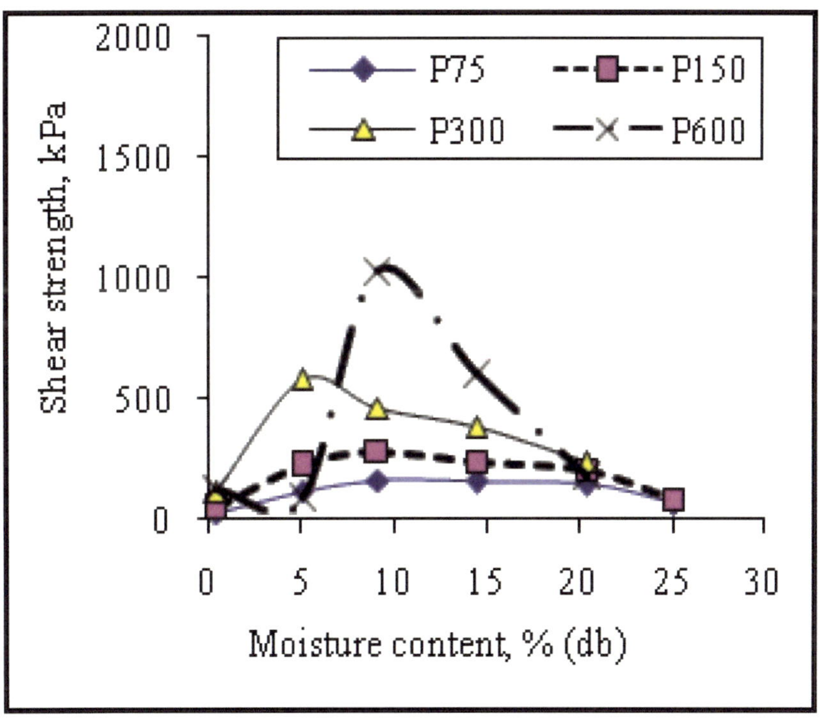

Figure 4. Effect of moisture content and applied pressure on shear strength

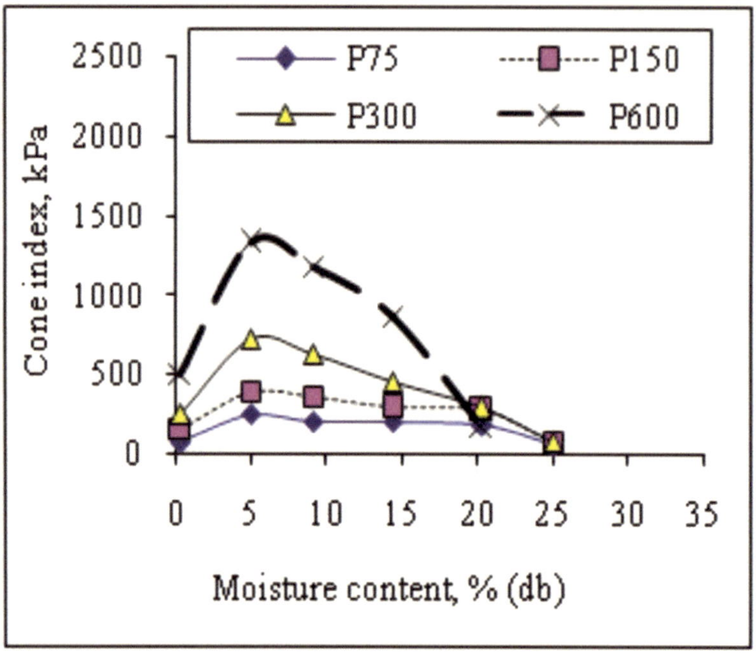

Figure 5. Effect of moisture content and applied pressure on cone index

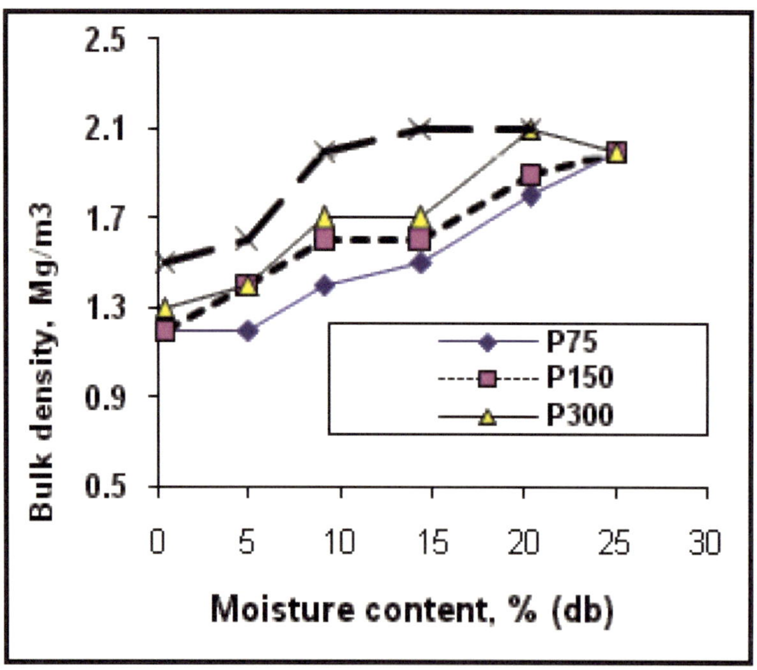

Figure 6. Effect of moisture content and applied pressure on bulk density

4.2. Unconfined compression

4.2.1. Sandy clay soil

The most common causes of soil compaction are agricultural machines such as tractors; harvesting equipment and implement wheels travelling over moist, lose soils [24]. Soils tend to be more compacted deeper into the soil profile due to the weight of overlaying soil. Heavily compacted soils contain few large pores, less total pore volume and consequently a greater density [25]. The experimental excavator can be categorized as light machinery in agreement with the report [26]. Also the characteristics of the experimental soil are shown in Table 9. This soil texture (sandy clay) is one of the dominant textures in the locality and even the state in general. Table 10 shows the rut that was produced as a result of the traffic frequency. This was considered to be high for the moisture content range (11 to 15% (db)) at which the experiment was carried out. A typical excavator track footprint (rut) is shown in Figure 7. For each traffic treatment, cone index, bulk density and rut depth were measured at 2 m intervals along a 48 m transect within the excavator tracks (left and right). All measurements were made in the centrelines of the tracks because that was where the compressive effects tend to concentrate [19]. Cone penetration resistance and bulk density were also measured in the control treatment where no tracks (excavator) had passed (Figure 8). Soil penetration resistance at the control site (zero excavator traffic) increased with depth due to shaft friction, and overburden pressure of the weight of soil above the specific depth (Figure 8). Also, lateral forces on the penetrometer cone increase with increasing depth so that more force was needed for the cone to displace soil [27]. Resistance can also increase with depth because of changes in soil texture, gravel content, structure and agricultural traffic if it had occurred. The variation of penetration resistance with depth along the excavator foot prints as a result of machinery traffic is presented in Figure 9. Generally, penetration resistance increased with depth up to a point and then decreased. This trend is consistent with the findings of many researchers. It was observed that the mean values of penetration resistance along the left and right foot prints of the excavator for each treatment were not significantly different from each other at 5% level of significance. It was reported that bulk density and penetration resistance increase with the number of passes but bulk density tended to be less responsive than penetration resistance [27]. Despite this, the measured changes in bulk density agreed with soil behaviour suggested by changes in penetration resistance. The result of this study agreed with this assertion. Examination of soil responses to traffic in deeper layers revealed that soil compaction increased as the traffic intensity increased. This also agreed with the findings of other researchers [28, 27]. Effect of traffic on bulk density is presented in Table 11 at two depths only due to the limitation imposed by the short (35 cm) shank of the penetrometer used in this study. At 9 and 11 passes, soil bulk density became significantly different than at lower traffic frequency.

Properties	Values
Sand (%)	51
Silt (%)	10
Clay(%)	39
Texture (%)	Sandy clay
Organic carbon (g/kg)	1.5
Organic matter (%)	2.58
C/N ratio	7.89
Total nitrogen g/kg)	0.19
pH in H_2O (1:2)	6.75
Ca^{2+} (cmol/kg)	3.30
Mg^{2+}(cmol/kg)	2.20
Na^{+}(cmol/kg)	0.14
K^{+} (cmol/kg)	0.26
P (mg/kg)	17.54

Table 9. Soil properties of sandy clay soil

Traffic treatments (No of passes)	Mean rut depth (mm)
1	27 a
3	45 b
5	70 c
7	99 d
9	132 e
11	148 f

Different letters within each traffic treatments show significant difference at 1% level of significance, Duncan's multiple range test)

Table 10. Effect of Excavator traffic frequency on rut depth

Depth (mm)	Control plot	Excavator traffic frequency (No of passes)					
		1	3	5	7	9	11
0-150	1.242	1.251 a	1.259 a	1.301 a	1.319 a	1.435 b	1.442 b
150-300	1.407a	1.422a	1.427a	1.431a	1.439a	1.506b	1.534b

Values with different letters (horizontally Bulk) are significantly different at each depth (P less than 0.001 Duncan's multiple range test).

Table 11. Bulk Density (Mg m^{-3}) at two depths under different degrees of Excavator traffic (1, 3, 5, 7, 9, 11)

http://dx.doi.org/10.5772/52774

Figure 7. Plan view of the rut produced by the excavator tracks during experimentation

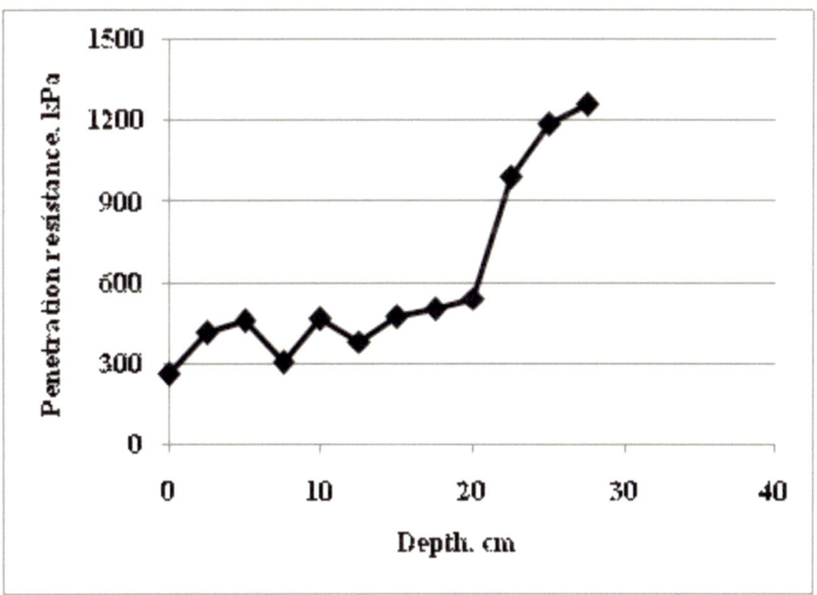

Figure 8. Variation of cone penetration resistance with depth at control plot

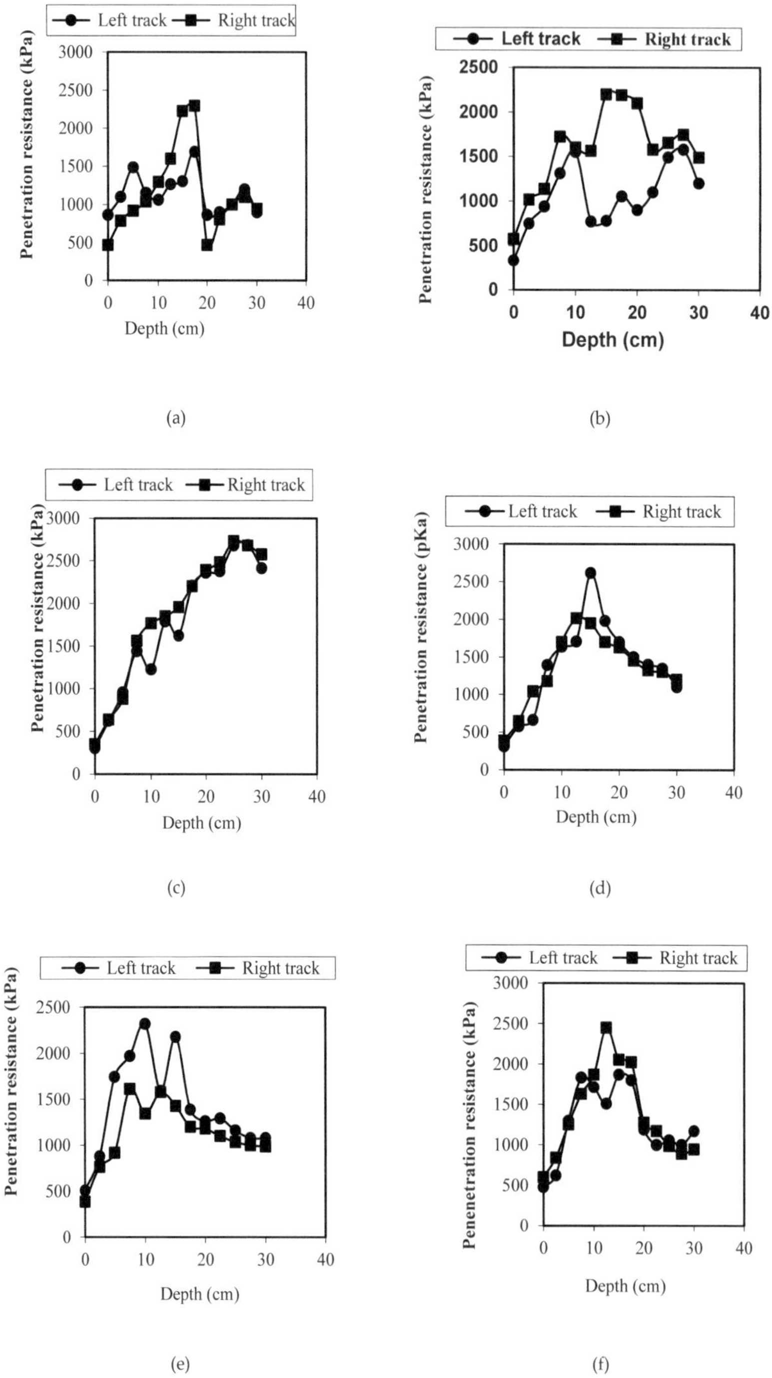

Figure 9. Cone penetration resistance in the centre lines of the excavator tracks- (a) 1 pass (b) 3 passes (c) 5 passes (d) 7 passes (e) 9 passes (f) 11 passes

5. Conclusion

2. The study shows that compaction behaviour of Akure sandy clay loam soil in a confined compression could be modelled after certain non-linear equations. The model derived from the data can be used to evaluate the relative importance of water content and applied pressure in the compaction process.

3. Similarly, the behaviour of Igbokoda loamy sand soil has been studied.The soil showed similar behaviours in the dry and near saturation conditions. Bulk density varied from 1.2389 to 1.6393 Mg/m^3 in the range of applied pressure of 0 to 618 kPa. Linear, intrinsically linear and quadratic models can be used to describe the behaviour of the soil for the purpose of predictions. In a multiple regression analysis, it was observed that the principal factor influencing compaction of the loamy sand soil was water content rather than applied pressure at the time of compaction.

4. The study also showed that compaction behaviour of Ilorin silt loam soil could be modelled after certain linear and non-linear regression equations. Cone index have good positive linear relationship with shear strength, but can also be fitted with polynomial (quadratic) function with higher coefficient of determination. The effect of moisture content and applied pressure on cone index, shear strength was best fitted with polynomial function of the second order. The effect of applied pressure and moisture content on bulk density of silt loam could be modelled after linear, exponential and polynomial regression functions. Cone index has good correlation with shear strength

5. The study of the compaction behaviour of Akure sandy clay in unconfined space (field) showed that Cone penetration resistance values under left and right centre lines of excavator tracks was not significantly different from one another. There were significant differences in the mean rut of traffic frequencies at 5% level of significance. There were also significant differences in the mean bulk density due to traffic frequencies at two depths, at 5% level of significance.

Acknowledgements

I hereby acknowledge the following that have made the various studies reported in this chapter possible: the Federal university of Technology, Akure (FUTA), Nigeria for the Senate Research Grant No (URG/MINOR/98/105); the National Centre for Agricultural Mechanisation (NCAM) for the privilege to use the facility of their Research Laboratory; World Bank for the STEP-B research Grant to the Institution (FUTA); and to my numerous undergraduate and post graduate students who have assisted in collecting useful data for the work reported in this chapter.

Author details

Seth I. Manuwa*

Address all correspondence to: sethimanuwa@yahoo.com

Department of Agricultural Engineering, School of Engineering and Engineering Technology, The Federal University of Technology, Akure, Nigeria

References

[1] Vanags, C, Minasny, B, & Mcbratney, B. and A. B. The Dynamic Penetrometer for Assessment of Soil Mechanical Resistance," (2004). http://www.regional.org.au/au/asssi/supersoil2004/s14/poster/1565_vanagsc.htm.

[2] Hillel, H. Soil and Water: Physical and processes,". Academic Press, New York, (1971).

[3] Gill, W. R. and G. E. Van den Berg, "Soil Dynamics in Tillage and Traction," Handbook 316, Agricultural Research Service, U. S. Department of Agriculture, Washington, D. C., (1968).

[4] Assouline, S. Modeling Soil Compaction under Uniaxial Compression," *Soil Science Society of America Journal*, (2002). , 66, 1784-1787.

[5] Koolen, A. J, & Kuipers, H. Agricultural Soil Mechanics," Springer- Verlag. New York, (1983). , 171-280.

[6] Koolen, A. J. A method of Soil Compatibility Determination," *Journal of Agricultural Engineering Research*, (1974). , 19, 271-278.

[7] Larson, W. E, Gupta, S. C, & Useche, R. A. Compression of Agricultural Soils from Eight Soil Orders," *Soil Science Society of America Journal*, (1980). , 44, 450-457.

[8] Saini, G. R, Chow, T. L, & Ghanen, I. Compatibility Indexes of Some Agricultural Soils of New Brunswick, Canada," *Soil Science*, (1984). , 137, 33-38.

[9] Gupta, S. C, & Allmaras, R. R. Models to Assess the Susceptibility of Soil to Compaction," *Advances in Soil Science*, (1987). , 6, 65-100.

[10] Oni, K. C. Traffic Compaction and Tillage Effects on the Performance of Maize in Sandy loam Soil of Nigeria," *Agricultural Mechanization in Asia, Africa and Latin America*, (1991). , 22(2), 27-31.

[11] Smith, C. W, Johnston, M, & Lorentz, A. Assessing the Compaction Susceptibility of South African Forestry Soils. 1. The Effect of Soil Type, Water Content and Applied Pressure on Uni-axial Compaction," *Soil Tillage & Research*, (1977). , 41, 53-73.

[12] Fritton, D. D. An Improved Empirical Equation for Uni-axial Soil Compression for a Wide Range of Applied Stresses, *Soil Science Society of America, Journal,* (2001). , 65, 678-684.

[13] Ohu, J. O, Raghavan, G. S. V, Mckyes, E, & Mehuys, G. Shear Strength Prediction of Compacted Soils with Varying added Organic Matter Contents,'' *Transactions of the American Society of Agricultural Engineers,* (1986). , 29(2), 251-255.

[14] Onwualu, A. P, & Anazodo, U. G. N. Soil Compaction Effect on Maize Production under Various Tillage Methods in a Derived Savanna Zone of Nigeria, '' *Soil & Tillage Research,* (1989). , 14, 99-114.

[15] Lambe, T. W. Soil Testing for Engineers,'' John Wiley, New York, (1951).

[16] Walkley, A, & Black, I. A. An Examination of the Different Methods for Determining Soil Organic Matter and a Proposed Modification of the Chromic Acid Titration Method,'' *Soil Science,* (1934). , 37, 29-38.

[17] Krzic, M, Bulmer, C. E, Teste, F, Dampier, L, & Rahman, S. Soil properties Influencing Comppactability of Forest Soils in British Columbia, *Can. J. Sci.* (2004). , 84, 219-226.

[18] Manuwa, S. I. (2009). Performance evaluation of tillage tines operating under different depths in a sandy clay loam soil. Soil and tillage research, , 103, 399-405.

[19] Sohne, W. (1958). Fundamentals of pressure distribution and soil compaction under tractor tyres. Agric. Eng. , 39, 276-281.

[20] Dexter, A. R, & Tanner, D. W. (1973). Response of unsaturated soils to isotropic stress. Journal of Soil Science, , 24, 491-502.

[21] Bertilsson, G. (1971). Topsoil reaction to mechanical pressure. Swede. Journal of Agricultural Research, , 1, 179-189.

[22] Fulton, J. P, Wells, L. G, Shearer, S. A, & Barnhisel, R. I. (1996). Spatial variation of soil physical properties: A precursor to precision tillage. ASAE Paper International Meeting, Phoenix, Arizona. July (961002), 14-18.

[23] Ssullivan, O, & , M. F. Uni-axial compaction effects on soil physical properties in relation to soil type and cultivation. Soil & Tillage Research, , 24, 257-269.

[24] [24]Alakuku, L, Weisskopf, P, Chamen, W. C. T, Tijink, F. G. J, Van Der Linden, J. P, Pires, S, Sommer, C, & Spoor, G. (2003). Prevention strategies for field traffic-induced subsoil compaction: A review. Part 1.Machine/soil interactions- *Soil and Tillage Res.,* , 73, 145-160.

[25] Horn, R, & Van Den, J. H. Akker and J. and Arvidsson. (Eds.) ((2001). Subsoil Compaction- Distribution, Processes and Consequences.- *Advances in Geology* 32, Catena-Verlag, Reiskirchen

[26] Botta, G. F, Jorajuria, D, Rosatto, H, & Ferrero, C. (2006). Light tractor traffic frequency on soil compaction in the rolling Pampa region of Argentina, Soil & Tillage Research, , 86, 9-14.

[27] Becerra, A. T, Botta, G. F, Bravo, X. L, Tourn, M, Melcon, F. B, Vazquez, J, Rivero, D, Linares, P, & Nardon, G. (2010). Soil compaction distribution under tractor traffic in almond (*Prunus amigdalus* L.) orchard in Almeria Espana, Soil & Tillage Research, , 107(1), 49-56.

[28] Botta, G. F, Jorajuria, D, & Draghi, L. M. (2002). Influence of the axle load, tyre size and configuration on the compaction of a freshly tilled clayey soil. Journal of Terramechanics, , 39, 47-54.

Chapter 8

Microaggregate Stability of Tropical Soils and its Roles on Soil Erosion Hazard Prediction

C.A. Igwe and S.E. Obalum

Additional information is available at the end of the chapter

http://dx.doi.org/10.5772/52473

1. Introduction

Soil aggregate stability influences a wide range of physical and biogeochemical processes in the agricultural and natural environments, including soil erosion [3]. The relative preponderance of aggregates of various sizes in the soil and their stability to external forces are, therefore, an issue of major concern to soil scientists. By definition, an aggregate is a composite body or granule of loosely bound mineral particles within a soil, the binding of which is characteristically mediated by a relatively minor amount of organic matter [Encyclopedia of Soil Science, ESS 2008]. The mineral and organic particles involved in such a natural conglomeration, otherwise known as aggregation, cohere to each other more than to the neighbouring particles and/or aggregates [ESS 2008]. Soil aggregates are therefore soil structural units of which classical soil research recognizes two major size-based categories, macroaggregates and microaggregates. Collapse of macroaggregates yields microaggregates. Thus, macroaggregates may be viewed as having microaggregates as their building blocks. Sometimes, external forces acting on a soil can also foster formation of aggregates from dispersed materials. It is the interplay of aggregate formation and breakdown that results in soil structure [54]. Although extremities in either of these structure-promoting processes are undesirable, they are considered an agronomic and environmental problem only in the case of breakdown. This is because it is much easier to break down over-sized aggregates into favourably sized ones than to achieve aggregation in structurally dilapidated soils. Consequently, studies on the responses of soil aggregates to natural and anthropogenic forces appear to tilt more towards stability or otherwise of soil aggregates than to their coalescence by these forces. Soil aggregation includes the processes of formation and stabilization, both of which occur continuously and concurrently [3]. Soil aggregate/structural stability may be defined as a measure of the ability of the soil structural units to resist change or the extent to which they remain intact when mechanically stressed by environmental factors [ESS 2008]. The environmental factors that become important in this regard generally depend on climate and soil characteristics related to the nature of the parent materials and age of the soil. Another important factor is the intensity of disturbance related to land use and management [1]. To understand the importance of climate, it will be good to first state that water is such an indispensable entity in the discussion of soil aggregate stability that the subject is sometimes referred to aggregate stability to water. Climate sets the limit of change in the state of water in the soil, whether the soil's response to the major climatic variables (rainfall and temperature) would be limited to mere wetting and drying or would also include freezing and thawing. In the tropical climate, soils are subject to frequent wetting and drying cycles in the short term during the rainy season and in the long term between the distinct rainy and dry seasons. Although freezing and thawing constitutes a greater stressor to soil aggregates, it rarely occurs in the tropics and therefore should be de-emphasized here for the sake of the scope of this chapter. In terms of inherent soil characteristics, tropicals soils generally show a higher aggregate stability compared to temperate soils [55], and this is due mainly to mineralogy of the former being characterized by dominance of oxides and kaolin clays [54]. Nevertheless, aggregate stability remains a valid topic in the tropics, especially in the broad area of environmental management, because many soils in the region are regarded as structurally fragile and unstable. This is due to other soil-related and certain climatic peculiarities of the region.The soil-related factors militating against the aggregate stability of the majority of tropical soils include the sandy nature of their parent materials, which often reflects in the texture of the soils. It has been reported that the resistance of soil aggregates to raindrop impact decreases with a decrease in clay content of the soil [40]. Most other soils occurring in areas with heavy rainfall, even when not originally sandy, have been so intensively washed by runoff and leaching that their texture tends toward coarseness [33]. It is perhaps because of the vast area occupied by soils in these categories, commonly referred to as 'tropical sandy soils' in the literature, and the effect of the sandy texture on their aggregate stability that considerable research goes into their management [FAO 2007]. The coarse texture of the soils, coupled with the low concentration and high mineralization rates of organic matter (the typical aggregating agent) in them, implies impaired aggregation. Again, most tropical soils have long weathering history as is often evident in their low silt content [33], and this also contributes to frustrating aggregation processes in the soils. Indiscriminate deforestation, inappropriate land use and non-sustainable soil management options are also a common feature of agriculture in the region. In terms of climate, the aforementioned long-term wetting and drying cycles in most of the tropical region can have important implications for the aggregate stability of the soils. Also, the characteristic rainstorms and the associated heavy raindrops in especially the humid tropics can have considerable splash effect [38] and, therefore, are a force to reckon with in soil aggregate destabilization. It is thus clear that most tropical soils are structurally fragile and susceptible to many forms of erosion including accelerated and catastrophic erosion. [3] noted that good soil structure, known by the presence of well formed and stable aggregates, is the most desirable of all soil attributes for sustaining agricultural productivity and for preserving environmental quality. In the above context, a good understanding of the aggregate stability of tropical soils and its relationship with their erodibility is needed to guide the management of these soils against erosive and similar degradative forces. In spite of the generally higher aggregate stability of tropicals soils compared to temperate soils [55], soil erosion remains a major threat to agricultural productivity in the tropical region. Proper management is necessary to position these soil resources for continued support of agricultural and allied activities while not compromising environmental quality. Soil aggregate stability has been shown to give some guide on the relative stability of Ultisols from sub-tropical China to externally imposed destructive forces and, hence, to be an appropriate indicator of the relative susceptibility of the soils to detachment, runoff and interrill erosion [63, 53]. Our focus here is on microaggregation and the relationship between microaggregate stability and erodibility of tropical soils.

2. Appropriate aggregate stability indices for assessing erosion hazards in tropical soils

The derivation of many aggregate stability indices involves all aggregate-size classes and, as a result, such indices provide information on the overall stability of the soil. A typical example and, perhaps, the mostly widely used of such indices is the mean-weight diameter (MWD) of aggregates. However, the MWD is often regarded as index of macroaggregate stability of soils, probably because of the preponderance of macroaggregate-size classes over microaggregate-size classes in its computation. Where authors of the papers reviewed in this chapter fail to specify which of macro- and microaggregate stability their indices represent, we regard the indices as representing macroaggregate stability rather than microaggregate stability, provided their determination did not involve dispersion. The MWD and indeed all such aggregate stability indices which integrate aggregate-size classes into one number are regarded as macroaggregate stability indices in this chapter. The use of such indices to assess erodibility may prove suitable in temperate soils, but may not in highly weathered tropical soils known for their oxyhydroxidic mineralogy and very stable microgranular structure [17]. The question remains which of macro- and microaggregate stability more closely relates to erodibility of the majority of tropical soils. To answer this question, we need to first understand the mechanisms that are generally responsible for the breakdown of macro- and microaggregates. The main mechanisms of aggregate breakdown for macro- and microaggregates are slaking and dispersion, respectively. Slaking is the initial break-up of macroaggregates into microaggregates when immersed in water, caused by pressure due to entrapped air [38] and/or by differential swelling [ESS 2008]. Unlike slaking, dispersion liberates the soil colloidal particles that are more transportable during erosion. Hence, microaggregate stability is often referred to as colloidal stability. This suggests that microaggregate stability may be a better indicator of potential soil erosion hazards. Some studies have related potential soil loss or, more specifically, the erodibility of tropical soils to their aggregate stability at both the macro and micro levels. These studies tend to support the view that erosion in the soils is related more to microaggregate stability than to macroaggregate stability. For instance, Igwe et al. [19] compared the predictability of soil loss by selected macro- and microaggregate stability indices for some soils from southeastern Nigeria. They found that all microaggregate stability indices predicted soil loss better than their macroaggregate stability counterparts. Some other researchers reported weak correlations between soil erodibility and macroaggregate stability indices for some Nigerian soils [30, 31]. The soils in question are by virtue of their parent materials dominated by quartz and, as is the case with many tropical soils, are at an advanced stage of weathering. Hence, such other minerals as Fe-oxyhydroxides and kaolinite abound in them, and these are the minerals that are known to cause highly stable aggregation [54]. Since these predominant minerals do not expand rapidly when immersed in water, slaking proceeds rather slowly in the soils. The implication is that the soils show fairly high macroaggregate stability which is a misrepresentation of their high erodibility and erosion status [33]. Considering the widely accepted role of soil organic matter in aggregate formation and stabilization/destabilization, the choice of microaggregate stability for the prediction of potential eroson hazards in tropical soils would also be explained by the relative influence of organic matter on macro- and microaggregate stability. Macroaggregates are generally considered more sensitive to soil organic matter concentrations–and hence are less stable–than microaggregates [58]. Whereas the theory of macroaggregates being less stable than microaggregates may hold true for tropical soils, that of macroaggregates being more sensitive to soil organic matter concentrations than microaggregates remains a controversial topic. It has been shown that the relationships between aggregate stability indices and organic matter concentrations in tropical soils are generally characterized by weak correlations [55], and these are thought to be due mainly to the relatively lower organic matter status of the soils. However, inconsistencies characterize the response of macro- and microaggregation to organic matter concentrations in tropical soils. The relationship between macroaggregate stability and soil organic matter concentration has been reported to be non-significant [17, 31, 64] or postively significant [7, 18, 26] or negatively significant [23]. There are indications that these relationships may depend on method of assessment of macroaggregate stability as well as on location. Soil clay content is another factor that may dictate the nature of organic matter effect on macroaggregate stability of tropical soils [61]. Similarly, the relationship between microaggregate stability and organic matter concentration in tropical soils has been reported to be non-significant [23] or positively significant [12, 42, 64, 51] or negatively significant [30, 33, 34]. There are indications that these relationships may depend on microaggregate stability index adopted by the authors as well as on the contents of organic matter in the soil relative to other microaggregating agents.

3. Soil microaggregation and microaggregate stability

There are a lot of inconsistencies in the literature regarding the appropriate size boundary between macro- and microaggregates. The placement of size boundary for the classification of aggregates into macro- and microaggregates and the delineation of their upper and lower limits, respectively appear to depend on the researcher's orientation and location. We adopt here the categorization scheme proposed by Oades and Waters [46], which specifies the boundary between macro- and microaggregates as 0.25 mm, and this is consistent with the use of 0.25 mm as the boundary between water-stable and water-unstable aggregates in aggregate stability studies. In the hierarchy of aggregate size order, the lower boundary of microaggregates is taken to be 0.02 mm [46]. However, these upper and lower boundaries may be exceeded in highly weathered tropical soils where the association between microaggregates and clay-sized granules often form a kind of continuum of very stable aggregates [59, 49]. The stable microgranular structure is often manifested in form of pseudo-sands composed of clay particles that are strongly cemented together by Fe oxides [31].

Microaggregates are formed in a number of ways, each influenced by a number of factors. The process of microaggregation combines break-up of aggregates due to slaking and aggregates due to subsequent attrition [14]. Factors that influence microaggregation may differ between the temperate and tropical regions. Some researchers working in a German temperate soil reported that microaggregation depended strongly on the size distribution of primary particles rather on land use [39]. Conversely, an assessment of microaggregate stability under different land use types in a Nigerian tropical soil revealed a strong dependence of the soil microaggregation on land use [51]. This implies that the agents of stabilization of microaggregates in tropical soils are sensitive to land use.

The high aggregate stability for which tropical soils are reputed is not limited to macroaggregates. As already noted, microaggregates formed in tropical soils at advanced stages of weathering are also of very high stability [59, 46, 28]. In spite of this, microaggregate stability may still be a good indicator of the erodibility of tropical soils because of its direct link with silt and clay dispersion. Mineralogy appears to have a great influence on microaggregate stability of soils [45]. In this regard, the major microaggregating agents in tropical soils are Fe and Al oxides [17, 64, 31, 2, 28, 57]. However, in hardsetting lowland soils with low organic matter concentration and which are prone to seasonal flooding, microaggregation may be achieved through practices that enhance the organic matter concentration in them, since the roles of Fe and Al oxides in such soils may be dispersive rather than microaggregating [27]. Also, the microaggregating effect of Fe_2O_3 has been reported to be masked in some soils with relatively high concentrations of organic matter (1.39-6.79%) while that of exchangeable Ca and Mg became evident due to the tie-up of these elements with organic matter and hence their minimal leaching [Opara 2009]. Closely related to the effect of Fe and Al oxyhydroxides on the microaggregate stability of tropical soils is that of the non-expanding clay types, which dominate the clay mineralogy of the soils [30, 2006, 61, 57].

In tropical soils, soil organic matter may act as a dispersing/deflocculating agent [31], as a microaggregating agent [26, 51] or as a facilitator to the microaggregating effect of Fe-Al oxides [28], depending on its relative abundance in the soils. By contrast, the effect of soil organic matter concentration on microaggregate stability of temperate soils appears not to be prounounced [39]. Apart from protecting the surface against raindrop impact, organic matter may impart hydrophobic characteristics to the soil, thereby reducing the slaking that usually precedes dispersion [38]. In some Fe-Al oxidic tropical soils from Malaysia, it was polysaccharide constituent of soil organic matter rather than total organic matter that influenced microaggregate stability [57]. Notably, the soil content of Fe and Al oxyhydroxides is not easy to manipulate by regular soil management practices [5]. The inference that can be drawn here is that the view that organic matter is not the main aggregating agent in tropical soils rich in Fe-Al oxyhydroxides [5] may not always apply to microaggregation, but the exact role of organic matter may depend on its concentration in the soil and on its chemical composition as may be determined by the prevailing land use and soil management.

4. Aggregate breakdown mechanisms and erosion processes

Some tropical soils occuring in high-intensity rainfall zones have the tendency to slake and form seals, thereby resulting in considerable runoff and soil erosion [50, 13]. Although rainfall impact and slaking cause much greater breakdown of macroaggregates than microaggregates, these two factors can also be important for microaggregate stability and soil erodibility in at least two ways. First, slaking precedes dispersion. And this is the reason why, even though slakability is different from dispersibility, soils with high slaking potential are at high risk of interill erosion [41]. Second, sealing and crusting often accompany slaking. Seal is defined as the orientation and packing, at the very surface of the soil, particles dispersed from soil aggregates due to the impact of rain drops, thereby rendering the soil relatively impermeable to water [44]. This is the first stage of seal formation. As the ponded water infiltrates or evaporates, the soil particles suspended in it get deposited on the soil surface, thereby increasing the thickness of the seal. This is the second stage of seal formation. The entire seal eventually dries out to become crust, a thin but much more compact and hard layer than the material directly beneath [44, 60]. Both seals and crusts are therefore formed in the same way and occur commonly in the semi-arid regions [44, 60]. Crusts formed due to the first and second stages of seal formation are called structural crusts and depositional or sedimentary crusts, respectively [44, 38].

Most tropical soils are highly weathered and lacking in expanding clay types. Where they occur, the associated shrink-swell hazard is mostly concentrated in the subsoil where there is increased content of clay particles due to translocation and illuivation or residual accumulation of clay [33]. Because of this, slaking is due more to compression of air entrapped inside aggregates during wetting than to swelling. In the absence of swelling, the intense rainstorms experienced in the tropical region may result in sedimentary sealing and crusting especially in soils with reasonably high clay content but with disproportionately low concentration of organic matter [62]. Surface sealing and crust formation are an important factor in erosion processes, for they influence detachability of soil particles from aggregates, as well as infiltration rate and surface roughness which determine runoff volume and speed, respectively [38].

For soil erosion in interrill areas, three generally recognized sub-processes completely define soil erosion; and they include detachment, transport and sedimentation [38]. Some researchers working with sandy-loam soils in the semi-arid tropics have obtained results which suggest that the erodibility of a soil depends on the relative proportion of aggregates in the soil, being higher when the aggregate size distribution shows a greater proportion of large-sized aggregates [35]. Others working with low-activity-clay tropical soils reported that the saturated hydraulic conductivity increased with an increase in structural stability of the soils [61, 48]. Increased saturated hydraulic conductivity implies reduced weakening and dispersion of the soil aggregates following rainfall and/or irrigation and, hence, less susceptibility to erosion. It appears therefore that, with respect to erosion, the predominance of large-sized aggregates in soils is not always an indicator of good soil structure, but the stability of the soil pore system is.

It has been shown that, in tropical soils, disruption of macroaggregates leaves them as microaggregates rather than as primary particles [17]. Disintegration of soil macroaggregates into microaggregates following rainfall, slaking, dispersion and sealing can decrease infiltration and saturated hydraulic conductivity of the soil [36, 37]. These effects which ultimately increase soil loss can be more severe in soils of low organic matter concentration [37], as those occuring in the tropical region. The main mechanism of microaggregate breakdown is dispersion into primary particles, and this is influenced by the electrolyte concentration of the soil solution and the applied water, exchangeable sodium percentage and mechanical disturbance [38]. Electrolyte concentration and the dispersion it induces can lead to a situation whereby re-deposition of the dispersed particles cause clogging of water-conducting pores in the soil, in which case the hydraulic conductivity becomes drastically reduced [10]. The roles of exchangeable sodium percentage and electrolyte concentration in microaggregate stability are also evident in tropical soils [31, 32], probably due to the effect of ions on the amount of aggregates cemented by Fe-Al oxyhydroxides.

Generally, polyvalent cations cause flocculation whereas the monovalent cations cause dispersion [38]. It appears, however, that in hardsetting tropical soils with low organic matter concentration and that are prone to seasonal flooding, the flocculating role of polyvalent cations and the dispersive role of monovalent cations are usually not evident [27]. On the other hand, polyvalent cations (Ca^{2+} and Mg^{2+}) are good microaggregating agents under upland soil conditions, provided there is sufficient organic matter in the soil to retain these cations against leaching [51]. For a range of tropical soils all from Nigeria, factors that have been identified to influence soil dispersion include presence and concentration of monovalent cations (mostly K^{+}) in prospective irrigation water [27], soil reaction (pH), sodium adsorption ratio, and soil properties related to cation exchange [32, 23, 26] In the same region, elemental contents in silt fraction were reported to influence microaggregate stability [24].

5. Assessment of soil microaggregate stability

Microaggregate stability is normally assessed by the extent of dispersion of microaggregates into granules and/or primary particles. This is difficult to do under field conditions where the dynamic nature of this soil property may not permit attainment of reliable data. Consequently, most methods of assessment of microaggregate stability are based either on conceptual model of microaggregation involving the finer and colloidal particles or on the response of isolated microaggregates to simulated dispersive force in the laboratory. Although the disintegration forces applied in the laboratory may attempt to simulate those found in the field, they do not fully duplicate field conditions [3]. Forces applied to achieve dispersion during microaggregate stability tests may even be bigger and too sudden compared to the ones that cause dispersion under field conditions. Results of such tests are, however, still useful for they allow for a discrimination between soils in accordance with field observations [3], thereby providing information that can guide management decisions. Some of the methods that have been applied to tropical soils are summarized (Table 1). The information presented in this table shows that all the indices have to do with clay and/or silt dispersion in water. Although either of the water-dispersible clay (WDC) and water-dispersible silt (WDS) can be used to do the assessment of microaggregate stability, most researchers prefer using indices that include both.

One observation that is noteworthy is the seemingly lack of agreement among the soil microaggregate stability indices included in this review. This lack of agreement is evident in the inconsistent pattern in which these indices relate to other soil properties, including soil contents of oxides and organic matter, both of which have been shown to be very important in microaggregation. For instance, WDC and clay dispersion ratio (CDR), both of which are indices of colloidal stability, have been reported to correlate with soil organic matter concentration in a contrasting manner [30]. It appears thus that, under certain conditions, some colloidal stability indices serve better, but under some other conditions, the same colloidal stability indices may not be suitable.

6. Soil erosion hazards in tropical soils and the need for prediction

The more widespread forms of erosion are rill and interrill erosion. Soil erosion can have both on-site and off-site effects which are the lowering of soil productivity and deposition of sediments, respectively. Crop yields are usually used as a proxy measure of soil productivity loss to erosion. Deposition of sediments, mostly colloidal particles detached from the soil by agents of erosion, occurs after they are transported by surface runoffs generated during rainfall (in the case of water erosion) and turbulent winds (in the case of wind erosion). Water erosion also results in the transport of runoff-laden solutes and dissolved contaminants and is thus a major source of land and water pollution. The problem is experienced more in the humid and sub-humid tropics where the rains often come as rainstorms. Here, soil loss to water erosion can be over 50 tons $ha^{-1}yr^{-1}$ [50, 15]. In contrast, the impact of wind erosion is felt more in the semi-arid and arid tropical climates, with soil loss rate that often surpasses that due to water erosion. In the West African Sahel, for instance, soil loss to wind erosion can be in the range of 58-80 tons $ha^{-1}yr^{-1}$ [34].

In those areas where water erosion is the bigger problem, taxonomically different soils can respond differently to erosion under similar conditions. For instance, Inceptisols and Entisols have been reported to be more erodible than Ultisols, due to higher Fe and Al contents of the latter [23]. With respect to crop yields, the productivity of adversely eroded soils can be restored through careful selection of appropriate soil management practices. However, except in a few cases where materials deposited by runoff are properly harnessed, the off-site effect of soil erosion always constitutes environmental problems. In contemporary agriculture where the emphasis is on not just achieving high yields but also on making agricultural enterprise environment-friendly, such environmental problems arising from soil erosion should be viewed as undermining agricultural productivity.

The problem of soil erosion and the associated negative impacts on agriculture and the environment is particularly severe in tropical sub-Saharan Africa, where it is a major cause of declining and stagnating soil productivity [48]. When considering appropriate soil conservation as an option, the first step is to try to understand the roles of microaggregate stability in checkmating soil erosion and in predicting soil erosion hazards. Prediction of soil erosion hazards involves a quantitative assessment of potential soil erosion in a land resource of an area. Such quantitative information is used in soil erosion hazard mapping for both short-term and long-term planning against erosion and associated deleterious effects, and this can have many agricultural and environmental benefits. Many atimes, erosion hazard maps are viewed as a tool for detailed farm planning and management [30]. Information on potential erosion hazards can also be used to embark on precautionary soil and water conservation measures. For instance, conservation specialists can use such information to select appropriate engineering designs and structures aimed at forestalling the occurrence of erosion in the first place, or controlling erosion in already eroded areas. Once started, rill and interill erosion need to be timely arrested, otherwise they may escalate into gully erosion, which is the more spectacular form of erosion that often threatens the integrity of the environment.

7. Microaggregate stability and erosion hazard prediction for tropical soils

Virtually all known methods of assessing microaggregate stability, discussed earlier, employ the extent of dispersion into primary particles. The relevance of microaggregate stability for assessing potential erosion hazards lies, therefore, on the effects of dispersion on soil hydrophysical processes. Dispersion generally induces processes that are related to soil erodibility such as very fast crusting, slow infiltrability, and great mobility of particles in water [38]. Soil erodibility may be defined as the degree or intensity of a soil's state or condition of being susceptible to erosion [56]. It is just one of the main parameters needed for erosion hazard prediction. The most commonly used index of erodibility is the erodibility factor (K-factor) of the revised universal soil loss equation (RUSLE). Although fragments/sediments detached by raindrops can be finer than the original soil, the detachment is often accompanied by mere displacement (splash effect); the actual transport and sedimentation involve silt- and clay-sized particles [38]. Therefore, microaggregates dispersion is a precondition for soil erosion to be complete. There is evidence from the United States that WDC and CDR can be good estimators of erodibility of some soils in Ohio [4].

Microaggregate stability, when used as a tool for predicting soil erosion hazards, takes into account only the aspect of such hazards that are due to the soil inherent erodibility. One would therefore expect researchers to relate microaggregate stability to only soil erodibility when assessing potential erosion hazards. However, because soil erodibility is a dynamic soil property, its accurate determination can sometimes be difficult. Acquisition of data for soil erodibility is particularly difficult in the case of the K-factor of the RUSLE, as this requires some basic land-use information as well as pre-measurement soil management specifications, actual practice of which is often tedious and time-consuming. Consequently, not all researchers relate microaggregate stability to soil erodibility; some often relate it directly to soil loss to natural or simulated erosion, while keeping constant such other factors that affect erosion as rainfall, topography, vegetation, and soil management and conservation practices. We reason that, unless the relationship between microaggregate stability indices and erodibility/soil loss are not established by statistical correlations, the effects of such methodological differences may be negligible.

Index	Derivation	Interpretation	References
Clay ratio, CR	$\frac{\% \text{ sand}}{\% \text{ silt} + \% \text{ clay}}$	A	Mbagwu (1986)
Degree of aggregation, DOA†	$\frac{w_a - w_b}{w_a}$	B	Zhang and Horn (2001)
Water dispersible clay, WDC	Clay after particle-size analysis with deionized water only	A	Mbagwu and Auerswald (1999); Igwe (2005)
Water dispersible silt, WDS	Silt after particle-size analysis with deionized water only	A	Igwe and Nkemakosi (2007)
Aggregated clay, AC or Clay aggregation, CA	Total clay – WDC	B	Mbagwu and Auerswald (1999); Igwe (2003)
Aggregated silt + clay, ASC	Total silt and clay – WDS and WDC	B	Igwe et al. (1999a)
Clay dispersion ratio, CDR or Clay dispersion index, CDI	$\frac{\% \text{ WDC}}{\% \text{ total clay}}$	A	Igwe and Nkemakosi (2007); Opara (2009)
Clay flocculation index, CFI	Total clay – WDC/total clay	B	Igwe and Nkemakosi (2007)
Dispersion ratio, DR or Water dispersible clay and silt, WDCS	$\frac{\% \text{ WDS} + \% \text{WDC}}{\% \text{ total silt and clay}}$	A	Mbagwu (1986); Igwe (2005); Igwe and Nkemakosi (2007); Sung (2012)

†w_a and w_b stand for the proportion of particles between 0.25 and 0.05 mm obtained by microaggregate size analysis and by particle size analysis, respectively.
A – The smaller the value, the more stable the microaggregates are.
B – The bigger the value, the more stable the microaggregates are.

Table 1. Indices of microaggregate stability commonly applied to tropical soils

Although a good number of studies have been conducted on aggregate stability of tropical soils, our survey of the literature reveals that not many of these studies related the erodibility of the soils or potential soil loss to aggregate stability. Soil aggregate stability or instability is such a critical factor in erosion processes that erosion is often the first thing that comes to the mind when pondering over usefulness of information on aggregate stability. It is thus surprising that the majority of studies on aggregate stability of tropical soils failed to describe its relationship with soil erosion. Again, among the few studies that did otherwise, only a small proportion used microaggregate aggregate stability indices in spite of the fact that, as we have been able to show earlier in this review, microaggregate stability more than macroaggregate stability corresponds to the dispersion and erodibility of tropical soils. We review in the preceding paragraph only those studies that related soil erodibility or potential soil loss to microaggregate stability in the tropical region.

In southeastern Nigeria, clay ratio (CR) and dispersion ratio (DR) were reported as being close substitutes to the K-factor in the prediction of soil loss [40]. Also in this region, Igwe et al. [29] related the K-factor to selected indices of microaggregate stability for soils from diverse geological formations. Their results showed a good correlation ($r = 0.53$) between K-factor and clay flocculation index (CFI), and they recommended that the CFI alone could be used to predict soil erosion hazard in the area. The stability and soil-loss response of a stony Nigerian tropical soil undergoing intensive cultivation to simulated tillage and stone removal was investigated [30]. This laboratory study revealed that tillage and stone removal led to increases in WDC, DR and CDR; and that this failure in microaggregate stability of the soil increased erosion of the soil. Still working with soils from southeastern Nigeria, Igwe [31] reported that any of DR, CDR and WDC could be used in predicting erodibility of some the soils. The CDR and DR were also found, in a separate study, to have significantly ($r = 0.44$ and 0.39, respectively) correlated with K-factor of the RUSLE and were therefore deemed good indices for predicting erodibility of soils of eastern Nigeria [32]. All these studies demonstrate the suitability of some microaggregate stability indices for assessing soil erodibility and potential soil loss in the tropical region.

8. Areas of further research

All the indices of microaggregate stability included in this review were developed based on silt and/or clay dispersion which occurs only in wet or submerged soils, and this limits their applicability to erodibility assessment to the case of water erosion [9]. In the semi-arid and arid tropics, wind erosion is a major source of soil and nutrient loss in agricultural soils. An index of microaggregate stability is therefore needed for such soils to also enable the assessment of potential erosion hazards in them. Similarly, there are indications that removal of gravels and stones from tropical soils characterized by high gravel content can confer higher erodibility to such soils [43, 25]. This implies that, for this category of soils, the use microaggregate stability indices determined from routine laboratory measurements as indicators of soil erodibility may be misleading. It may therefore be necessary to correct microaggregate stability results for gravel content, especially when they are intended for use in the assessment of soil erodibility. Research is needed on the best method of doing such a correction as may be confirmed by a good agreement between the ensuing results and field-measured erodibility of the soil. Also, some researchers have reported good correlations between their microaggregate stability indices and soil contents of silt [57] or clay [26, 51], just as others have reported that elemental contents in silt fraction affected microaggregation [24]. Silt is known to be the soil particle that is most suspectible to loss during erosion [52], and the data presented by Igwe and Ejiofor [2005] for a severely gullied tropical soil support this assertion. This suggests that paying attention to soil texture, especially variations in silt content, may benefit microaggregate stability studies in relation to erodibility.

It is known that oxides which abound in many tropical soils are a major promoter of their colloidal stability. The role of particularly Fe oxides in microaggregate stability may not be limited to the promotion of microaggregate formation. A study conducted in a mediterranean environment revealed that Fe oxides also acted to decrease dispersion of clay [6]. The possibility of this phenomenon and the factors promoting it in Fe-oxide-rich soils in the core tropics need to be explored. This review reveals that there are conflicting reports on the effect of organic matter concentration on soil microaggregation and microaggregate stability of tropical soils. Forms of oxides in the soil can influence not only their aggregating potential but also that of organic matter [11], and this has been demonstrated specifically for microaggregation of tropical soils [28]. On the other hand, the chemical composition of organic matter and its distribution in the aggregate-size classes (whether it is physically protected within microaggregates or not) may also contribute to determining how it influences microaggregation in the soil. More studies are therefore suggested on the role of organic matter in microaggregate stability of tropical soils, with emphasis on soils differing in both contents and forms of oxides.

In erosion processes, field capacity is expected to be an important factor because of its direct link with infiltration and runoff. It has been shown that slaking potential of a soil decreases with an increase in its field capacity [8], suggesting that the tendency for dispersion may also decrease with an increase in field capacity. However, in severely gullied soils in eastern Nigeria showing silt content of not more than 1% and mean organic matter concentration of 0.18% (both on weight basis), CDR was shown to increase (i.e. decrease in colloidal stability) with an increase in field capacity [23]. Recently, Abrishamkesh et al. [1] reported higher field capacity under a condition of higher structural stability than lower structural stability in a temperate environment. Similarly, Obalum et al. [48] reported that the lower the structural stability of some coarse-textured tropical soils, the higher the pressure at which they attain field capacity. They attributed the observation to reduced dispersion and hence increased internal drainage of the soils as their stability increased. It appears therefore that the field capacity represents a structural index related to both dispersability and stability of soil aggregates. Research is needed to fully explore the relationships among field capacity, microaggregate stability and erodibility of tropical soils differing in degree of past erosion.

9. Conclusion

The majority of tropical soils show high microaggregate stability irrespective of their low organic matter concentration. This is due mainly to their high contents of Fe and Al oxides which are known to promote microaggregation in soils of low organic matter concentration. However, there are some conflicting reports on the effects of the various players, especially oxides and organic matter, on microaggregation in tropical soils. So many natural and anthropogenic factors can lead to dispersion of the soils, but the factors tend to vary from study to study. A number of agricultural and environmental problems can arise from the dispersion of clay especially in sandy soils characterized by low concentration of organic matter [9], like the ones predominants in the tropics. The most important of these problems is soil erosion. Although only few studies have related soil erosion hazard (assessed either by soil erodibility or by soil loss) to selected indices microaggregate stability, these studies show that microaggregate stability is a useful tool for predicting erosion hazards in tropical soils. However, comparisons of results of erosion hazard prediction would be meaningful only when the same index of microaggregate stability is used. We suggest some areas for further research on microaggregation in tropical soils and the relationship between colloidal stability and soil erodibility.

Author details

C.A. Igwe and S.E. Obalum

Department of Soil Science, University of Nigeria, Nsukka, Nigeria

References

[1] Abrishamkesh S, Gorji M, Asadi H. Long-term effects of land use on soil aggregate stabilty. International Agrophysics 2011; 25 103-108.

[2] Alekseeva TV, Sokolowska Z, Hajnos M, Alekseev AO, Kalinin PI. Water stability of aggregates in subtropical and tropical soils (Georgia and China) and its relationships with the mineralogy and chemical properties. Eurasian Soil Science 2009; 42 415-425.

[3] Amezketa E. Soil aggregate stability: a review. Journal of Sustainable Agriculture 1999; 14 83-151.

[4] Bajracharya RM, Elliot WJ, Lal R. Interrill erodibility of some Ohio soils based on field rainfall simulation. Soil Science Society of America Journal 1992; 56 267-272.

[5] Barthes BG, Kouakoua E, Larre-Larrouy M, Razafimbelo TM, de Luca EF, Azontonde A, Neves CSVJ, de Freitas PL, Feller CL. Texture and sesquioxide effects on water-stable aggregates and organic matter in some tropical soils. Geoderma 2008; 143 14-25.

[6] Calero N, Barron V, Torrent J. Water dispersible clay in calcareous soils of southwestern Spain. Catena 2008; 74 22-30.

[7] Chappell NA, Ternan JL, Bidin K. Correlation of physicochemical properties and sub-erosional landforms with aggregate stability variations in tropical Ultisol disturbed by forestry operations. Soil and Tillage Research 1999; 50 55-71.

[8] De Boodt M. West European Methods for Soil Structure Determinations. International Society of Soil Science Communique 1, West European Group 2005; Part VI 17-18, 36-38.

[9] Dexter AR, Cyzz EA. Effects of soil management on the dispersibility of clay in a sandy soil. International Agrophysics 2000; 14 269-272.

[10] Dikinya O, Hinz C, Aylmore G. Dispersion and re-deposition of fine particles and their effects on saturated hydraulic conductivity. Australian Journal of Soil Research 2006; 44 47-56.

[11] Duiker SW, Rhoton FE, Torrent J, Smeck NE, Lal, R. Iron (hydr)oxide crystallinity effects on soil aggregation. Soil Science Society of America Journal 2003; 67 606-611.

[12] Dutartre Ph, Bartoli F, Andreux F, Portal JM, Ange A. Influence of content and nature of organic matter on the structure of some sandy soils from West Africa. Geoderma 1993; 56 459-478.

[13] Ekwue EI. A simple technique for measuring soil infiltration rates during simulated rainfall. Journal of Arid Agriculture 1994; 3-7 95-104.

[14] Encyclopedia of Soil Science, ESS. Aggregate stability to drying and wetting. *In* Chesworth C (Ed). *Encyclopedia of Soil Science,* Springer 2008; pp. 28-33.

[15] FAO. Land and environmental degradation and desertification in Africa. FAO Corporate Document Repository, 1995. http://www.fao.org/docrep/X5318E/X5318E00.htm.

[16] FAO. Management of tropical sandy soils for sustainable agriculture. FAO Corporate Document Repository, 2007. 536 pp. http://www.fao.org/docrep/010/ag125e/ag125e00.htm.

[17] Folly A. Estimation of erodibility in the savanna ecosystem, northern Ghana. Communications in Soil Science and Plant Analysis 1995; 26 799-812.

[18] Idowu OJ. Relationships between aggregate stability and selected soil properties in humid tropical environment. Communications in Soil Science and Plant Analysis 2003; 34 695-708.

[19] Igwe CA, Akamigbo FOR, Mbagwu JSC. Application of SLEMSA and USLE erosion models for potential erosion hazard mapping in southeastern Nigeria. International Agrophysics 1999b; 13 41-48.

[20] Igwe CA, Akamigbo FOR, Mbagwu JSC. Chemical and mineralogical properties of soils in southeastern Nigeria in relation to aggregate stability. Geoderma 1999a; 92 111-123.

[21] Igwe CA, Akamigbo FOR, Mbagwu JSC. Physical properties of soils of southeastern Nigeria and the roles of some aggregating agents in their stability. Soil Science 1995b; 160 431-441.

[22] Igwe CA, Akamigbo FOR, Mbagwu JSC. The use of some soil aggregate indices to assess potential soil loss in soils of southeastern Nigeria. International Agrophysics 1995a; 9 95-100.

[23] Igwe CA, Ejiofor N. Structural stability of exposed gully wall in central eastern Nigeria as affected by soil properties. International Agrophysics 2005; 19 215-222.

[24] Igwe CA, Nkemakosi JT. Nutrient element contents and cation exchange capacity in fine fractions of southeastern Nigerian soils in relation to their stability. Communications in Soil Science and Plant Analysis 2007; 38 1221-1242.

[25] Igwe CA, Okebalama CB. Soil strength of some central eastern Nigeria soils and effect of potassium and sodium on their dispersion. International Agrophysics 2006; 20 107-112.

[26] Igwe CA, Udegbunam ON. Soil properties influencing water-dispersible clay and silt in an Ultisol in southern Nigeria. International Agrophysics 2008; 22 319-325.

[27] Igwe CA, Zarei M, Stahr K. Clay dispersion of hardsetting Inceptisols in southeastern Nigeria as influenced soil components. Communications in Soil Science and Plant Analysis 2006; 37 751-766.

[28] Igwe CA, Zarei M, Stahr K. Colloidal stability in some tropical soils of southeastern Nigeria as affected by iron and aluminium oxides. Catena 2009; 77 232-237.

[29] Igwe CA. Changes in water-dispersible clay and erosion of stony slope soil under tillage in Nigeria. Polish Journal of Soil Science 2002; 35 31-36.

[30] Igwe CA. Clay dispersion of selected aeolian soils of northern Nigeria in relation to sodicity and organic carbon content. Arid Land Research and Management 2001; 15 147-155.

[31] Igwe CA. Erodibility in relation to water-dispersible clay for some soils of eastern Nigeria. Land Degradation and Development 2005; 16 87-96.

[32] Igwe CA. Erodibility of soils of the upper rainforest zone, southeastern Nigeria. Land Degradation and Development 2003; 14 323-334.

[33] Igwe CA. Tropical soils, physical properties. *In* J Glinski, J Horabik, J Lipiec (Eds.), *Encyclopedia of Agrophysics* (1st ed.), Springer 2011; 934-937.

[34] Ikazaki K, Shinjo H, Tanaka U, Tobita S, Funakawa S, Kosaki T. Field-scale aeolian sediment transport in the Sahel, West Africa. Soil Science Society of America Journal 2011; 75 1885-1897.

[35] Kukal SS, Manmeet-Kaur, Bawa SS. Erodibility of sandy loam aggregates in relation to their size and initial moisture content under different land uses in semi-arid tropics of India. Arid Land Research and Management 2008; 22 216-227.

[36] Lado M, Paz A, Ben-Hur M. Organic matter and aggregate-size interactions in saturated hydraulic conductivity. Soil Science Society of America Journal 2004a; 68 234-242.

[37] Lado M, Paz A, Ben-Hur M. Organic matter and aggregate-size interactions in infiltration, seal formation, and soil loss. Soil Science Society of America Journal 2004b; 68 935-942.

[38] Le Bissonnais Y. Aggregate breakdown mechanisms and erodibility. *In* Lal R (Ed.), *Encyclopedia of Soil Science* (2nd ed.), Taylor and Francis 2006; 40-44.

[39] Leifeld J, Kogel-Knabner I. Microaggregates in agricultural soils and their size distribution determined by X-ray attenuation. European Journal of Soil Science 2003; 54 167-174.

[40] Mbagwu JSC, Bazzoffi P. Soil characteristics related to resistance of breakdown of dry soil aggregates by water-drops. Soil and Tillage Research 1998; 45 133-145.

[41] Mbagwu JSC, Auerswald K. Relationship of percolation stability of soil aggregates to land use, selected properties, structural indices and simulated rainfall erosion. Soil and Tillage Research 1999; 50 197-206.

[42] Mbagwu JSC, Piccolo A, Mbila MO. Water stability of aggregates of some tropical soils treated with humic substances. Pedologie 1993; XLIII-2 269-284.

[43] Mbagwu JSC. Erodibility of soils formed on a catenary toposequence in southeastern Nigeria as evaluated by different indexes. East African Agricultural and Forestry Journal 1986; 52 74-80.

[44] Morin J. Soil crusting and sealing. In: Soil Tillage in Africa: Needs and Challenges. FAO Soils Bulletin; 1993, FAO Corporate Document Repository.

[45] http://fao.org/docrep/T1696E/T1696E00.htm.

[46] Nwadialo BE, Mbagwu JSC. An analysis of soil components active in microaggregate stability. Soil Technology 1991; 343-350.

[47] Oades JM, Waters AG. Aggregate hierarchy in soils. Australian Journal of Soil Research 1991; 29 815-828.

[48] Obalum SE, Buri MM, Nwite JC, Hermansah, Watanabe Y, Igwe CA, Wakatsuki T. Soil degradation-induced decline in productivity of sub-Saharan African soils: the prospects of looking downwards the lowlands with the *sawah* eco-technology. Applied and Environmental Soil Science; Volume 2012, Article ID 673926, 10 pages. doi: 10.1155/2012/673926.

[49] Obalum SE, Igwe CA, Hermansah, Obi ME, Wakatsuki T. Using selected structural indices to pinpoint the field moisture capacity of some coarse-textured agricultural soils in southeastern Nigeria. Journal of Tropical Soils 2011; 16 151-159.

[50] Obi ME, Salako FK, Lal R. Relative susceptibility of some southeastern Nigeria soils to erosion. Catena 1989; 16 215-225.

[51] Obi ME. Runoff and soil loss from an Oxisol in southeastern Nigeria under various management practices. Agricultural Water Management 1982; 5 193-203.

[52] Opara CC. Soil microaggregates stability under different land use types in southeastern Nigeria. Catena 2009; 79 103-112.

[53] Richter G, Negedank JFW. Soil erosion processes and their management in the German area of the Moselle river. Earth Surface Proceedings 1977; 2 261-278.

[54] Shi Z, Yan F, Li L, Li Z, Cai C. Interrill erosion from disturbed and undisturbed samples in relation to topsoil aggregate stability in red soils from subtropical China. Catena 2010; 81 240-248.

[55] Six J, Bossuyt H, Degryze S, Denef K. A history of research on the link between (micro)aggregates, soil biota and soil organic matter dynamics. Soil and Tillage Research 2004; 79 7-31.

[56] Six J, Feller C, Denef K, Ogle SM, de Moraes-SA JC, Albrecht A. Soil organic matter, biota and aggregation in temperate and tropical soils – Effects of no-tillage. Agronomie 2002; 22 755-775.

[57] Soil Science Society of America. *Glossary of Soil Science Terms*, American Society of Agronomy, Crop Science Society of America, Soil Science Society of America; 2001, Wisconsin, USA.

[58] Sung CTB. Aggregate stability of tropical soils in relation to their organic matter constituents and other soil properties. Pertanika Journal of Tropical Agricultural Science 2012; 35 135-148.

[59] Tisdall JM. Formation of soil aggregates and accumulation of soil organic matter. In: Structure and Organic Matter Storage in Agricultural Soils (eds. Carter MR, Stewart BA). Advances in Soil Science 1996. CRC Press, Boca Raton, FL. pp. 57-96.

[60] Trapnell CG, Webster R. Microaggregates in red earths and related soils in east and central Africa, their classification and occurrence. Journal of Soil Science 1986; 37 109-123.

[61] Valentin C. Soil crusting and sealing in West Africa and possible approaches to improved management. In: Soil Tillage in Africa: Needs and Challenges. FAO Soils Bulletin; 1993, FAO Corporate Document Repository. http://fao.org/docrep/T1696E/T1696E00.htm.

[62] Wuddivira MN, Camps-Roach G. Effects of organic matter and calcium on soil structural stability. European Journal of Soil Science 2007; 58 722-727.

[63] Wuddivira MN, Stone RJ, Ekwue EI. Clay, organic matter, and wetting effects on splash detachment and aggregate breakdown under intense rainfall. Soil Science Society of America Journal 2009; 73 226-232.

[64] Yan F, Shi Z, Li Z, Cai C. Estimating interrill soil erosion from aggregate stability of Ultisols in subtropical China. Soil and Tillage Research 2008; 100 34-41.

[65] Zhang B, Horn R. Mechanisms of aggregate stabilization in Ultisols from subtropical China. Geoderma 2001; 99 123-145.

Chapter 9

Soil Physical Properties and Nitrous Oxide Emission from Agricultural Soils

Natalya P. Buchkina, Elena Y. Rizhiya,
Sergey V. Pavlik and Eugene V. Balashov

Additional information is available at the end of the chapter

http://dx.doi.org/10.5772/53061

1. Introduction

1.1. N_2O in agriculture and microbial processes of nitrification and denitrification

N_2O is an important greenhouse gas as it contributes to global warming and the depletion of the stratospheric ozone layer (Bouwman, 1990; Crutzen, 1979; Houghton et al., 1990). At the present time the atmospheric concentration of N_2O is rising linearly at a rate of 0.3% per year (IPCC, 2007). Its current concentration in the atmosphere is equal to 319 ppbv (IPCC, 2007). A warming potential of 1 kg of N_2O is 310 times greater than 1 kg of CO_2 over a 100-year period (IPCC, 2007).

It is now widely accepted that agriculture is the main source of anthropogenic N_2O. The agriculture contributes to 60% of the global N_2O emissions (OECD. 2000). Agricultural soils are recognized as the major source of atmospheric N_2O, globally contributing 1.7-4.8 Tg N yr^{-1} (IPCC, 2007; Mosier, 1998).

N_2O is formed in two microbiological processes – nitrification and denitrification (Bremner, 1997; Firestone and Davidson, 1989) (Figure 1).

N_2O N_2O

↑ ↑

$NH_3 \rightarrow NH_2OH \rightarrow NO_2^- \rightarrow NO_3^-$ (a)

Figure 1. Production of N_2O in soils during (a) nitrification and (b) denitrification processes. (Natalya P. Buchkina, Elena Y. Rizhiya, Sergey V. Pavlik, Eugene V. Balashov)

The contribution of these processes to N_2O emission will vary with climate, soil management and properties (Granli, Bockman, 1994; Skiba, Smith, 2000). During nitrification N_2O is formed as a by-product in the reactions of oxidation of NH_3 to NH_2OH and NH_2OH to NO_2^- under strictly aerobic conditions. These two reactions are catalysed by such enzymes as ammonia monooxigenase and hydroxylamine oxidoreductase (McCarty, 1999; Wood, 1986). The reaction of NO_3^- production from NO_2^- is catalysed by nitrite oxidoreductase (Bock et al., 1986).

In contrast to nitrification, N_2O is an obligatory intermediate product of denitrification. Denitrification is a stepwise reduction of NO_3^- to N_2 by facultative anaerobic microorganisms. Enzymes catalysing these above-mentioned four reactions are nitrate reductase, nitrite reductase, nitric oxide reductase and nitrous oxide reductase (Hochstein, Tomlinson, 1988).

Despite denitrification is considered as a strictly anaerobic process, it is now accepted that N_2O can be produced during nitrifier denitrification (as a pathway of nitrification) under aerobic conditions. During a first part of nitrifier denitrification an oxidation of NH_3 to NO_2^- (nitrification) occurred, whereas a reduction of NO_2^- to N_2O and N_2 is a second part of the microbial process (Poth, Focht, 1985; Ritchie, Nicholas, 1972).

The rates of these processes and of N_2O production are controlled by several soil factors: soil water-filled pore space (WFPS), oxygen availability, pH, mineral nitrogen (NO_3^-, NH_4^+), available soil organic carbon (SOC), and temperature (Dobbie et al., 1999; Dobbie, Smith, 2003; Khalil et al., 2004., Wlodarczyk et al., 2003). Nitrification is the main process of N_2O emission at 35-60% WFPS. Denitrification becomes more important at soil water contents greater than 60% WFPS due to a decreased oxygen supply (Dobbie, Smith, 2003; Drury et al., 2003).

N_2O flux from agricultural soils depends on a complex interaction between climatic factors, soil properties and soil management. Management practices such as tillage systems and fertilizer applications can significantly affect the production and consumption of N_2O because of alteration in soil physical, chemical, and biochemical properties. The only way to reduce N_2O emissions from agricultural soils is to affect the soil properties through: 1. application of appropriate soil tillage system, 2. improving efficiency of N fertilizers with better placement, timing and rates of required N for growing relevant crops with high N uptake efficiency (Chatskikh, Olesen, 2007).

1.2. Role of soil tillage and soil physical properties in N_2O emission from soils

Tillage systems change soil structural, air, and water status (Beare et al., 1994; Green et al., 2007; Simansky et al., 2008). Nevertheless, continuous use of conventional moulboard ploughing in traditional farming systems can lead to degradation of soil structure and to losses in soil organic matter through erosion and oxidation (Lal, 2004; Lopez-Garrido et al., 2011; Norton et al., 2006; Reicosky, 2002). H. Riley et al. (2008) showed that a 15-year conventional ploughing of loamy soil caused a significant increase in bulk density from 1.29 to 1.43 Mg m^{-3} and a significant decrease in total porosity from 49.1 to 44.8%. E. Balashov and N. Buchkina (2011) showed that a 75-year use of conventional mouldboard ploughing resulted in a significant decline in a concentration of SOC (from 43.1 to 30.2 g C kg^{-1} soil) as well as in an amount of water-stable aggregates (from 90.1 to 70.2%) and clay content (from 30.5 to 26.9%), compared to the fallow land in a clayey loam Haplic Chernozem.

There is a trend for the adoption of no-tillage and reduced tillage around the world, because these systems have been shown to improve soil organic matter status, structural state and water regime (Green et al., 2007; Simansky et al., 2008; Angers et al., 1997; Fernandez et al., 2010; Nyakatawa et al., 2000; Six et al., 2002; West, Post, 2002). V. Simansky (2008) reported that reduced tillage to a depth of 10-12 cm, compared to conventional tillage to the depth of 22-25 cm, led to a significant increase in the average amount of water-stable macro-aggregates of Orthic Luvisol. Long-term studies also suggested that the reduced tillage resulted in an essential change in soil structure that increased soil porosity and decreased bulk density (Zhang et al., 2007).

The conversion from mouldboard ploughing to no-tillage can result in a reduced N_2O emission (Almaraz et al., 2009) as well as in an increased N_2O emission from soils (Lemke et al., 1999). This unfavorable increase in N_2O emission can be caused by higher WFPS and bulk density of soils recently converted to no-tillage system (Liu et al., 2007).

Direct measurements of N_2O flux from soils with different tillage systems have shown that their effects on the greenhouse gas emission are governed by climatic conditions. According to the results reported in (Six et al., 2002), a conversion of conventional tillage to no-tillage resulted to an increase in soil organic matter level in temperate and tropical soils in a wide range of agroecosystems. In the temperate soils an increase in N_2O emissions and CH_4 uptake was observed under no-tillage compared with conventional tillage. In general, light-textured, well-aerated soils under drier climate conditions do not produce more N_2O when under no-tillage or reduced tillage system compared to conventional tillage. However, poorly aerated soils in wet climate can produce more N_2O when conventional tillage is converted for no-tillage or reduced tillage system (Chatskikh, Olesen, 2007; Ball et al., 1999; Choudhary et al., 2002).

Changes in the total porosity and the amount of water-stable aggregates may result in a formation of favorable conditions for denitrification as it can take place in soils with high availability of organic matter and low availability of oxygen. Oxygen is regarded as a key determinant of denitrification rates, while the importance of available SOC and NO_3^- (or other nitrogen oxides) will vary in the ecosystems (Robertson, Groffman, 2007). A proportion of the total gaseous products of denitrification that is actually emitted to the atmosphere as N_2O depends on the soil aggregation and water content (Smith et al., 2003). A destruction of soil aggregates by tillage can lead to an increase in availability of SOC for soil denitrifying microorganisms (Gregorich et al., 1989; Petersen et al., 2008).

Despite positive effects of no-tillage and reduced tillage systems on soil aggregation and porosity, denitrification can occur even in the soil aggregates under soil aerobic conditions (Smith, 1980). K. Khalil (2004) reported that significant denitrification rates (i.e. greater than 0.2 mg N kg^{-1}) from 2-3-mm aggregates were observed in the 0 and 0.35 kPa O_2 pressure treatments, whereas nitrification (both NH_4^+ and NO_2^- oxidation rates) was reduced by a factor of 4–10 when O_2 concentration decreased from 20.4 to 0.35 kPa. According to the results reported by Uchida et al. (2008), the highest N_2O-N fluxes of 9920 μg m^{-2} h^{-1} occurred in the 0–1.0 mm aggregates, while the respective values in the 1.0–2,0; 2.0–4.0 and 4.0–5.6 mm aggregates were 3985, 2135 and 2750 μg m^{-2} h^{-1}.

An absence of tillage can lead to soil compaction and, as a result, to a reduced air-filled porosity and availability of oxygen (Rasmussen, 1999). At high soil water content, the effects of increasing soil bulk density may contribute to a formation of anaerobic conditions (Douglas, Crawford, 1998). R. Ruser et al. (1998) reported that an increase in bulk density of fine silty soil only by 20% led to increasing N_2O emission. According to the results reported by M. Beare et al. (2009), there was a strong exponential increase in N_2O production above 0.60 cm^3 cm^{-3} WFPS of a compacted fine loamy soil where denitrification was a dominant pathway of N_2O production.

Therefore, tillage systems can have contradictory effects on soil physical status and on N_2O emissions from soils. If it is necessary to obtain valid information on N_2O emissions from soils differing in soil physical status and tillage systems the site-specific measurements need to be carried out in various agroecosystems with different agricultural uses.

1.3. Role of increasing nitrogen rates in N_2O emission from soils

Application of mineral N-fertilizers into agricultural soils usually results in increasing N_2O emissions (Smith et al., 2003; Jones et al., 2007; MacKenzie et al., 1998; Rizhiya et al., 2011). However, there is contradictory information on linearity between applied N rates and N_2O emissions from soils. According to results reported by Gregorich et al. (2005), N_2O emission from agricultural soils increased linearly with the applied amount of mineral N fertilizer. At N rates not exceeding or equal to those required for maximum yields, N rates tended to create a linear response in N_2O emissions, with approximately 1% of applied mineral N lost as N_2O (Bouwman, 1996; Halvorson et al., 2008).

However, there are threshold N rates, which can exceed the crop and ecosystem uptake capacity (Grant et al., 2006). At N rates exceeding crop requirements, N_2O emissions are more variable and often rise exponentially with increasing N fertilization (Hobet et al., 2011; Snyder et al., 2009). C.P. McSwiney and G.P. Robertson (2005) reported on nonlinear increase between N_2O emission and N rates. N_2O emission was moderately low (20 g N_2O-N ha^{-1} day^{-1}) at N rate of 101 kg N ha^{-1}, but at N rate above 134 kg N ha^{-1}, N_2O emissions sharply increased to about 450 g N ha^{-1} day^{-1}. The greatest percentage of fertilizer N lost as N_2O (7%) occurred at the N rate of 134 kg N ha^{-1}. The authors also (1) found that the proportion of applied N lost as N_2O decreased to 2–4% (as emission factor) with N rates exceeding 134 kg N ha^{-1} and (2) stated that this threshold of N_2O response to N fertilization could be reduced with no or little yield penalty by reducing N fertilizer inputs to levels that just satisfy crop needs.

According to J.W. Van Groenigen et al. (2010), agricultural management practices to reduce N_2O emissions should focus on optimizing fertilizer-N use efficiency under median rates of

N input, rather than on minimizing N application rates. The authors proposed to assess the N_2O emissions as a function of crop N uptake and crop yield.

1.4. Role of crop roots in N_2O emission from soils

Living roots of crops can play an integral role in soil N_2O production by providing their labile organic exudates and mineral N to microbial community involved into nitrificarion and denitrification. J.A. Bird (2011) reported that living roots of *Avena barbata* increased the turnover and loss of belowground ^{13}C compared with unplanted soils. The rhizosphere of *Avena barbata* shifted the microbial community composition, resulting in greater abundances of gram-negative markers and lower abundances of gram- positive, actinobacteria and cyclopropyl PLFA markers compared to unplanted soil. According to the results of Khalid et al. in (Khalid et al., 2007), the presence of grass species significantly increased the size of a number of dissolved nutrient pools in comparison to the unplanted soil (e.g. dissolved organic carbon, total phenolics in solution) but had a little affect on other pools (e.g. free amino acids).

The plant roots change such soil physical properties as air porosity, pore distribution, penetrometer resistance, permeability and bulk density (Williams, Weil, 2004; Mitchel et al., 1995; Balashov, Bazzoffi, 2003). Balashov and Bazzoffi (2003) showed that the resilience capacity of wheat root system at the boot stage was sufficient for the formation and stabilization of water-stable aggregates in a sandy loam Typic Udorthents and a clayey loam Vertic Xerorthent compacted with ground contact pressures of 51 and 103 kPa. After the compaction with ground contact pressure of 154 kPa, the root system was unable to maintain the water-stable aggregation of soils even at its initial level. Therefore, roots of plants can improve soil physical status of compacted soils and can reduce a risk of formation of favorable conditions for denitrification, which are observed at WFPS exceeding 60%.

A rather high attention has been given to the dynamics of N_2O profile concentration and its subsequent diffusion to the soil surface. The concentrations and fluxes of N_2O are dependent on WFPS, mineral N content, microbial activity, temperature, and macro-porosity in the soil profiles (Burton, Beauchamp, 1994; Clough et al., 2005; Jacinthe, Lal, 2004; Kusa et al., 2010; Li et al., 2002; Muller et al., 2004; Phillips et al., 2012; Van Groenigen et al., 2005; Velthof et al., 1996; Yoh et al., 1997).

J.W. Van Groenigen et al. (2005) found that maximum N_2O concentrations were observed at depths of 48 and 90 cm at WFPS > 70% in a sandy soil. The authors reported that during the summer the maximum N_2O fluxes from the soil surface were coinciding with high subsoil N_2O concentrations suggesting that denitrification in the subsoil was a main determinant of N_2O formation. In winter with low air and soil temperatures as well as with low soil microbial activity high N_2O concentrations in the subsoil did not lead to corresponding high fluxes of N_2O from the soil surface.

K. Kusa et al. (2010) showed that the contribution ratios of the N_2O produced in the top soil (0–0.3 m depth) to the total N_2O emitted from the soil to the atmosphere in the treatments with the Gray Lowland soil and the Andosol were 0.86 and 1.00, respectively, indicating that the N_2O emitted from the soils to the atmosphere was mainly produced in the top soil. Variations in the profile N_2O concentrations between two soils were caused by those in the soil structure, in particular, because of the presence of macro-pores and cracks in the soil structure, which resulted in different production and movement of N_2O in the soils. The higher N_2O production in the subsoil of the Gray Lowland soil could have been activated by NO_3^- leaching through macro-pores and cracks, and subsequently the N_2O produced in the subsoil could have been rapidly emitted to the atmosphere through the macro-pores and cracks.

The aim of this study was to quantify direct N_2O emissions from a light-textured arable soil under several crops grown on ridges, in North-Western Russia, that was typical of the region, during four growing seasons, with special attention to soil physical properties.

2. Materials and methods

The measurements of N_2O emission were conducted at the Menkovo Experimental Station of the Agrophysical Research Institute in the St. Petersburg region of Russia (59°34'N, 30°08'E) in 2004–2007. Average annual rainfall in the area for the previous 10 years was 1,109 mm, with 50–60% falling during the growing season (May–September). Average annual air temperature was +4.5°C, and average air temperature in the growing season was +13.6°C. The studied soil was the most typical soil of the area: a sandy loam Spodosol containing 7–8% clay, 17–23 g C kg^{-1} soil, and with bulk density ranging from 0.95 to 1.5 g cm^{-3}. Three 0.5-ha fields containing this soil were used to conduct the experiment. The first of them received 160 t ha^{-1} (656 kg of total N ha^{-1} or 219 kg of available N ha^{-1}), the second – 80 t ha^{-1} (328 kg of total N ha^{-1} or 108 kg of available N ha^{-1}) of farmyard manure (FYM) in spring of 2003 and early summer of 2004, and the third received no FYM.

Part of the field with no FYM was allocated for potato (*Solanum tuberosum* L.) in 2004 and 2005. Part of the field with the highest application of FYM was allocated for potato in 2005. All the three fields were used to grow cabbage (*Brassica oleracea capitat* L.) in 2006 and carrot (*Daucus carota* L.) in 2007. By 2005 the SOC content in the soils of the plots receiving 160 t ha^{-1} of FYM, 80 t ha^{-1} of FYM and no FYM at the beginning of the experimental studies was equal to 21, 19 and 17 g C kg^{-1} soil, respectively.

In 2004-2005 large single plots were used for the experiment while smaller but randomized replicate plots were used in 2006 and 2007. The use of single plots earlier in the experiment was a necessary consequence of the cropping arrangements adopted by the manager of the experimental station. The experimental study described here was therefore the only available way in which to conduct this project. Mineral N fertilizer (ammonium nitrate, AN) was applied to the soil at different rates (Table 1). The mineral N was applied (broadcast) to the soil in granular form mechanically: before the furrows and ridges were formed (1st application) and the second and third time (for potato crop only) the fertilizer was applied to furrows. The depth of mouldboard ploughing for all the plots was 22–24 cm. Most of the plots were ploughed in spring between 13 and 20 May. Crops were planted within a day or two after mouldboard ploughing and harvested at the end of August/early September.

http://dx.doi.org/10.5772/53061

The closed chamber technique was used to measure direct N_2O fluxes from the soil. Gas samples were collected two-three times a week (between noon and 2 pm) throughout the growing seasons (May-September) of 2004–2007. Chambers were made of inverted cylindrical plastic buckets, 18.9 cm in diameter and 11 cm high, and put into the field every time before gas samples were collected (Figure 2). Eight replicate chambers were used at all the plots — four in the furrows and four on the ridges.

Year	Crop	SOC, g C kg^{-1} soil	Ammonium nitrate fertilizer application					
			1st	Date	2nd	Date	3rd	Date
2004	Potato	17	72	14/05	48	17/07	0	
2005	Potato	17	72	13/05	38	10/06	10	27/07
		21	72	13/05	38	10/06	10	27/07
2006	Cabbage	17	0		0		0	
		17	70	15/05	0		0	
		19	0		0		0	
		19	90	15/05	0		0	
		22	0		0		0	
		22	110	15/05	0		0	
2007	Carrot	17	40	20/05	0		0	
		17	110	20/05	0		0	
		19	40	20/05	0		0	
		19	120	20/05	0		0	
		23	60	20/05	0		0	
		23	130	20/05	0		0	

SOC – soil organic carbon

Table 1. Crops grown and rates of mineral fertilizers used during the growing seasons of 2003-2007 (Natalya P. Buchkina, Elena Y. Rizhiya, Sergey V. Pavlik, Eugene V. Balashov).

Figure 2. Chamber placement on the ridges and in the furrows of the potato (*Solanum tuberosum* L.), cabbage (*Brassica oleracea capitat* L.) and carrot (*Daucus carota* L.) plots. (Natalya P. Buchkina, Elena Y. Rizhiya, Sergey V. Pavlik, Eugene V. Balashov)

Chambers were pressed into the soil to a depth about 2 cm and the soil outside the chambers was compacted against the chamber walls to make them gas-tight. A three-way tap on the top of each chamber was closed only after the chamber was fixed into the topsoil, to avoid extra air pressure inside the chamber. After 60 min gas samples were collected via the three-way tap. Similar sampling at the end of the closure period has been employed elsewhere (Ball et al., 2007; Hergoualc'h, 2008).

Gas samples were taken with a 60-ml syringe; 50 ml were flushed through a 10 ml vial that had been previously flushed with air, then the remaining 10 ml were forced into the vial to over-pressurise it. 3-ml subsamples from the vials were analysed for N_2O in a Carlo Erba 4130 gas chromatograph (GC) fitted with an electron capture detector. The GC was calibrated with standard gas mixtures. The detection limit for the sampling/analytical system was 0.05 ppm. Daily fluxes were calculated using the arithmetic mean of the four replicate chambers. Cumulative N_2O fluxes for different months as well as for whole growing seasons of 2004–2007 were calculated by plotting daily fluxes through time, interpolating linearly between them, and integrating the area under the curve. Cumulative standard errors were also determined by the same way (Dobbie, Smith, 2003; Buchkina et al., 2010).

The climatic factors most likely to affect the amount of N_2O produced in the soil were monitored during the experiment. Data on daily rainfall and air temperatures (maximum, minimum, and average) were collected from the Institute's meteorological station, situated c. 100 m from the experimental plots. Soil temperature at a 10-cm depth was measured with a digital thermometer on each sampling occasion. Soil bulk density and water content were measured (in three replicates) 1–2 times a month by standard methods (Rastvorova, 1988; Soil Survey, 1996) and the results were used to calculate water-filled pore space (WFPS). Soil samples for mineral nitrogen (NH_4^+ and NO_3^-), were also collected 1–2 times a month from the 0–10 cm layer. At least 10 subsamples were collected from each plot and combined. The composite samples were dried at room temperature and then amounts of NH_4^+ and NO_3^- were measured in water extractions using ion-selective electrodes in three replicates (Bankin et al., 2005). SOC content was measured by a wet combustion method using potassium dichromate ($K_2Cr_2O_7$) solution in sulfuric acid (Soil Survey, 1996).

All the measurements of the soil properties were done in three replicates. Means and standard deviations were calculated for each parameter within each treatment. Significance of differences between treatments was estimated by analysis of variance (one-way ANOVA) at $p \leq 0.05$. Relationships between the soil parameters were assessed with a linear regression analysis using computer statistical package.

3. Results and discussion

3.1. Weather conditions

The amount of precipitation and its distribution over the growing season (1 May–30 September) varied between the studied years significantly (Figure 3). The wettest growing season was 2004 and the driest 2006, with 845 and 456 mm of rain falling, respectively, compared to 575 mm in 2005, and 555 mm in 2007. During the growing season there were 75 days with rain in 2004, 55 in 2005, 41 in 2006 and 52 in 2007. Almost 75% of the rain fell on days when precipitation was greater than 10 mm: there were 25 such occasions in 2004, 15 in 2005, 13 in 2006 and 20 in 2007. The longest dry spells were observed in 2006: 27 days in late April-May, 15 days in July and 11 days in August. Even the wettest growing seasons of 2004 had a 10-day dry spell in May. The growing seasons of 2005 and 2007 had their longest dry spells in July (9 and 11 days, respectively) and in August-September (12 and 16 days, respectively). Amount of rainfall exceeded 100 mm in June, July, August and September of 2004, in May, June, and July of 2005, in May of 2006, in May and July of 2007.

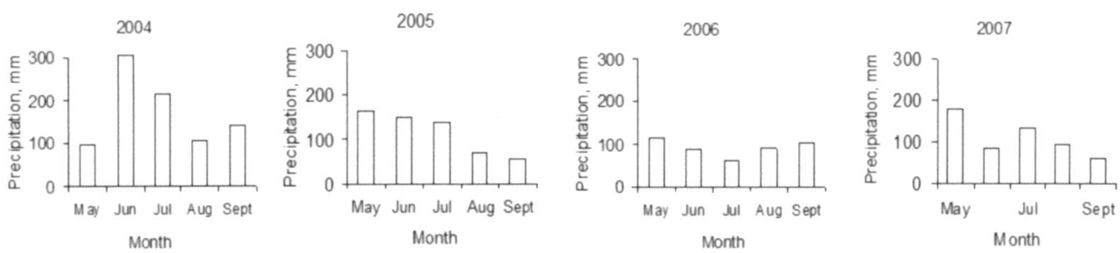

Figure 3. Monthly rainfall during the growing seasons of 2004-2007. (Natalya P. Buchkina, Elena Y. Rizhiya, Sergey V. Pavlik, Eugene V. Balashov)

Average air temperature for the studied growing seasons was more stable than the precipitation. July and August were always the warmest months, (16–20°C). In May, June and September the average was always between 10 and 15°C. The warmest growing season overall was 2006, averaging 15°C, compared to 14.4°C in 2005 and 2007, and to 13.9°C in 2004. In the growing season of 2005 the minimum air temperature never fell below zero but in all the other growing season it did: to -0.5°C at the end of May of 2004 and to -2.1°C in early May of 2006 and 2007. Maximum air temperatures were 27.2°C in 2004, 29.2°C in 2005, 33.8°C in 2006 and 31.4°C in 2007.

3.2. Soil properties

Soil bulk density varied during the growing seasons of 2004-2007 from 0.9 to 1.4 g cm^{-3} being the highest in potato furrows during the wettest growing season of 2004 and the lowest on cabbage ridges during the driest growing season of 2006 (Figure 4). The differences in soil bulk density between ridges and furrows were higher in the wetter growing seasons of 2004, 2005 and 2007 than in dryer growing season of 2006. That must be a result of soil compaction in the furrows during crop management operations. It was shown that the studied soils were much more easily compacted when wet than when they were relatively dry. In wetter growing seasons the differences in soil bulk density between ridges and furrows were greater in spring and early summer time when the ridges were formed for crop planting and mechanical cultivation of soil was regularly done (as a weed control operation). It also coincided with higher monthly rainfalls at the beginning of the wetter growing seasons. Later in the season the differences in soil bulk density between ridges and furrows were smaller. During the driest growing season of the four – 2006 – the difference in soil bulk density between ridges and furrows was less pronounced and did not change much during the season. The difference in soil bulk density between soils with the low (19 g C kg^{-1} soil) and high (22 g C kg^{-1} soil) SOC content was also often significant with more fertile soils having lower bulk density than the less fertile soils. Still, the differences in bulk density between soil ridges and furrows were almost always more significant than the differences in that for ridges (or furrows) between soils with different SOC content. The significantly higher bulk density of soil furrows, compared to that of soil ridges, could result in a formation of soil physical conditions more favorable for microbial process of denitrification in the studied soil (Beare et al., 2009).

The soil water content during the four studied growing seasons changed from 8 to 36% (of weight) and was higher than 20% for most of the wettest growing season of 2004, varied from 8 and 36% and from 8 and 31% during the growing seasons of 2005 and 2007, respectively, and never exceeded 24% during the driest growing season of 2006 (Figure 5). The differences in soil water content between ridges and furrows were greater in wetter growing seasons of 2004, 2005 and 2007 than in the driest of the four growing seasons of 2006. A combination of greater water content and higher bulk density in the soil of furrows as compared to the soil of ridges could lead to decreasing oxygen availability and to increasing N_2O fluxes as the intermediate of accelerated process of denitrification under such soil physical conditions (Drury et al., 2003; Beare et al., 2009).

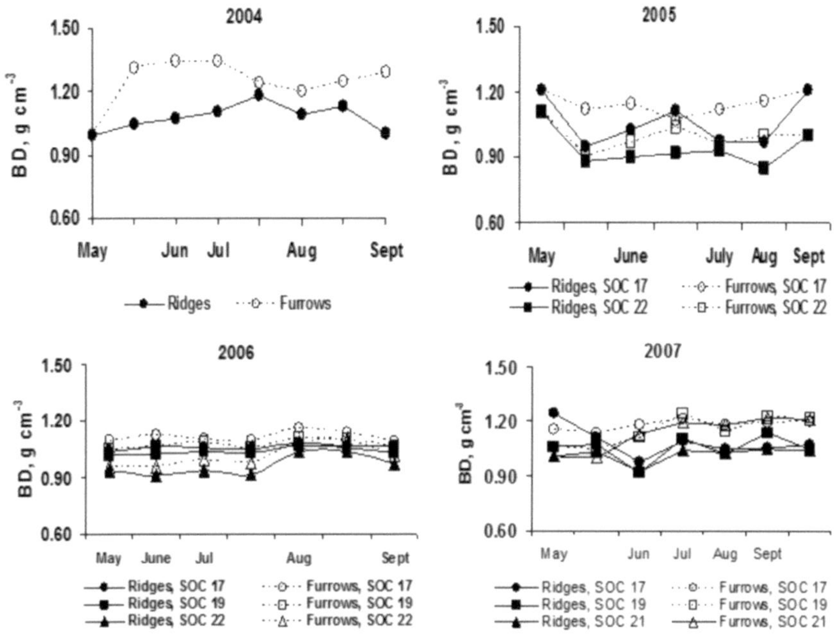

Figure 4. Soil bulk density (BD) in furrows and on ridges at the plots with potato (2004, 2005), cabbage (2006) and carrot (2007) at different soil organic matter (SOC) content (17, 19 and 21 g C kg^{-1} soil). (Natalya P. Buchkina, Elena Y. Rizhiya, Sergey V. Pavlik, Eugene V. Balashov)

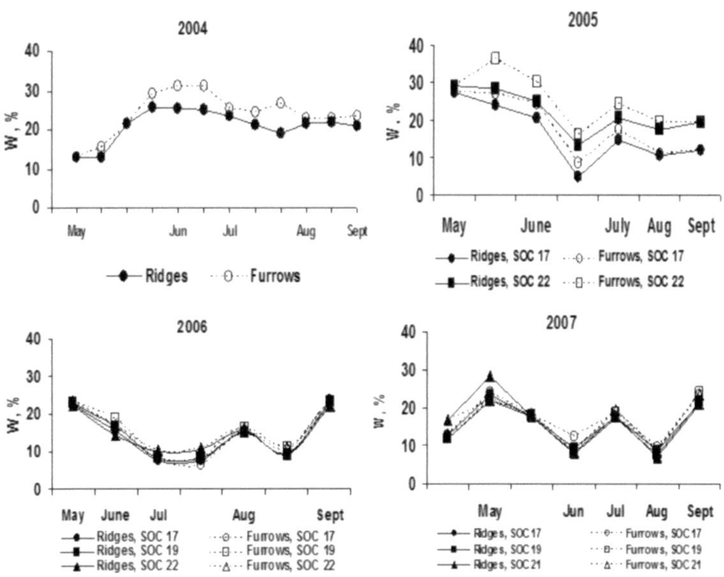

Figure 5. Soil water content (W) in furrows and on ridges at the plots with potato (2004, 2005), cabbage (2006) and carrot (2007) at different soil organic matter (SOC) content (17, 19 and 21 g C kg^{-1} soil) (Natalya P. Buchkina, Elena Y. Rizhiya, Sergey V. Pavlik, Eugene V. Balashov).

The relationship between soil water content (average for a month) and monthly precipitation is shown on Figure 6 and 7. A linear correlation ($p < 0.05$) between the two parameters was higher for the soil with the lower SOC content than for the soil with the higher SOC (especially for furrows). The water-holding capacity of the studied soils was not very high as they were light- textured and should mainly depend on rainfall and SOC content. The higher SOC of a light-textured soil the more water it would be able to hold and would be less dependent upon rainfall if the dry spells were not very long.

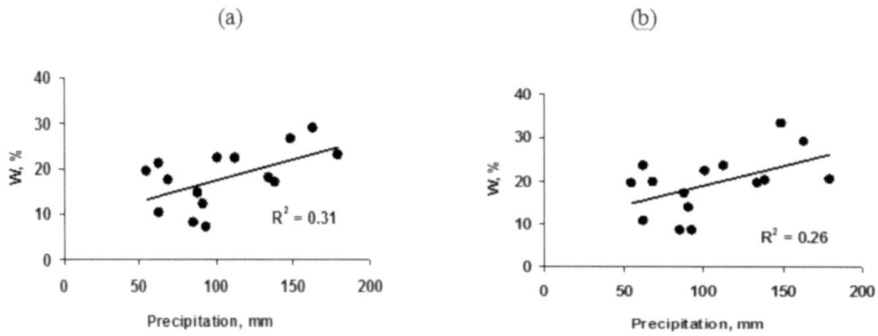

Figure 6. Relationships between monthly precipitation and average (for a month) soil water content (W, % of weight) for (a) soil ridges and (b) soil furrows with the highest (22 g C kg^{-1} soil) soil organic carbon content during the growing seasons of 2004-2007 (Natalya P. Buchkina, Elena Y. Rizhiya, Sergey V. Pavlik, Eugene V. Balashov)

During the studied growing seasons the soil WFPS varied widely within and between years (Figure 8). In general, the soils of all plots were wetter in 2004 than in 2005, 2006, and 2007 with WFPS for most of the growing seasons varying from 35 to 70%, compared to 15–50% in 2005, 15–40% in 2006, and to 20–45% in 2007. In 2004, in contrast to all the other studied seasons, maximum values of WFPS were observed in June and July (40–85%). During the wettest part of the growing season soil in furrows was 20–40% wetter than soil on ridges (60–85% versus to 40–45%, respectively). In May and August of 2004 the soil WFPS was lower than in the middle of the summer, changing from 20 to 40% and from 38 to 58%, respectively. The difference in soil water content between ridges and furrows during the drier part of the season was 10–20%. In 2005, the lowest values of soil WFPS were observed at the end of July, August and September, when WFPS on potato ridges went down to 10–22%, being the lowest of the year. During this dry spell the soil WFPS in potato furrows varied between 19 and 38%. The highest values of soil WFPS for the whole growing season of 2005 were observed in May and June but even then they were varying between 38 and 62%, and the difference between furrows and ridges was 11-13%. In 2006, the driest growing season of all, soil WFPS was very low in July (15-20%). The highest values of soil WFPS, but still quite low compared to the highest values of the other growing seasons, were observed in May and September (33-43%). The difference in soil WFPS between furrows and ridges during the growing season of 2006 was never higher than 8%. In 2007, the lowest values of the soil WFPS were found at the end of May-early June and at the end of June – early August (15-25%). Early May and August of 2007 were wetter but the highest values of the soil WFPS were not exceeding 53%. The difference in soil WFPS on ridges and furrows in 2007 was 7-10% during dry spells and 10-15% during wetter spells of the season. As was indicated above, nitrification is the main process of N_2O emission at 35-60% WFPS, but denitrification is more important at soil water content greater than 60% WFPS due to a decreased oxygen supply (Drury et al., 2003; Dobbie, Smith, 2003). Therefore, the observed soil moisture conditions were often more favorable for nitrification rather than denitrification during the growing seasons of 2004-2007.

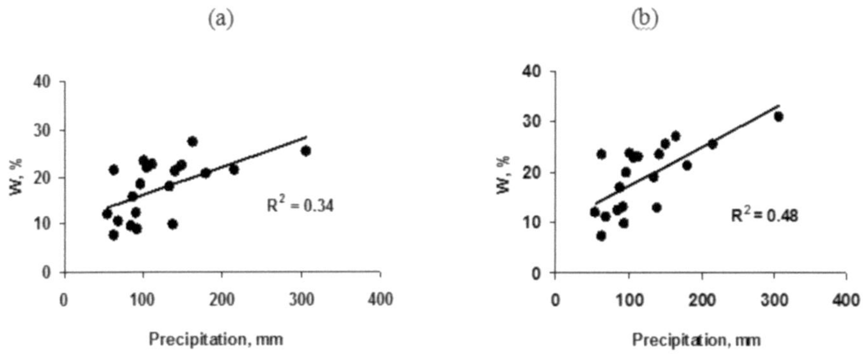

Figure 7. Relationships between monthly precipitation and average (for a month) soil water content (W, % of weight) for (a) soil ridges and (b) soil furrows with the lowest (19 g C kg^{-1} soil) soil organic carbon content during the growing seasons of 2004-2007 (Natalya P. Buchkina, Elena Y. Rizhiya, Sergey V. Pavlik, Eugene V. Balashov).

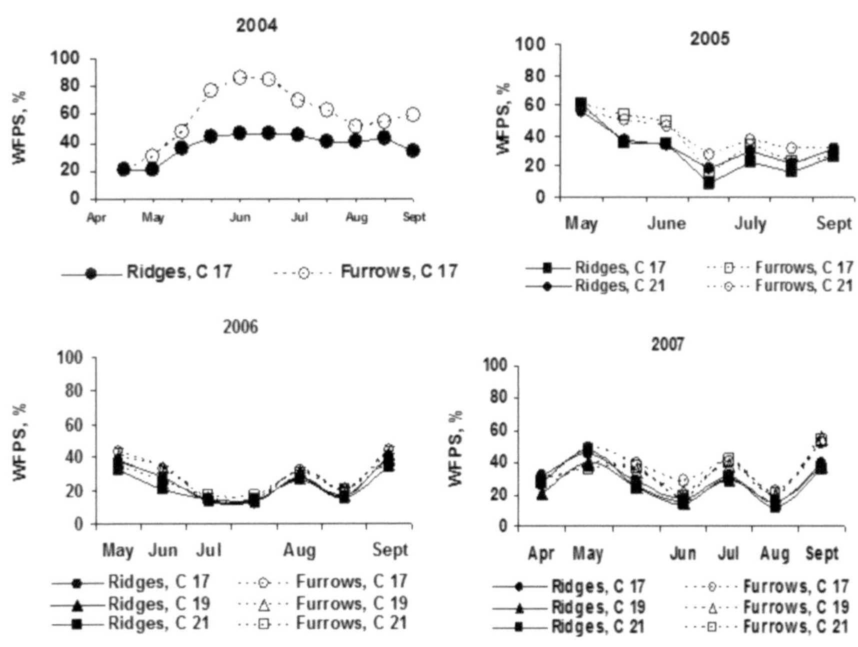

Figure 8. Soil water-filled pore space (WFPS) for furrows and ridges for potato (2004 and 2005), cabbage (2006) and carrot (2007) at different soil organic matter (SOC) content (17, 19 and 21 g C kg^{-1} soil) (Natalya P. Buchkina, Elena Y. Rizhiya, Sergey V. Pavlik, Eugene V. Balashov).

The linear correlation between monthly rainfall and soil WFPS (average for a month) was strong for soil furrows as well as for soil ridges only during the driest growing season of 2006 (R^2=0.80-0.88, p<0.05). For the growing season of 2007 this relationship was strong only for soil ridges (R^2=0.92-0.94, p < 0.05) but not for furrows (R^2 = 0.65-0.81, p = 0.1-0.19) while in 2004 it was strong for soil furrows (R^2=0.95, p < 0.01) but not for ridges (R^2=0.57, p = 0.14). The relationship between monthly rainfall and soil WFPS (average for a month) for all 5 growing seasons is shown on the Figure 9. In average, the soil WFPS in furrows is more sensitive to monthly rainfall (R^2=0.64, p < 0,001) than that on ridges (R^2=0.37, p < 0,001). Our results corresponded with the data reported by L. Meng et al. (2005) where the strong correlations between WFPS and precipitation were also observed.

According to T. Granli and O.C. Bockman (1994) optimum soil temperatures for N_2O emissions vary between 25-40°C. During the growing seasons soil temperatures at 10 cm depth varied between 3 and 26°C, and were highest (15–26°C) always at the end of July and in early August. Average soil temperature was almost 1°C lower during the growing season of 2004 (14.3°C) and 2007 (14.4°C) than of 2003 (15.1°C), 2005 (15.2°C) and 2006 (15.0°C). Over the 4 years, higher soil temperatures were always found in those plots where soil was more exposed to the sun. The soil temperature was often higher in the ridges than in the furrows.

The experimental soil was very low with mineral N content when N-fertilizers were not used and contained 2–9 mg N kg^{-1} soil at the beginning of all studied growing seasons. Soil mineral N concentrations always increased after N-fertilizer was applied to the soil to 20-100 mg N kg^{-1} soil but were never high for longer than 3 weeks and were always back to below 10 mg N kg^{-1} soil at the end of the growing seasons.

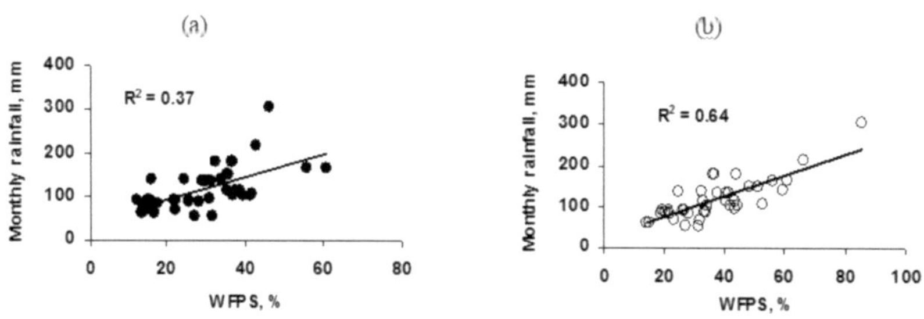

Figure 9. Relationships between monthly rainfall and soil water filled pore space (WFPS, average for a month) for (a) soil ridges and (b) soil furrows during the growing seasons of 2004-2007 (Natalya P. Buchkina, Elena Y. Rizhiya, Sergey V. Pavlik, Eugene V. Balashov).

K. Dobbie and K.A. Smith (2003) reported that high N_2O emission could be even observed at WFPS (65%), soil temperature (4.5°C) and NO_3^--N content (5 mg kg^{-1} soil). The results of our studies showed that these threshold limits had been observed or exceeded during the growing seasons of 2004-2007.

http://dx.doi.org/10.5772/53061

3.3. N_2O emissions

3.3.1. *Daily fluxes*

During the growing seasons 0f 2004-2007 daily N_2O fluxes varied highly between plots differed in fertilization, soil physical conditions and crops. Agricultural soils without any fertilization can produce N_2O during the growing season because of mineralization of soil organic matter and N rich plant residues. In our study the daily N_2O fluxes for the four studied growing seasons were never higher than 10 g N_2O–N ha^{-1} on the unfertilized plots (both in furrows and on ridges). Conversely, daily N_2O fluxes from the soil could be very low even on fertilized soils (especially on ridges), indicating that other parameters, for example, soil mineral N content, temperature, soil WFPS and oxygen availability were main determinants controlling production of N_2O as the by-product of nitrification (Khalil et al., 2004; Meng et al., 2005; Conen et al., 2000).

Our data corresponded with the results published by H. Flessa and P. Dorsch (1995) and also by B. Ball et al. (2002) who found that application of N-fertilizers by itself did not always enhance N_2O emission and that weather conditions at the time of fertilizer application influenced N_2O emissions substantially. Nevertheless, C. Henault et al. (1998) reported that N_2O emissions were higher 20 g N ha^{-1} day^{-1} from soils with low water content as a result of domination of nitrification.

Possible differences in N_2O emissions from soils with different crop types depend on the management practices and climatic conditions. Tillage, sowing and other cultivations for agricultural plants can disturb soil aggregates, increase soil organic matter mineralization and deteriorate soil physical status and, as a result, can lead to an increased N_2O emissions (Elder, Lal, 2008; Ruser et al., 1998; Beare et al., 2009; Six et al., 1998). Crops can (1) demonstrate different rates of soil water uptake, (2) prevent the accumulated soil water against evaporation under different weather conditions, and (3) remove more N from the soil that will result in different availability of mineral N for nitrifying and denitrifying microorganisms.

During the growing seasons of 2004-2007 daily N_2O fluxes from the soil with crops grown on ridges (potato, carrot, cabbage) were often higher than those from the soil with cereals, grasses, grass-clover mixture and grass with oats (the latter crops grown in the same experiment are not being discussed in this paper but were described earlier in our paper - Buchkina et al., 2010).

For the latter crops daily N_2O fluxes were never exceeding 20 g N_2O-N ha^{-1} while for the crops grown on ridges N_2O fluxes from the soil of furrows could be as high as 115 g N_2O-N ha^{-1} day^{-1} (Figure 10). Higher N_2O emissions from ridges can be explained by a greater mineralisation of soil organic matter that resulted to an increased N_2O emission from a bare soil subjected to a higher warming (Maljanen et al., 2002). According to the results reported by M. Maljanen et al. (2002), higher N_2O fluxes were observed from soils kept bare by tillage or cutting presumably from lack of competition for nitrate (NO_3^-) between microbes and plants.

Daily N_2O fluxes during the growing seasons of 2004-2007 were never affected by WFPS if no N had been applied with the mineral fertilizers as the original soil was very low in available N. However, in the N-fertilized plots the greatest N_2O fluxes were found when WFPS exceeded 40% (Figure 11) and that happened more often in the soil of furrows with the higher soil bulk density rather than in the soil of ridges with the lower soil bulk density. The highest N_2O emission in the cultivated soils generally are observed at 70–90% WFPS due to denitrification which rapidly increased when WFPS exceeded 60%, whereas at 30–70% WFPS nitrification was the main source of N_2O (Granli, Bockman, 1994; Smith et al., 2003; Maljanen et al., 2002).

It was shown for field experiments in Scotland (Dobbie et al., 1999; Conen et al., 2000) that high N_2O fluxes actually occurred more often when WFPS was higher than 60%. In our experiment the soil WFPS quite rarely exceeded 60% (the longest period for the four growing seasons was about 20 days in 2004 and only for potato furrows) as the rainfall of the area was less than it was in Scotland and also the soil of the experimental plot was light-textured with relatively easy drainage down the soil profile. The work of E.A. Davidson (1991) suggests that denitrification played its role in N_2O emissions from our soil only occasionally, when there were anaerobic conditions in some parts of the soil, while the main source of N_2O emission most of the time was nitrification. The same results were reported in our earlier work (Buchkina et al., 2010) for other crops (cereals, grasses, and grass-legume mixtures) at our experimental station.

Soils in the ridges and furrows could differ in bulk density, air and total porosity, available SOC, mineral N and oxygen availability. For instance, the average bulk density of the soil on ridges and in furrows was equal to 0.9 and 1.4 g cm^{-3}, respectively. Therefore, the soil furrows could show more favorable conditions for denitrification before and after a rainfall. Our results corresponded with the data of Beare et al. (2009) who showed that most (88%) of the total N_2O production from compacted soil occurred after soil rewetting, at a time when there was ample NO_3^--N and dissolved organic carbon available and the air-filled porosity was low (0.22 cm^3 cm^{-3}), rather than during the drying phase when compacted soil was accumulating NO_3^-N and air filled porosity was high (0.62 cm^3 cm^{-3}). P. Merino et al. (2012) also reported that the interaction of WFPS and soil NO_3^- content was statistically significant ($p < 0.001$), indicating that the response of N_2O emission from a loamy clay soil to changes in NO_3^- content was very dependent on WFPS.

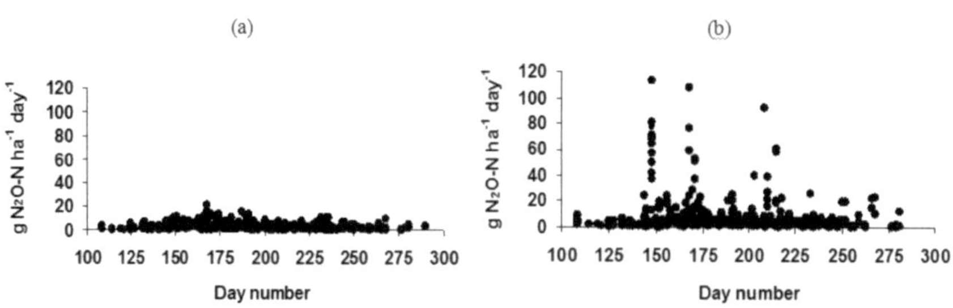

Figure 10. Daily N_2O fluxes from the soil under (a) cereals, grasses, grass-legume mixtures, grass-cereal mixtures during the growing seasons of 2003-2005 (Buchkina et al., 2010) and (b) crops grown on ridges (potato, cabbage, carrot) during the growing seasons of 2004-2007 (Natalya P. Buchkina, Elena Y. Rizhiya, Sergey V. Pavlik, Eugene V. Balashov).

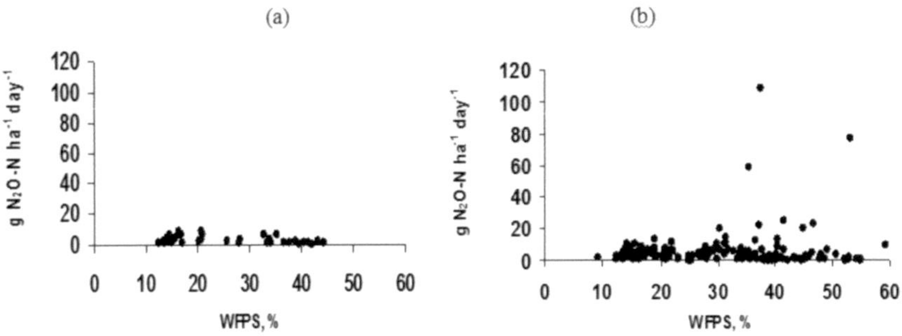

Figure 11. Soil water-filled pore space (WFPS) and daily N_2O fluxes (both from furrows and ridges) from the soil with potato (2004, 2005), cabbage (2006) and carrot (2007): (a) unfertilized plots, (b) fertilized plots (Natalya P. Buchkina, Elena Y. Rizhiya, Sergey V. Pavlik, Eugene V. Balashov).

3.3.2. Cumulative fluxes and emission factors

Cumulative N_2O fluxes for the growing seasons of 2004-2007 from the studied soil for different crops varied from 0.34±0.09 to 1.83±0.45 kg N_2O-N ha^{-1} (Table 2). Within the same growing season N_2O cumulative fluxes were higher from the soils of furrows than from the soils of ridges (especially, on the plots where mineral nitrogen was applied with fertilizers) but the difference was not significant when all the four growing seasons were taken into account: the cumulative N_2O fluxes from soils in furrows varied from 0.37±0.10 to 1.83±0.45 kg N_2O-N ha^{-1} while from soils on ridges – from 0.36±0.011 to 1.25±0.29 kg N_2O-N ha^{-1}.

The lowest cumulative N_2O fluxes for both positions were measured for the growing season of 2006 – the driest of the four. During this growing season cumulative N_2O fluxes from soils of furrows were often even lower than those from soils of ridges in wetter growing seasons (Table 2, Figure 12). During the dry growing season of 2006 cumulative N_2O fluxes were not affected by N-fertilizer rates on the soils having SOC content of 17 and 19 g C kg^{-1}. On the plot containing 23 g C kg^{-1} only the soil in furrows emitted more N_2O for this growing season when higher amount of N-fertilizer was applied to it. During the dry growing season the compacted soil in the furrows demonstrated greater soil water content than the uncompacted soil on the ridges. Hence, the first one had a higher amount of smaller pores than the latter one with a higher amount of larger pores. Therefore, during this growing season the compacted furrow soil with smaller pores could (1) hold more water, (2) have more favorable conditions for nitrification and denitrification, and (3) response in the higher values of N_2O emissions to the mineral N fertilization than the uncompacted ridge soil.

The cumulative N_2O fluxes from the soils showed the higher values during the wetter growing season of 2007 compared to the driest growing season of 2006 when there was no relationship between fertilizer rates and cumulative N_2O fluxes. In 2007, an increase in a rate on mineral N-fertilizer from 40 kg N ha^{-1} to 110 kg N ha^{-1} did not result in an increase of cumulative N_2O fluxes from the soil on ridges under carrot (Table 2). Only the rates of 120 and 130 kg N ha^{-1} caused a significant increase of cumulative N_2O fluxes from this soil. Thus, the relationship of the cumulative N_2O fluxes with the mineral N rates was nonlinear for soil on ridges for the growing season of 2007. The cumulative N_2O fluxes from the soil in furrows under carrot were significantly higher than those from the ridges, and demonstrated a non-linear and unstable increase with increasing N rates from 40 kg N ha^{-1} to 130 kg N ha^{-1}. However, the highest cumulative N_2O fluxes from the soil both on ridges and in furrows for the growing season of 2007 were observed at the N rate of 130 kg N ha^{-1}.

Year	Crop	Crop yield, kg ha^{-1}	SOC, g kg^{-1} soil	N applied, kg N ha^{-1}	Position	Cumulative N_2O flux, kg N_2O-N ha^{-1}	Emission factor
2004	Potato	17000	17	120	Ridges	1.13±0.39	0.94
					Furrows	1.51±0.69	1.26
2005	Potato	18000	17	120	Ridges	0.60±0.21	0.50
					Furrows	0.91±0.35	0.76
	Potato	21000	21	120	Ridges	1.25±0.29	1.04
					Furrows	1.83±0.45	1.52
2006	Cabbage	41400	17	0	Ridges	0.37±0.08	-
					Furrows	0.45±0.11	-
	Cabbage	59900	17	70	Ridges	0.36±0.05	0.51
					Furrows	0.42±0.08	0.60
	Cabbage	45040	19	0	Ridges	0.43±0.06	-
					Furrows	0.53±0.07	-
	Cabbage	75900	19	90	Ridges	0.45±0.07	0.50
					Furrows	0.56±0.08	0.62
	Cabbage	81100	22	0	Ridges	0.34±0.09	-
					Furrows	0.37±0.10	-
	Cabbage	103320	22	110	Ridges	0.36±0.11	0.33
					Furrows	0.60±0.12	0.54
2007	Carrot	54830	17	40	Ridges	0.62±0.02	1.55
					Furrows	0.65±0.07	1.63
	Carrot	61220	17	110	Ridges	0.64±0.03	0.58
					Furrows	0.93±0.07	0.85
	Carrot	68110	19	40	Ridges	0.71±0.01	1.78
					Furrows	0.92±0.10	2.3
	Carrot	70170	19	120	Ridges	0.74±0.01	0.62
					Furrows	0.99±0.11	0.83
	Carrot	74570	23	60	Ridges	0.65±0.03	1.08
					Furrows	1.09±0.05	1.82
	Carrot	70040	23	130	Ridges	0.80±0.04	0.62
					Furrows	1.45±0.14	1.12

SOC – soil organic carbon

Table 2. Crops, amounts of mineral N-fertiliser applied, cumulative N_2O fluxes and emission factors.

http://dx.doi.org/10.5772/53061

According to several authors, there is the threshold of N rates which can exceed the N requirements of crops (Granli, Bockman, 1994) and if the N rates exceed crop requirements, N_2O emissions can become more variable and increase exponentially with increasing N fertilization (Hoben et al., 2011; Snyder et al., 2009). Our data supported the results received by these authors. In our opinion, the N rates of 120 and 130 kg N ha^{-1} exceeded the threshold of N requirements of carrot during the growing season of 2007. Moreover, there was only a weak correlation (r=0.16) between the yields of carrot and the mineral N-fertilizer rates for the growing season of 2007.

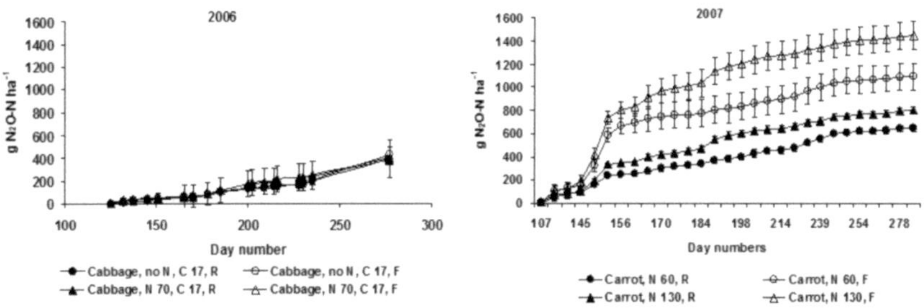

Figure 12. Differences in cumulative N_2O fluxes between soil in ridges and on furrows (F) in relatively dry (2006) and wet (2007) growing seasons at different N-fertilizer rates (0, 60, 70, 130 kg N ha^{-1}) (Natalya P. Buchkina, Elena Y. Rizhiya, Sergey V. Pavlik, Eugene V. Balashov)

It was recommended by J.W. Van Groenigen et al. (2010) that to obtain more valid information on N_2O emissions from agricultural soils it is better to assess N_2O emissions as a function of crop yield. According to our results, the ratios of N_2O cumulative fluxes (average for furrows and ridges) to carrot yields were equal to $1.16*10^{-5}$, $1.17*10^{-5}$, $1.27*10^{-5}$, $1.20*10^{-5}$ and $1.61*10^{-5}$ kg N_2O-N ha^{-1} kg^{-1} yield at the N-fertilizer rates of 40, 60, 110, 120 and 130 kg N ha^{-1}, respectively. These results also demonstrated that the N rate of 130 kg ha^{-1} exceeded the N requirements of carrot during the growing season of 2007.

When all the data for all the growing seasons were taken into account we observed that the significant positive correlation between amount of N applied into the soil with mineral fertilizers and cumulative N_2O flux for a growing season was higher in the soils of furrows (r=0.66, p =0.01) than in the soil of ridges (r=0.56, p=0.01) (Figure 13). The soil of furrows contained the same amount of mineral N and carbon as the soil of ridges but often had higher water content, bulk density and WFPS and, as a result, demonstrated more favorable soil physical conditions for microbial process of denitrification.

The emission factor of 1% is recommended by (IPCC, 2006) for evaluating the efficiency of direct N_2O emissions from agricultural soils. The emission factor (calculated for 5 months' cumulative fluxes) for the different crops varied during the four growing seasons from 0.33 to 2.30%. The emission factors were expectedly higher in wetter growing seasons of 2004, 2005 and 2007 (0.50-2.30%) than in drier growing season of 2006 (0.50-0.62%) The emission factors were often higher than 1% for the soil of both furrows and ridges only in wetter growing seasons of 2004, 2005 and 2007 but not in dry growing season of 2006.

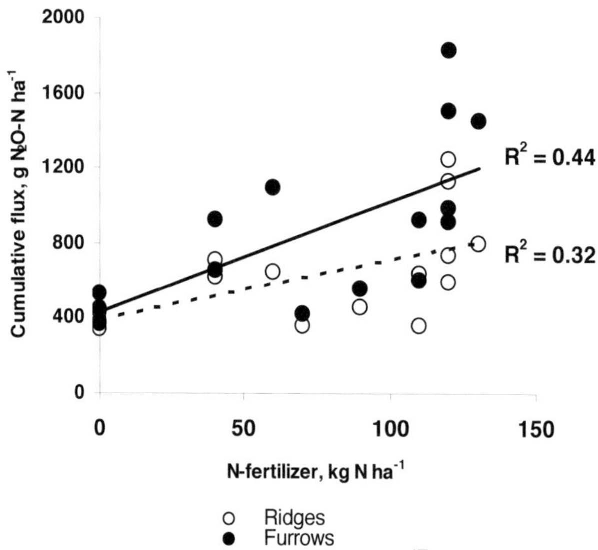

Figure 13. Relationship between the rates of mineral nitrogen fertilizer and cumulative N_2O flux from a sandy loam Spodosol in furrows (black dots and line) and on ridges (empty dots and dotted line) for the growing seasons of 2004-2007 (Natalya P. Buchkina, Elena Y. Rizhiya, Sergey V. Pavlik, Eugene V. Balashov)

4. Conclusions

Cumulative N_2O fluxes from a sandy loam Spodosol for the growing seasons of 2004–2007 varied between 0.34±0.09 and 1.83±0.45 kg N_2O-N ha^{-1} for different crops studied in this experiment. Seasonal N_2O cumulative fluxes were higher from the soils of furrows than from those of ridges (especially on the plots where mineral nitrogen was applied with fertilizers) only in wetter growing seasons of 2004, 2005 and 2007. This difference was not significant for the drier growing season of 2006 or when all the four growing seasons were taken into consideration as climatic conditions affected the relationship and made it much weaker.

The N application contributed to a higher N_2O emission from soil in furrows, where the soil was more compacted with higher water-filled pore space, than from soil on ridges but only in wetter growing seasons. During the dry growing season of 2006 there was no significant difference between N_2O emissions from soils on ridges and in furrows. There was a nonlinear N_2O response to increasing N fertilizer rates from the soil on ridges and in furrows for carrot during the wet growing season of 2007. The mineral N-fertilizer rate of 130 kg N ha^{-1} could exceed the N requirements of carrot during the growing season of 2007.

http://dx.doi.org/10.5772/53061

Soil water-filled pore space affected N_2O emission from the soil only if mineral N-fertilizer was applied into the soil. Plots receiving no extra N never emitted much N_2O whatever the soil water-filled pore space.

Acknowledgements

The authors are very grateful to Prof. V.A. Semenov, the founder, and to Dr. E.A. Olenchenko, the main research manager, of the field experiment where all the described data were collected. Data on the crop yields in 2006 and 2007 were provided by Dr. E.A. Olenchenko (not published). The Royal Society of London supported the project (ref no 15336) in 2003-2005. The project would have been impossible without the donation of the gas chromatograph by the University of Edinburgh to the Agrophysical Research Institute. We thank the staff of the Menkovo Experimental Station of the Agrophysical Research Institute for providing the information on daily weather conditions.

Author details

Natalya P. Buchkina*, Elena Y. Rizhiya, Sergey V. Pavlik and Eugene V. Balashov

*Address all correspondence to: buchkina_natalya@mail.ru

Department of Soil Physics, Physical Chemistry and Biophysics, Agrophysical Research Institute, St. Petersburg, Russia

References

[1] Almaraz JJ, Mabood F, Zhou X, Madramootoo C, Rochette P, Ma B, Smith DL. Carbon dioxide and nitrous oxide fluxes in corn grown under two tillage systems in Southwestern Quebec. Soil Science Society of America Journal 2009;73 113–119.

[2] Angers DA, Bolinder MA, Carter MR, Gregorich EG, Drury CF, Liang BC, Voroney RP, Simard RR, Donald RG, Beyaert RP, Martel J. Impact of tillage practices on organic carbon and nitrogen storage in cool, humid soils of eastern Canada. Soil and Tillage Research 1997;41 191–201.

[3] Balashov E, Bazzoffi P. Aggregate water stability of differently compacted sandy and clayey loam soils with and without wheat plants. International Agrophysics 2003;17 151-155.

[4] Balashov E, Buchkina N. Effect of short- and long-term agricultural use of chernozem on its quality indicators. International Agrophysics 2011;25 1-5.

[5] Ball BC, McTaggart IP, Watson CA. Influence of organic ley-arable management and afforestation in sandy loam to clay loam soils on fluxes of N2O and CH4 in Scotland. Agriculture, Ecosystems and Environment 2002;90 305–317.

[6] Ball BC, Scott A, Parker JP. Field. N2O, CO2 and CH4 fluxes in relation to tillage, compaction and soil quality in Scotland. Soil and Tillage Research 1999;53 29–39.

[7] Ball BC, Watson CA, Crichton I. Nitrous oxide emissions, cereal growth, N recovery and soil nitrogen status after ploughing organically managed grass/clover swards. Soil Use and Management 2007;23 145–155

[8] Bankin MP, Bankina TA, Korobeinikova LP. Physicochimicheskiye metody v agrochimii i biologii pochv (Physico-chemical methods in agrochemistry and soil biology). Izdatelstvo Sankt-Peterburgskogo Universiteta, Sankt-Peterburg; 2005.

[9] Beare MH, Hendrix PF, Coleman DC. Water-stable aggregates and organic matter fractions in conventional- and no-tillage. Soil Science Society of America Journal 1994;58 777–786.

[10] Beare MH, Gregorich. EG, St-Georges P. Compaction effects on CO2 and N2O production during drying and rewetting of soil. Soil Biology and Biochemistry 2009; 41 611-621.

[11] Bird JA, Herman DJ, Firestone MK. Rhizosphere priming of soil organic matter by bacterial groups in a grassland soil. Soil Biology and Biochemistry 2011;43 718-725.

[12] Bock E, Koops HP, Harms H. Cell biology of nitrifying bacteria. In: Prosser JI. (ed) Nitrification. Special Publications of the Society for General Microbiology; 1986. 20 p17–38.

[13] Bouwman AF. Exchange of greenhouse gas between terrestrial ecosystems and atmosphere. In: Bouwman AF. (ed) Soil and the greenhouse effects. Wiley, Chichester, 1990. p.61–127.

[14] Bouwman AF. Direct emission of nitrous oxide from agricultural soils. Nutrient cycling in agroecosystems 1996;46 53-70.

[15] Bremner, JM. Sources of nitrous oxide in soils. Nutrient Cycling in Agroecosystems 1997;49 7-16.

[16] Buchkina NP, Balashov EV, Rizhiya EY, Smith KA. Nitrous oxide emissions from a light-textured arable soil of North-Western Russia: effects of crops, fertilizers, manures and climate parameters. Nutrient Cycling in Agroecosystems 2010;87 429-442.

[17] Burton DL, Beauchamp EG. Profile nitrous oxide and carbon dioxide concentrations in a soil subject to freezing. Soil Science Society of America Journal 1994;58 115–122.

[18] Chatskikh D, Olesen JE. Soil tillage enhanced CO2 and N2O emissions from loamy sand soil under spring barley. Soil and Tillage Research 2007;97 5–18.

[19] Choudhary MA, Akramkhanov A, Saggar S. Nitrous oxide emission from a New Zealand cropped soil: tillage effects, spatial and seasonal variability. Agriculture, Ecosystems and Environment 2002;93 33–43.

[20] Clough TJ, Sherlock RR, Rolston DE. A review of the movement and fate of N2O in the subsoil. Nutrient Cycling in Agroecosystems 2005;72 3–11.

[21] Conen F, Dobbie KE, Smith KA. Predicting N2O emission from agricultural land through related soil parameters. Global Change Biology 2000;6 417–426.

[22] Crutzen PJ. The role of NO and N2O in the chemistry of the troposphere and stratosphere. Annual Review of Earth and Planetary Sciences 1979;7 443-472.

[23] Davidson EA. Fluxes of nitrous oxide and nitric oxide from terrestrial ecosystems. In: Rogers JE, Whitman WB (eds) Microbial production and consumption of greenhouse gases: methane, nitrogen oxides and halo-methanes. Washington, DC: American Society of Microbiology; 1991; p219–235.

[24] Dobbie KE, Smith KA. Nitrous oxide emission factors for agricultural soils in Great Britain: The impact of soil water-filled pore space and other controlling variables. Global Change Biology 2003;9 204–218.

[25] Dobbie KE, Smith KA. Impact of different forms of N fertilizer on N2O emissions from intensive grassland. Nutrient Cycling in Agroecosystems 2003;67 37–46.

[26] Dobbie KE, McTaggart IP, Smith KA. Nitrous oxide emissions from intensive agricultural systems: variations between crops and seasons; key driving variables; and mean emission factors. Journal of Geophysical Research 1999;104 26891- 26899.

[27] Douglas JT, Crawford CE. Soil compaction effects on utilization of nitrogen from livestock slurry applied to grassland. Grass and Forage Science 1998;53 31-40.

[28] Drury CF, Zhang TQ, Kay BD. The non-limiting and least limiting water ranges for soil nitrogen mineralization. Soil Science Society of America Journal 2003;67 1388–1404.

[29] Elder JW, Lal R. Tillage effects on gaseous emissions from an intensively farmed organic soil in North Central Ohio. Soil and Tillage Research 2008; 98 45–55.

[30] Fernandez R, Quiroga A, Zorati C, Noellmeyer E. Carbon contents and respiration rates of aggregate size fractions under no-till and conventional tillage. Soil and Tillage Research 2010;109 103-109.

[31] Firestone MK, Davidson EA. Microbiological basis of NO and nitrous oxide production and consumption in soil, In Andreae MO, Schimel DS. (eds.) Exchange of Trace Gases between Terrestrial Ecosystems and the Atmosphere. New York: John Wiley and Sons; 1989. p7-21.

[32] Flessa H, Dorsch P. Seasonal variation on N2O and CH4 fluxes in differently managed arable soils in southern Germany. Journal of Geophysical Research 1995;100(D11):23115–23124.

[33] Granli T, Bøckman OC. Nitrous oxide from agriculture. Norwegian Journal of Agricultural Science 1994;12 128 pp.

[34] Grant RF, Pattey E, Goddard TW, Kryzanowski LM, Puurveen H. Modeling the effects of fertilizer application rate on nitrous oxide emissions. Soil Science Society of America Journal 2006;70 235–248.

[35] Green VS, Stott DE, Cruz JC, Curi N. Tillage impacts on soil biological activity and aggregation in a Brazilian Cerrado Oxisol. Soil and Tillage Research 2007;92 114–121.

[36] Gregorich EG, Kachanoski RG, Voroney RP. Carbon mineralization in soil size fractions after various amounts of aggregate disruption. Journal of Soil Science 1989;40 649:659.

[37] Gregorich EG, Rochette P, VandenBygaart AJ, Angers DA. Greenhouse gas contributions of agricultural soils and potential mitigations practices in Eastern Canada. Soil and Tillage Research 2005;83 53-72.

[38] Halvorson AD, Del Grosso SJ, Reule CA. Nitrogen, tillage, and crop rotation effects on nitrous oxide emissions from irrigated cropping systems. Journal of Environmental Quality 2008;37 1337–1344.

[39] Henault C, Devis X, Page S, Justes E, Reau R, Germon JC. Nitrous oxide emissions from different soil and land management conditions. Biology and Fertility of Soils 1998;26 199–207.

[40] Hergoualc'h K, Skiba U, Harmand J-M, Henault C. Fluxes of greenhouse gases from Andosols under coffee in monoculture or shaded by Inga densiflora in Costa Rica. Biogeochemistry 2008;89 329–345.

[41] Hoben JR, Gehl RJ, Millar N, Grace PR, Robertson GP. Nonlinear nitrous oxide (N2O) response to nitrogen fertilizer in on-farm corn crops of the US Midwest. Global Change Biology 2011;17 1140-1152.

[42] Hochstein LI, Tomlinson GA. The enzymes associated with denitrification. Annual Review of Microbiology 1988;42 231-261.

[43] Houghton JT, Jenkins GJ, Ephraums JJ., editors. Climate Change: The IPCC Scientific Assessment. Cambridge: Cambridge University Press; 1990.

[44] IPCC. Changes in atmospheric constituents and in radiative forcing. Cambridge: Cambridge University Press; 2007.

[45] Intergovernmental Panel on Climate Change, IPCC Guidelines for National Greenhouse Gas Inventories, prepared by the National Greenhouse Gas Inventories Programme. In: Eggleston, HS, Buendia L, Miwa K, Ngara T, Tanabe K. (eds.) N2O Emissions from Managed Soils, and CO2 Emissions from Lime and Urea Application, vol. 4. Hayama: IGES (Chapter 11); 2006.

[46] Jacinthe PA, Lal R. Effects of soil cover and land-use on the relations flux-concentration of trace gases. Soil Science 2004;169 243–259.

[47] Jones SK, Rees RM, Skiba UM, Ball BC. Influence of organic and mineral N fertiliser on N2O fluxes from a temperate grassland. Agriculture Ecosystems and Environment 2007;121 74–83.

[48] Kaiser EA, Ruser R. Nitrous oxide emissions from arable soils in Germany – an evaluation of six long-term field experiments. Journal of Plant Nutrition and Soil Science 2000;163 249–259.

[49] Kavdir Y, Hellebrand HJ, Kern J. Seasonal variations of nitrous oxide emission in relation to nitrogen fertilization and energy crop types in sandy soil. Soil and Tillage Research 2008;98 175–186.

[50] Khalid M, Soleman N, Jones DL. Grassland plants affect dissolved organic matter and nitrogen dynamics in soil. Soil Biology and Biochemistry 2007;39 378-381.

[51] Khalil K, Mary B, Renault P. Nitrous oxide production by nitrification and denitrification in soil aggregates as affected by O2 concentration. Soil Biology and Biochemistry 2004;36 687–699.

[52] Kusa K, Sawamoto T, Hu R, Hatano R. Comparison of N2O and CO2 concentrations and fluxes in the soil profile between a Gray Lowland soil and an Andosol. Soil Science and Plant Nutrition 2010;56 186–199.

[53] Lal R. Soil carbon sequestration to mitigate climate change. Geoderma 2004;123 1–22.

[54] Lemke RL, Izaurralde RC, Nyborg M, Solberg ED. Tillage and N source influence soil-emitted nitrous oxide in the Alberta Parkland region. Canadian Journal of Soil Science 1999;79:15–24.

[55] Li X, Inubushi K, Sakamoto K. Nitrous oxide concentrations in an andisol profile and emissions to the atmosphere as influenced by the application of nitrogen fertilizers and manure. Biology and Fertility of Soils 2002;35 108–113.

[56] Liu Hui, Zhao Ping, Lu Ping, Wang Yue-Si, Lin Yong-Biao, Rao Xing-Quan. Greenhouse gas fluxes from soils of different land-use types in a hilly area of South China. Agriculture, Ecosystems and Environment 2008;124 125–135.

[57] Liu XJ, Mosier AR, Halvorson AD, Reule CA, Zhang FS. Dinitrogen and N2O emissions in arable soils: Effect of tillage, N source and soil moisture Soil Biology and Biochemistry 2007;39 2362–2370.

[58] Lopez-Garrido R, Madejon E, Murillo JM, Moreno F. Soil quality alteration by mouldboard ploughing in a commercial farm devoted to no-tillage under Mediterranean conditions. Agriculture, Ecosystems and Environment 2011;140 182–190.

[59] MacKenzie AF, Fan MX, Cadrin F. Nitrous oxide emission in three years as affected by tillage, corn-soybean-alfalfa rotations, and nitrogen fertilization. Journal of Environmental Quality 1998;27 698–703.

[60] Maljanen M, Martikainen PJ, Aaltonen H, Silvola J. Short-term variation in fluxes of carbon dioxide, nitrous oxide and methane in cultivated and forested organic boreal soils. Soil Biology and Biochemistry 2002;34 577–584.

[61] McCarty GW. Modes of action of nitrification inhibitors. Biology and Fertility of Soils 1999;29 1-9.

[62] McSwiney CP, Robertson GP. Nonlinear response of N2O flux to incremental fertilizer addition in a continuous maize (Zea mays L.) cropping system. Global Change Biology 2005;11 1712–1719.

[63] Meng L, Ding W, Cai Z. Long-term application of organic manure and nitrogen fertilizer on N2O emissions, soil quality and crop production in a sandy loam soil. Soil Biology and Biochemistry 2005;37 2037–2045.

[64] Merino P, Artetxe A, Castellon A, Menendez S, Aizpurua A, Estavillo JM. Warming potential of N2O emissions from rapeseed crop in Northern Spain. Soil and Tillage Research 2012;123 29–34.

[65] Mitchell AR, Ellsworth TR, Meek BD. Effect of root systems on preferential flow in swelling soil. Communications in Soil Science and Plant Analysis 1995;26 2655-2666.

[66] Mosier A, Kroeze C, Nevison C, Oenema O, Seitsinger S, Van Cleemput O. Closing the global N2O budget: nitrous oxide emissions through the agricultural nitrogen cycle. Nutrient Cycling in Agroecosystems 1998;52 225–248.

[67] Muller C, Stevens R, Laughlin R, Jager HJ, Microbial processes and the site of N2O production in a temperate grassland soil. Soil Biology and Biochemistry 2004;36 453–461.

[68] Norton LD, Mamedov AI, Huang C. Soil aggregate stability as affected long-term tillage and clay mineralogy. In: Horn R, Fleige H, Peth S, Peng X. (eds.) Soil Management for Sustainability: Advances in Geoecology. Catena Verlag, Reiskirchen; 2006.

[69] Nyakatawa EZ, Reddy KC, Lemunyon JL. Predicting soil erosion in conservation tillage cotton production systems using the revised universal soil loss equation (RUSLE). Soil and Tillage Research 2000;57 213–224.

[70] OECD. 2000. Environmental indicators for agriculture, methods, and results, executive summary. Paris, France (www.oecd.org/dataoecd/0/9/19166629.pdf)

[71] Petersen SO, Schjonning P, Thomsen IK, Christensen BT. Nitrous oxide evolution from structurally intact soil as influenced by tillage and soil water content. Soil Biology and Biochemistry 2008;40 967–977.

[72] Phillips RL, Wick AF, Liebig MA, West MS, Lee DW. Biogenic emissions of CO2 and N2O at multiple depths increase exponentially during a simulated soil thaw for a northern prairie Mollisol. Soil Biology and Biochemistry 2012;45 14-22.

[73] Poth M, Focht DD. 15N kinetic analysis of N2O production by Nitrosomonas europaea: an examination of nitrifier denitrification. Applied and Environmental Microbiology 1985;49 1134–1141.

[74] Rasmussen KJ, Impact of ploughless soil tillage on yield and soil quality: a Scandinavian review. Soil and Tillage Research 1999;53 3–14.

[75] Rastvorova OG. Soil physics: practical use. Leningrad: Leningrad University Press (in Russian). 1988.

[76] Reicosky DC. Long-term effect of mouldboard ploughing on tillage-induced CO2 loss. In: Kimble JM, Lal R, Follet RF. (eds.) Agricultural Practices and Policies for Carbon Sequestration in Soil. Florida: CRC/Lewis, Boca Raton; 2002. p.87–97.

[77] Riley H, Pommeresche R, Eltun R, Hansen S, Korsaeth A. Soil structure, organic matter and earthworm activity in a comparison of cropping systems with contrasting tillage, rotations, fertilizer levels and manure use. Agriculture, Ecosystems and Environment 2008;124 275–284.

[78] Ritchie GAF., Nicholas DJD. Identification of the sources of nitrous oxide produced by oxidative and reductive processes in Nitrosomonas europaea. Biochemistry Journal 1972;126 1181–1191.

[79] Rizhiya EY, Boitsova LV, Buchkina NP, Panova GG. The influence of plant recidues with different C/N ratio on nitrous oxide emission from derno-podzolic loamy-sand soil. Eurasian Journal of Soil Science 2011;10 1251-1259.

[80] Robertson GP, Groffman PM. Nitrogen Transformation. In: Paul EA. (ed.) Soil Microbiology, Biochemistry, and Ecology. New York: Springer; 2007. p341-364.

[81] Ruser R, Flessa H, Schilling R, Steindl H, Beese F. Soil compaction and fertilization effects on nitroius oxide and methane fluxes in potato fields. Soil Science Society of America Journal 1998;62 1587-1595.

[82] Simansky V, Tobiasova E, Chlpik J. Soil tillage and fertilization of Orthic Luvisol and their influence on chemical properties, soil structure stability and carbon distribution in water-stable macro-aggregates. Soil and Tillage Research 2008;100 125–132.

[83] Six J, Elliott ET, Paustian K. Aggregate and SOM dynamics under conventional and no-tillage systems. Soil Science Society of America Journal 1998;63 1350–1358.

[84] Six J, Feller Ch, Denef C, Ogle SM, de Morales Sa JC, Albrecht A. Soil organic matter, biota and aggregation in temperate and tropical soils – effects of no-tillage. Agronomie 2002;22 755–775.

[85] Skiba U, Smith KA. The control of nitrous oxide emissions from agricultural and natural soils. Chemosphere 2000;2 379–386.

[86] Smith KA. A model of the extent of anaerobic zones in aggregated soils, and its potential application to estimates of denitrification. Journal of Soil Science 1980;31 263–277.

[87] Smith KA, Ball T, Conen F, Dobbie KE, Massheder J, Rey A. Exchange of greenhouse gases between soil and atmosphere: interactions of soil physical factors and biological processes. European Journal of Soil Science 2003;54 779–791.

[88] Snyder CS, Bruulsema TW, Jensen TL, Fixen PE. Review of greenhouse gas emissions from crop production systems and fertilizer management effects. Agriculture, Ecosystems and Environment 2009;133 247-266.

[89] Soil Survey. Laboratory methods manual, soil survey investigations report N42, version 3.0, January 1996. Washington: US Department of Agriculture, Natural Resources Conservation Service, National Soil Survey Center; 1996.

[90] Uchida Y, Clough TJ, Kelliher FM, Sherlock RR. Effects of aggregate size, soil compaction, and bovine urine on N2O emissions from a pasture soil. Soil Biology and Biochemistry 2008;40 924–931.

[91] Van Groenigen JW, Kuikman PJ, de Groot WJM., Velthof GL. Nitrous oxide emission from urine-treated soil as influenced by urine composition and soil physical conditions. Soil Biology and Biochemistry 2005;37 268–274.

[92] Van Groenigen JW, Velthof GL, Oenema O, van Groenigen KJ, van Kessel C. Towards an agronomic assessment of N2O emissions: a case study for arable crops. European Journal of Soil Science 2010;61 903–913.

[93] Velthof GL, Brader AB, Oenema O. Seasonal variations in nitrous oxide losses from managed grasslands in the Netherlands. Plant and Soil 1996;181 263–274.

[94] West TO, Post WM. Soil organic carbon sequestration rates by tillage and crop rotation: a global data analysis. Soil Science Society of America Journal 2002;66 1930–1946.

[95] Williams SM, Weil RR. Crop cover root channels may alleviate soil compaction effects on soybean crop. Soil Science Society of America Journal 2004;68 1403-1409.

[96] Wlodarczyk T, Stepniewska Z, Brzezinska M. Denitrification, organic matter, and redox potential transformations in Cambisols. International Agrophysics 2003;17 219-227.

[97] Wood PM. Notrification as a bacterial energy source. In: Prosser J.I. (ed.) Nitrification. Special Publications of the Society for General Microbiology 1986;20 39-62.

[98] Yoh M, Toda H, Kanda K, Tsuruta H. Diffusion analysis of N2O cycling in a fertilized soil. Nutrient Cycling in Agroecosystems 1997;49 29–33.

[99] Zhang GS, Chan KY, Oates A, Heenan DP, Huang GB. Relationship between soil structure and runoff/soil loss after 24 years of conservation tillage. Soil and Tillage Research 2007;92 122–128

Section 3

Physical Properties of Plants and Their Products

Chapter 10

Identification of Wheat Morphotype and Variety Based on X-Ray Images of Kernels

Alexander M. Demyanchuk, Stanisław Grundas and
Leonid P. Velikanov

Additional information is available at the end of the chapter

http://dx.doi.org/10.5772/52236

1. Introduction

High quality seeds of wheat are necessary for fast biological and technological progress. As the proverb says, "what you sow that you will reap". High quality of seeds is ensured by its variety and the extent of damage to the kernels. First of all, the seeds should be of a variety characterised by high productivity and suitable for the climate zone where it is cultivated. Secondly, the seeds should have a high level of germination capacity and be free of damage of biotic and abiotic character. To generate a variety and to obtain seed zoning requires a minimum of about ten years. The potential of variety, on average, decreases by 20% a year, and a few years later it is replaced. Transition to hybrids reduces the time of obtaining new seeds to five years, but the agricultural economy is dependent on the activities of breeding centres. Harvested grain cannot be used as seed. This state of affairs with relation to seeds stimulates the search for new approaches to an improvement of their quality.

In this chapter we propose to discuss one approach to the obtainment of seeds with improved properties. The approach proposed provides a more detailed description of the study of morphological types in the process of ontogenesis for the identification and selection of seeds on the basis of established indicators. The time required to obtain high quality seeds, ready for mass sowing, can thus be reduced from several years to hours, with multiple-fold reduction of the cost of their obtainment.

In the organism, in one form or another, "all is reflected in everything". In practical breeding selection, economically useful properties of plants are most often associated with their morphological features. Ideally, one would like to see these features in the kernels, and based on these features to select them as seeds. Suggestions about such a connection were made long ago, but there were no arguments to substantiate such a connection, nor methods for mathematical description of the morphology of kernels, allowing identifying characteristic traits or indicators allocating particular batches of kernels to the morphological type, variety, or cultivar. Improvement of measurement techniques allows us to address anew previously known problems which have not yet matured to the point of solution, and to set new ones which previously could not even be thought about.

On the basis of the study of morphology in ontogenesis and characteristics of the wheat genome, algorithms of the kernel forming were obtained (Batygin & Demyanchuk, 1995; Demyanchuk, 1997) which allows to establish a relation between its morphology and the growth conditions of the plant, kernel position in the ear, etc. As the work conducted indicated, all of those internal and external factors determine, in ontogenesis, a specific morphological portrait of the kernel, classifying it to a specific variety under the conditions of its cultivation. Based on such a "portrait" one can identify the variety, as well as some of its properties, for example, the duration of its vegetation period.

In the next part of this chapter the required definitions necessary for the selection of morphological characters, and the obtained effects are described. To date, there exist mature technical possibilities of rapid identification of the form of kernels and of high-speed separation or grain flow on the basis of selected feature or property. The selection of kernels whose shape conforms to the specific form of a variety will ensure permanent maintenance of that variety.

The application of X-rays in this technology is due to the fact that X-rays can clearly identify the silhouette of kernels and - most importantly - identify the characteristics of their inner structure, which can be of decisive importance. For example, the presence of insects in kernels preserving their natural shapes or infestation with the sunn pest. The X-ray method can reveal hidden germination, or symptoms of germ necrosis or, alternatively, indications of increased potential productivity of seeds.

2. Description of the study object

Wheat - one of the most widely cultivated plants. It was grown over many millennia BCE. Archaeological evidence suggests that in many parts of Asia, Europe, and also in Egypt, wheat was grown for 5-7 thousand years BC. Wheat grain was found in Egyptian pyramids, the pile-buildings of Switzerland, and at many sites of ancient man. The exceptional ability of wheat to synthesise gluten, with high baking quality of flour, guarantees its monopolistic position among other crops. Wheat (genus *Triticum* L.) belongs to the order *Poales,* family *Poaceae*. In cultivation it is represented by a great many varieties adapted to growing conditions, constantly updated with new cultivars, more productive, more adaptable to local conditions, more than meeting the requirements of modern manufacturing. No other cereal species has so many varieties or cultivars as wheat. As a rule, the countries that grow wheat, apart from commonly occurring types, have their own local varieties and cultivars. Centuries of experience of growing wheat in different soil and climatic conditions contributed to the formation of a large diversity of species and varieties.

Belonging to a variety is primarily determined by the main parameters of the vegetative organs – the stem and the ear, by the size, shape and colour of the grains, as well as by their chemical composition.

The real or proper wheat cultivars are characterised by elastic and flexible straw that does not get fragmented during threshing, the ear is set firmly on the straw, and the kernels are naked and during threshing get easily separated from the fitting scales.

The second group, known as spelt, is characterised by opposite features, namely, spelt cultivars have very brittle straw, easily broken during threshing, the ear is also easily separated from the straw, while the kernels are strongly attached to the scales and get separated from them only with great difficulty.

Morphology is one of the most important characteristics of plants. Despite centuries of observation of morphological features, the process of their formation in ontogeny cannot be considered well understood. If some of them can be considered as signs of species, genus or even variety, others remain highly volatile, retaining, however, in their variety, things in common which, however, are not an easy object of scientific description.

Researchers involved in the study of wheat, depending on their position, apply such a classification as is the closest to them. The most common classifications include the botanical, genetic, economic, and morpho-physiological. Let us focus on the last one. It is the one that is always used in classical breeding which has given and still gives humanity more and more productive and valuable varieties. The breeder selects plants that have some or other morphological features. Their combinations, based on his experience, knowledge and intuition, are related to the productivity, stability or some technological properties of kernel or plant as a whole, such as for example the length of straw and ear type, the size, colour and number of spikelets and flowers in the spikelet, etc., characteristics of kernels. The kernel is an integral representative of the plant. In fact, it is the whole plant in miniature.

Leading morpho- and physiologists, who left the most significant mark on the study of wheat, indicated that the kernel should carry signs of belonging to a variety. Because "everything is reflected in everything", the uniqueness of a variety should be reflected in the uniqueness of some of the morphological proportions of its kernels. The significance of this idea is implicitly recognised in the classification of wheat varieties used for admission to the collection of the Vavilov Institute of Plants (VIP) in St. Petersburg, Russia. According to the existing methods of morphology, wheat kernels are assessed by three linear dimensions ($a \times b \times c$) and their multiplication with a coefficient equal to 0.52. It is clear that the approximation of the complex-shaped kernel by means of a simple parallelepiped is an extreme simplification. In such a description both the shape of the object and its biological characteristics are completely lost. However, that was a step forward and corresponded with the level of knowledge on kernels and methods of description of their forms accumulated by that time. At the same time it was accepted that "the study of morphological characters and their relationships with the physiological functions and biochemical parameters - one of the most effective ways to improve the selection of high-protein wheat cultivars" (Henkel, 1969).

Morphophysiological systematization (Henkel, 1969) relates the type of wheat with the duration of the growing season and the weight of one thousand kernels. The characteristic features of morphotype include also the building of apexes at the moment of transition from stage of growth to the stage of development. This set of features, combination of which characterise varieties, presents a difficulty in the identification of varieties, and brings us back to the kernel. It appears to be a tempting challenge is to find such a set of morphological traits of wheat kernel that would characterise both the variety and the quality of seeds. There is a known non-formalised relationship of the form of the embryo with plant productivity. It's widely known, for example, that kernels from the upper, central and lower parts of the ear are somewhat different in shape, and that the difference in moisture content of these kernels characterises the drought resistance of the variety. In kernels most variable parameter is length. With increasing thickness of kernels (Larikova, 2007; Larikova & Kondratyev, 2002) their length and width increase as well. Compared to slender kernels, in well-filled kernels the embryo and endosperm are bigger as a rule. Depending on the degree of kernel filling, the differences in the mass of endosperm are larger than the differences in the mass of embryos.

Yielding properties of kernels as seeds are largely dependent on their place in the ear. High yield properties are typical to kernels formed in the outer flowers of the central ear. These kernels, compared to other kernels in the ear, have the greatest width. It is believed that this indication is the preferred morphological indicator of kernels for selection as the most productive seeds.

In wheat there is an interrelated system of right hand-left hand features, both for an individual plant and for the cultivar, variety or species.

In the morphology of wheat it was noted that the filling of the particular elements of the embryo (i.e. filling of leaves, roots, and other formed elements of the embryo) may be a variety-specific indicator (Henkel, 1969). Identification of these indicators with the help of a microscope was not widely developed in practice, because this involves the destruction of the kernel and therefore makes it impossible to compare its structure with the properties of the plant which would grow out of it.

In this chapter it is proposed to apply a mathematical description of the "assembly technology" of the embryo and kernel throughout the successive strictly directed cell divisions (Demyanchuk, 1997). This description is related with the morphotype of the plant, with its characteristic length of the growing season and all morphometric characteristics. This permitted mathematical description of the characteristic features of the shape of the kernel of a given morphotype and variety, and the presentation of an algorithm for the identification of kernels in complex mixture of grain in bulk. The application of X-ray techniques for non-destructive analysis of the internal structure of an object opens new perspectives in obtaining morphometric characteristics of kernels. It permits estimation of the shape of kernels as a whole, in various projections, as well as of the morphometry of the embryo. The exact location of various defects of the internal structure, both of biotic (for example, number of insects) and abiotic nature (endosperm cracks), can be determined. Only the X-ray method can reveal the peculiarities of the geometry and fine structure of the embryo and then associate them with the properties of the plant obtained from the seed, beginning with its external appearance and ending with the structure of its yield (number of elements, their mass, and other technological characteristics, including sensory features). Thus, selection of kernels can be conducted on the basis of their X-ray projections, characteristic for variety and for high yielding capacity. In this way it will be possible to maintain a variety over indefinite periods of time and protect it against "degeneration", while achieving increased productivity and at the same time freeing the seed material obtained from kernels that are disease-stricken, damaged, infested by insects (Varshalowich, 1958), fungal diseases, and from contamination with foreign material.

3. Peculiarity of kernel formation

The specifics of wheat kernel formation in ontogenesis consists in the combination of the growth processes of the parent organism and the kernel. Compared to the normal quadruple complication of the organisation of the parental structures, including those that ensure nutrition for the new structures of the embryo, at this stage radical changes take place in the organism. After the dual fertilisation, in the seven-cell eight-nucleus germ sac of the *Poligonum* type three different processes develop:

1. Process of direct growth and development of the embryo, also characterised by quadruple complication compared to the structures of the preceding stage;

2. Process of the formation of endosperm, whose formation ensures such a configuration of intercellular bonds that accelerates the exchange processes many times. The intensity of those processes is not less than 21-fold greater than that of the exchange processes in the embryo;

3. Process of the formation of cells of the gametophyte (antipodes), whose intensity decreases, producing a total of slightly over one hundred cells. As a rule, the cells of the gematophyte (antipodes) form approximately (as standard) a structure of 108 cells that undergo lysis (degradation) in the course of development.

The developing caryopsis is covered with growing structures of the parental organism. The increasing complexity and rate of acceptance of nutrition supply from the parental structure of covers is considerably less than the consumption requirements of the internal processes taking place within the caryopsis. The nutrition requirements during the formation of endosperm are nearly 100-fold greater than the level of requirement for the development of a kernel in the ear, and eventually the endosperm loses contact with the mother plant.

For analytical estimates of acceptable changes of the structures under study, we will now consider an algorithmic description of wheat kernel development.

4. Modelling of kernel development

The huge number of factors that need to be combined into a coherent picture of the growth and development reduce this problem to the area of either a description of results already available or the credibility of that result. With all the grand results of genetics, we are left only with a probability of an incomplete number of possibilities. Almost all of the designs of a biological experiment can be modelled by the study of the "black box". In practically any experiment we are left with the possibility of facing the unpredictable. This dramatically reduces the possibility of wider application of existing knowledge about the nature of things and events.

Complex systems operate under conditions of large numbers of random factors. The sources of random factors are the external environment, as well as errors and reactions that occur within the organism. In plants, growth and differentiation are so arranged as to allow talking about ontogenesis as a purposeful process. Irrespective of the diversity of conditions and ways of realisation of ontogenesis, only one and the same final result of development is constantly observed (determination).

The inductive-deductive construction allows collecting new ideas about the processes and the behaviour of an object.

Let us introduce a number of assumptions about the morphogenesis of kernel as a developing biological system. These assumptions can then be removed if necessary.

1. The zygote in conjunction with its surroundings constitute a system which includes the functions of defining and securing the structure during development;
2. The elements that determine the genetic program of implementation are in the cell;
3. The location and condition of the surrounding cells are defined by the terms of running the program of hereditary realisation, i.e. by the conditions of the program of the genome;
4. The environment meets the conditions of existence of the organism and contains everything needed to perform the processes of growth and development;
5. The organism realises itself in accordance with the provision of metabolites and following a genetic program as a response of its condition to the surrounding environment (comfort, safety);
6. The cell contains elements that are already capable of forming and eliminating connections;
7. If not enough of these elements initially, the cell system receives all that is necessary during the cycle of division as a result of exchange with neighbouring cells from the accessible environment;
8. Energy elements necessary for the process of cell division accumulate in the cell through the process of exchange with the environment via the developed connecting channels of the cell;
9. If during a cycle the necessary structural changes in the cell did not take place, and/or sufficient energy for cell division has not been accumulated, the division does not occur.

Schmalhausen, (1968), and Schrödinger, (1944) argued that the entire genetic information is transferred from the zygote to all the cells of the developing organism through the mechanism of cell division. Methods of encoding and transmission of information affect the organisation of connections between elements. Therefore, as the basis of the construction we used the concepts of systems theory and the assumption that information necessary for the implementation of the genetic programs is contained in the ancestral cell chromosomes. Our system (cell, cellular system) is defined as an ordered set, a connection pool which, through interaction, leads the system to a particular purpose. Hence an important interaction among the three fundamental components of the concept of the system: elements, relationships and operations. The algorithm usually represents the method for computing functions, and in our case it determines the sequence of actions to be performed by the organism in the process of morphogenesis.

The solutions for such an algorithm represent the chromosome sets of cells, groups of cells, organs of plants, forming a connected hierarchical structure. In formulating the problem, the study of morphogenesis in ontogeny, the very "purpose" of computing the algorithm, is secondary. The main task is to follow the scheme of calculation (self-construction) of a developing system. That scheme, in our case, defines the parameters of morphogenesis of the system. The defined, repeatable sequence of steps and fulfilment of the conditions lead to cell division.

Operations on the elements and relationships of the system should take into account:

1. Direction of the development process;
2. Obligatory character of hierarchy of structures, typical for each stage;
3. What elements trigger the development process;
4. Multiplicity of choice of pathways between the stages;
5. Oscillatory (cyclic) nature of the processes in the system being a form of its existence.

The algorithmic construction describing the development of the system allowed to achieve agreement between the object and the formal concepts of systems theory. The large number of factors of different nature that determine the development and growth of the organism does not allow the selection of a dominant biophysical and biochemical interpretation of the processes. At the initial stage, the algorithm we propose for the description of development is of a formal character. In the case of the construction being effective, there will be a most credible systemic, biophysical and biochemical, interpretation of events. If our starting hypotheses are able to substantiate the main morphological properties of the organism, not previously combined into a whole in ontogenesis, we will assume that our algorithmic construction is suitable for the prediction of the morphological features of the organism.

Therefore, let us consider the behaviour of the carriers of genetic information, genes in the chromosomes. For wheat, the basic chromosome number is a multiple of 7 ($2n = 14, 28, 42$). The first step to construct our algorithm is to determine the features of two functions, each of which is also a multiple of 7, and namely:

1. Function defining the conditions of the organism building cycle;

2. Function directly describing organism building in the cycle.

As far back as in the mathematics of the ancient Egyptians it was believed that a mathematical operation is determined by the two entities over which and with which it is carried out. Pythagoras, defining the concept of number, compared it to a sphere and endowed it with four dimensions (three spatial dimensions and density). From the formal point of view, the cell is similar to that concept of number, but it undergoes modification in the cycle. Our number – cell – establishes a connection with its environment, exchanges and controls connection channels whose number equals that of the chromosomes. An “operation” in the cycle, by means of the connecting channels conforming to the program of the preceding stage of development, forms cells, each of which has the same number of channels. Here ends the process of configuration of “numbers” combined into a specific form. The operation defines the process on these numbers, each of which communicates with another with the same number of communication channels, that number being also equal to the basic chromosome number. The duration of such a process is defined by the conditions of cycle completion and by the method of creation of the “cycle body”. The time required for the computation is defined by the temporal complexity of the algorithm and the computing power of the computer. The temporal complexity of computation of the target function of development is defined by the duration of the vegetation period (e.g. in the case of wheat – the time of its life).

Same examples of interpretation. In the algorithmic model, the direction of metabolite exchange between the maternal and the developing organism is determined by the orientation of chromosomes in the metaphase plate. Biophysical processes (opening up of the helixes) initiate an electromagnetic pulse. The resultant force at the time of the burst pulse determines the orientation of the metabolism, and simultaneously occurring biochemical processes "fix" the structure in its current form. The "directions" (communication channels) of the processes in the cell cycle, cell division and location, are defined as the "operations" of the cell cycle. We will define the algorithm of constructing on the basis of cell division as the purpose function of development.

To clarify the composition and sequence of operations, let us consider the cycle of cell division. The cell cycle is usually divided into four periods: pre-synthetic (G1), period of DNA synthesis (S), post-synthetic (G2) and mitosis (M). Actually, mitosis accounts for 1/7-1/10 of the cell cycle (Table 1):

The phases of the cycle will be juxtaposed with operations, the implementation of which leads to doubling of the chromosomes, the direction of metabolic processes and the necessary conditions of the cell cycle (Demyanchuk, 1997). Cell division is more convenient to consider in the phase of arrangement of chromosomes in the metaphase plate. Technically speaking this phase formed a stable non-equilibrium system of implementation of a sequence of hereditary factors among which the most important are the following (Fig. 1): 1) orientation of metabolic fluxes through "channels" in the cellular environment of a developing structure, 2) biophysical processes in chromosomal band; and 3) biochemical processes that fix the shape.

Item	Cell division cycles	Interpretation
1.	In phase Gap 1 (G1) there takes place transcription (first step leading to gene expression) and translation (transfer of genes from one chromosome to another) in both cells which are a result of the preceding division. Plastids and mitochondria are multiplied. At this stage the cells of a multicellular organism perform all functions necessary for the organism.	Synchronization of metabolic processes with the surrounding cells. Setting up exchange in the plane of the perimeter of the future metaphase plate. In selected areas the "waves of incoming and outgoing" metabolic fluxes are formed.
2.	S-phase - a period when the DNA in the nucleus doubles. DNA replication begins at many but exactly defined locations. By the end of S-phase, each molecule of DNA is doubled in full. Along with the DNA the amount of histones and non-histone proteins of chromatin should double at the same time. In S-phase also centrosome doubles, the place of microtubule formation. In interphase microtubules grow from the centrosome toward the whole periphery to the cell. In late G1 phase of the centrioles move apart by a few microns, and in the S-phase next to each centriole a second centriole is built, and centrosome doubles.	Exchange. Doubling of structures responsible for development: synthesis of DNA, doubling of the chromosomes (comparative operation "×2", i.e. doubling of elements at points indicated by communication channels, so that the new elements are in agreement with the cellular environment). "The wave of incoming metabolic flux" from the surrounding cells provides a process in the cell, and the "wave of the outgoing flux" specifies the location of the cell structure which should be formed in this cycle of development.
3.	The next phase, G2 - preparation for division. At this time, the formation of the two centrosomes ends, and the system of interphase microtubules begins to break down, releasing tubulin from microtubules. The chromosomes at this time are beginning to further condense, but that is not visible under the microscope.	Formation of the "motor system of chromosomes."
4.	Actually mitosis (M phase) is also divided into several stages. The stages of mitosis - prophase, prometaphase, metaphase, anaphase and telophase.	
4.1.	In the prophase there is an additional packing (condensation) of chromosomes to the extent that they become similar to first tangled filaments, visible in the light microscope. Depolymerisation takes place in the cytoplasm present in the microtubule cell. At this point the cell, as a rule, loses its special form and becomes rounded. Around the centrosome, there appears a so-called star - a system of radial microtubules, which are gradually extended. In the process of mitosis, microtubules start renewing 20 times faster than in the interphase, and the small number of long, stable microtubules get replaced with a lot of short and unstable ones. When the microtubules extending from two poles (cell centres) meet each other, they come into contact and get connected to each other by means of certain proteins that stabilise them and	The process in the surrounding cell space. Accumulation and transformation of energy in the form required for cell division.

Item	Cell division cycles	Interpretation
	prevent them from depolymerisation. These microtubules form the spindle of division. Microtubules from the star growing in other directions, either become ultimately destroyed or establish connections near the poles.	
4.2.	In the prometaphase the membrane of the nucleus gets defragmented into vesicles and the nucleus disappears as a structure. The contents of the nucleus are combined with the cytoplasm. A condition similar to the prokaryotic is established. During the division the nucleus disappears. In the prometaphase chromosomes condense, and finally take the form of pair formations. Each pair becomes connected at the point of crossing. In the prometaphase chromosomes, led by microtubules, get arranged in the equatorial plane perpendicular to the spindle. Microtubules act as springs. These forces are balanced when the microtubules emanating from opposite poles are the same length.	Compaction of chromosomes. Orientation of the chromosomes in accordance with the cycle of exchange flows of the cellular environment. Let us imagine that the directions relative to each other in the metaphase plate are oriented roughly with a shift at the angle of (2π / 7). See the circle in central part of Fig. 1.
4.3.	In the metaphase, all the processes in the cell freeze. Chromosomes formed in the metaphase plates take part only in vibrational motion.	Location of bivalents in the equatorial plane. Process cycle (rhythm?), when the exchange stopps.
4.4.	The next stage - Anaphase – is started by a sudden and simultaneous separation of centromeres of the two chromatids from of each other. This is in response to a rapid tenfold increase in the concentration of calcium ions in the cell. They are released from the membrane vesicles surrounding the cell centre. Led by the attraction of microtubules, the chromosomes begin to diverge to the poles of the cell, each of the two sister chromatids to its pole.	The divergence of homologous chromosomes to the poles. Independent divergence of chromosomes included in various bivalents.
4.5.	In the next stage, telophase, a new nucleus envelope begins to form around the chromosomes gathered around each centrosome. A double membrane is recreated from the vesicles, nuclear lamina proteins are dephosphorylated and then form a proper lamina, nuclear pores are assembled again from component parts. And thus, we have considered the stages of mitosis consisting in the doubling of the nucleus. It begins with hidden from the eyes doubling of chromosomes in the interphase, and continues through its self-destruction as a structure during mitosis. When the nucleus has doubled, it is necessary to divide the cytoplasm - to carry out cytokinesis.	The formation of two haploid nuclei in the cell, which may differ genotypically. To perform the "division operation," it is believed that one of the nuclei forms structures in the cell, and the other defines the conditions and the number of cycles of division of the cell system.

Table 1. The phases of the cell cycles.

http://dx.doi.org/10.5772/52236

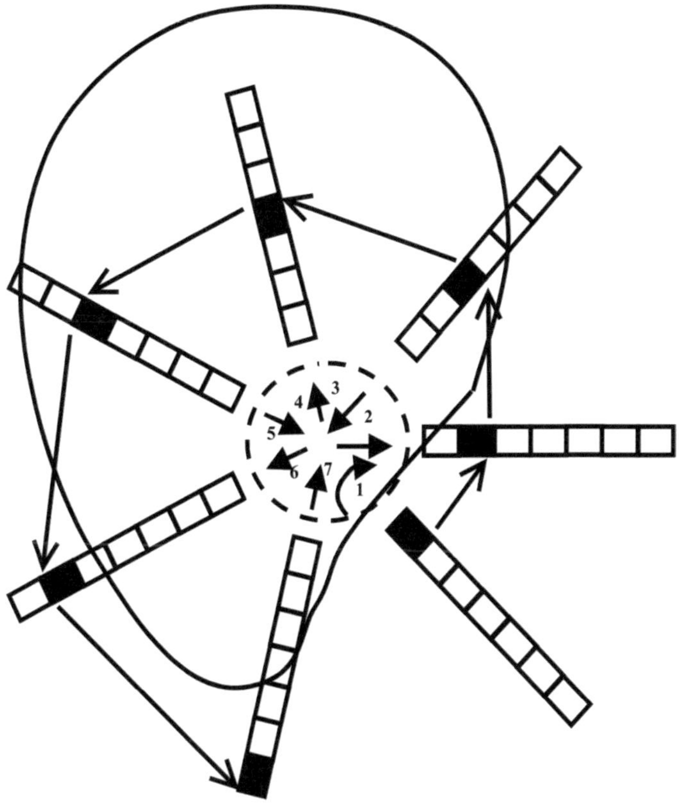

Figure 1. Scheme of the formation of embryo cells of cereals with the base chromosome numbers multiple of "7" at the initial stage of development, with the critical number of 588 cells. The arrows in a circle indicate the direction of flows, the shaded rectangles - the location of the cells, and the shape of the envelope lines around the perimeter of the shaded rectangles - the shape of the embryo.

Electromagnetic pulses of the breaking valence bonds of chromosomes determine the selectivity of resonance in the metabolism of functionally related groups of cells that have no direct contact.

During preparation for zygote division, in the surrounding space there appears a scheme of the general number and arrangement of cells which should be formed in the first phase of development. For wheat it is - ($2 \times 3 \times 2 \times 7 \times 7 = 588$). Thus, the growing cell structure then forms a shape when the "operation of cycle completion" in space will set the future structure of the embryo formation phase. Only then each of its dividing cells will be able to take the position specified in this process. To ensure such positioning of the cells, in each step of the exchange there should be a "link" of each cell with each.

The total number of cycles of the creation of structures according to the algorithm coincided with the number of stages in the development of wheat. The list of structures constituting the plant, the critical number of cells in the initial forms, the achievement of which is necessary for the completion of a stage of development and transition to the next stage of growth in the experiment, also fully coincide with the calculated ones. The angles of displacement in the formation of metameres on the stalk (leaves, buds) and the placement of other organs of the plant also "obey" the rule of displacement "points" of their formation introduced in this manner (Fig. 2.). In Fig. 3. X-ray image of barley seed embryo is presented.

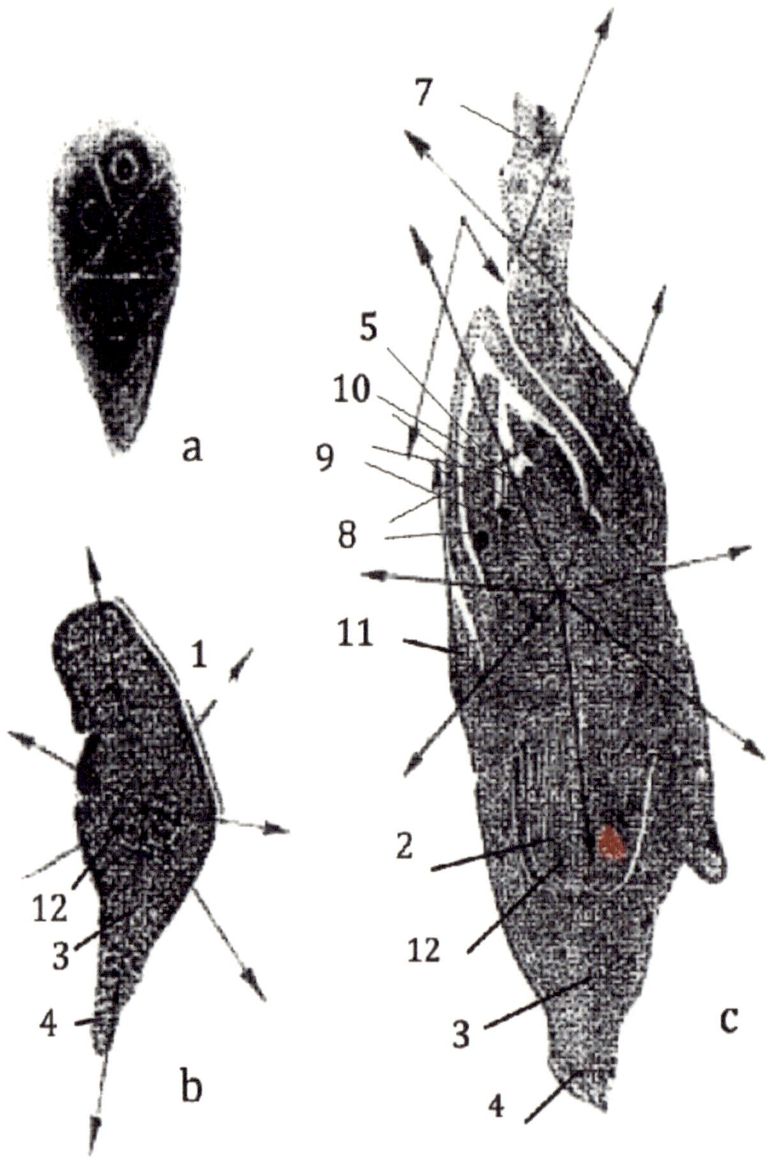

Figure 2. Development of wheat germ (Batygina, 1974; Batygina, 1987): a - four-cell embryo in the dorsal-ventral section; b and c - subsequent stages of embryonic development. 1 - plate, 2 - embryonic root, 3 – coleorhizae, 4 - suspensor, 5 - point of growth, 6 - coleoptiles, 7 - ligulas, 8, 9, 10 - first leaves, 11 - epiblast, 12 - root cap. Directions of the formation of structures and of the "organism as a whole," according to the algorithm of wheat morphogenesis (Demyanchuk, 1997), are indicated by arrows in the figures.

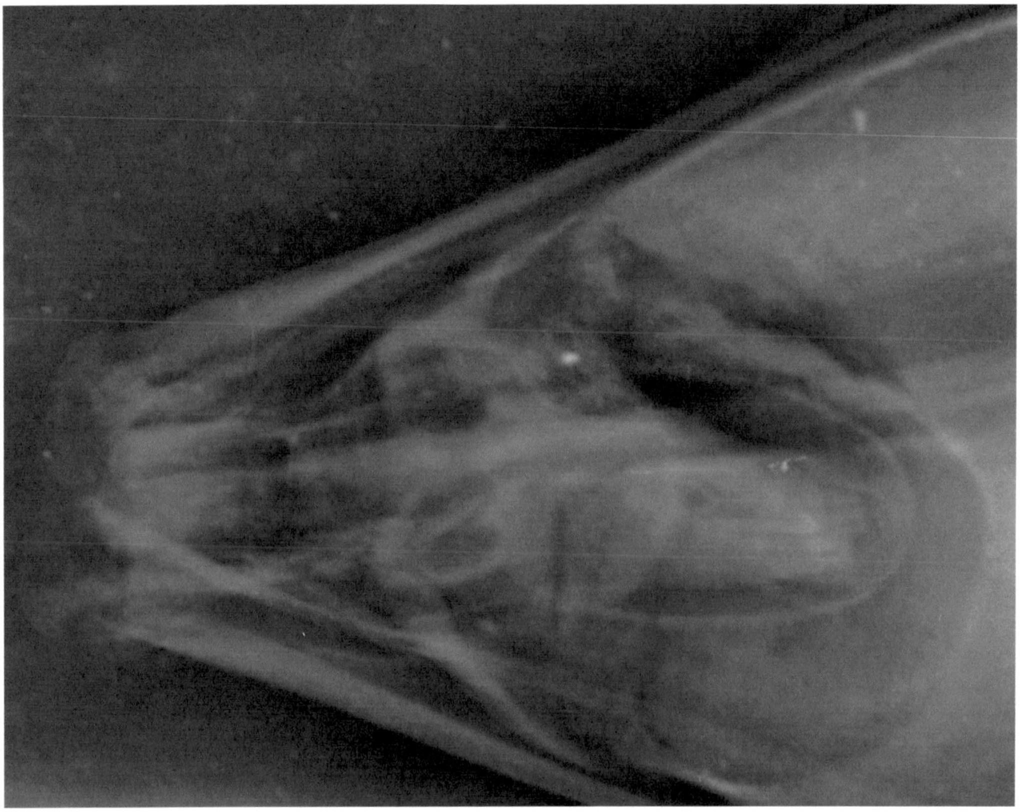

Figure 3. X-ray image of barley seed embryo. Magnification of 30x.

In fact, the method allowed specifying all the critical numbers that must be achieved in all the transitions in the development. Deviations from the common "standard" of the formation of cells of cereals with the basic chromosome number equal to seven (7), are characterised by stable formation in the species- and variety-related additional functionally related groups of cells, in multiples of 49. Depending on the location within the organism, the ways of placement of cells into connected groups differ from one another. For example, the laying of cells according to the scheme, with a constant shift of the direction of the location of the next cell along the line of arrows (7 - 4), indicated in Fig. 1, with a shift at the end of the circle in the perpendicular plane and the angle of $2\pi /7$, forms the shape of the sprout. Analysis of subgroups of cells, corresponding to the phase of establishment of metameres (leaf, shoot, and bud) showed also that in the structure of leaf the observed proportions, necessary for its construction, were maintained (Demyanchuk, 1997). Each variety has its own stable scheme of the formation of cells which constitutes its morphological specificity. The formation of the endosperm is based on interactions of haploid groups of chromosomes of the three nuclei. Algorithmic description of the development of specific forms of cereal grains will correspond to reality if the process is presented in a species (Fig. 4).

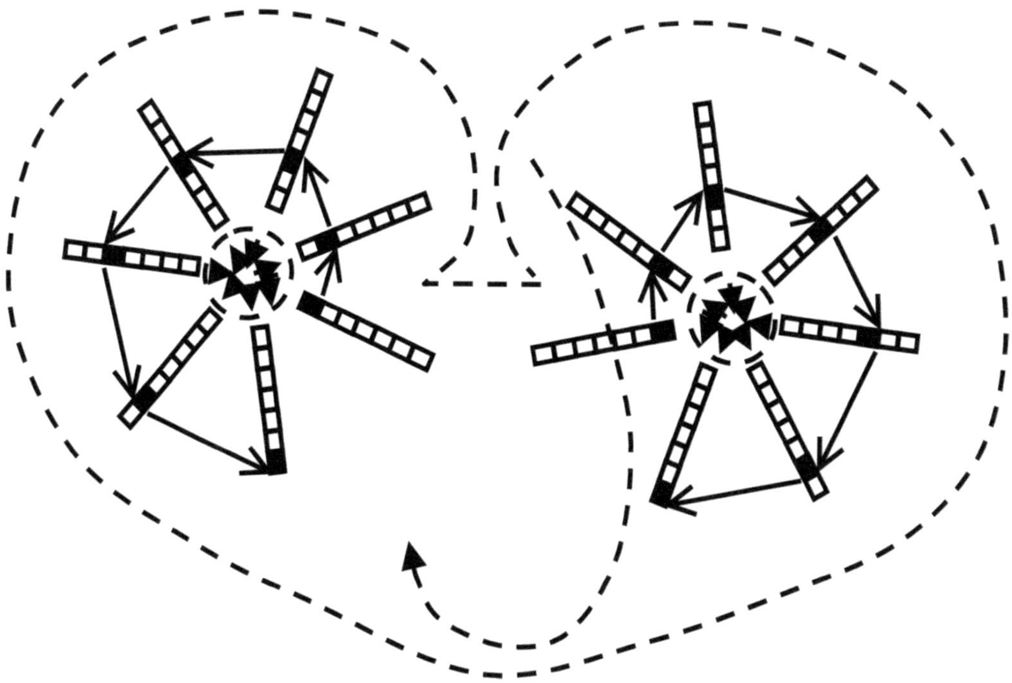

Figure 4. Diagram describing the development of the endosperm by the "closing" of the three groups of connection channels.

The form assumed by the cell system is one of possible choices from a list of assemblages of cells at specific angles to each other. The characteristic angles of deviation in the construction of the embryo are shown in Fig. 2a. The most distinct combination of cells under these angles is observed in the primary divisions of the embryo, during the period when the cells are initially placed in a certain plane of division. As can be seen from Fig. 2b, in a formed embryo, the angles formed by the elements of the embryo, as well as their orientation relative to each other, are also located at a deflection angle equal to or a multiple of $2\pi /7$, i.e. approximately 51.4 degrees.

Of key importance for the specificity of form is the condition of the formation of a critical number of cells. If the conditions for transition to the next stage of development are not met, the laying of the critical number of cells is initiated again. In such a case, the groups of cells initiated in the preceding attempt at establishing the critical conditions of transition remain in the organism and continue to grow. The external manifestation of the process will be an increase in the "life cycle" and addition of a certain number of functionally related groups of cells, multiples of 49, of a given stage, which will change the shape of the specifics.

The development of the endosperm corresponds to a system that defines the formation of cells under the control of three groups of channels. Two of those groups indicate the distribution of cells in conformance with Archimedes spirals (Fig. 4.). Deviation from the conditions for achieving the critical number of endosperm cells will also lead to a repetition of the full cycle of formation of the previous critical number.

In the current morphophysiological classification there are 10 major types of soft and hard wheat. Some of them are presented as 2-5 subtypes. Let us now consider the morphological characteristics of kernels of the known morphophysiological types. Unfortunately, the material available at the time did not permit to process data for all known morphophysiological types, however, the results showed that the geometric characteristics of shape have distinct specific symptoms that can be used both in the practical and the theoretical aspects.

4.1. The first morphophysiological type

This type includes very early and early maturing varieties of spring wheat - North Circumpolar, East Siberian and Far East selections and is divided into subtypes - a, b, c (Henkel, 1969). The length of their growing season (from germination to maturity) equals 68 - 85 days, at least 75-100 days, respectively. Wheat cultivars of this type are characterised by the ability of seed germination at 0 °C, resistance to relatively low temperatures in spring and even to weak frosts, short first stage of development; they are adapted to develop under the conditions of a short summer and an early fall. They accelerate their development at daytime length of 18-24 hours and under predominance of light flux in the long wavelength region of the spectrum (red-orange). They do not have high heat requirements in stages X-XII of organogenesis. Maturation can take place even at +12, +14^{0}C.

Due to the rapid passage of stages II-V of their organogenesis, they form 5-6 leaves. The leaves are short (8 - 10 cm), narrow (0.7 - 0.8 cm), light green, with a slight pubescence. Leaf sheaths are usually smooth. Nodes not pubescent. Short internodes, plant height of about 70 cm. Stem is thin, relatively strong. Long day accelerates their development in stages VII-VIII of organogenesis, therefore they form short ears under such conditions (4.5 - 6.5 cm). High transparency of the air and lower temperature do not contribute to increased length of segments of the spike in stages VII-VIII of organogenesis, therefore, even at small size the ears are usually dense. The kernels are small. Weight of 1000 kernels varies from 14 to 18 g.

Varieties included in the first morphophysiological type are represented, to a considerable extent, by red-colour, non–pubescent cultivars.

Subtype "a" of the first morphological type is represented by the soft spring wheat variety 'Alenkaya' (Fig. 5-7.). The second subtype, "b", is represented by variety 'Balaganka' (Fig. 8-10.). The third subtype, "c", is represented by variety 'Amurskaya 77' (Fig. 11-13.).

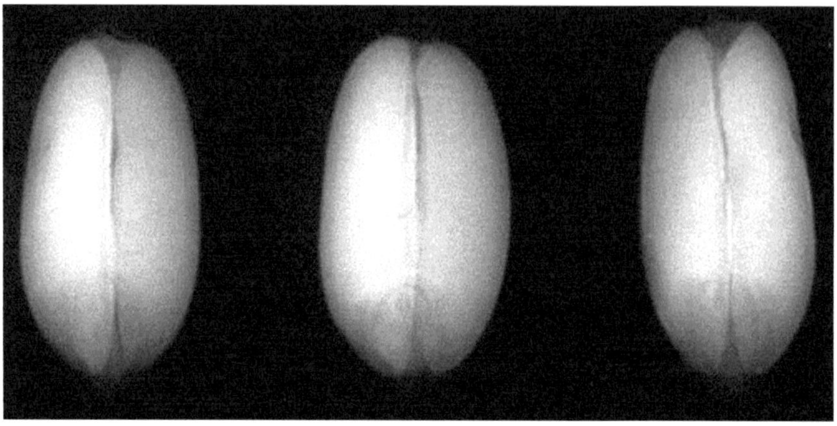

Figure 5. Var. Alenkaya. Front view: Kernels extended, i.e. significant predominance of length over width. In the embryo half of the kernel visible varying thickening at the apex of the embryo (little "chubby cheeks"). The radii of curvature of both ends of the kernels are approximately the same, large enough (blunt ends). The projections of the grooves along the length of the kernels are narrow. At the top end a little shadow in the form of an equilateral triangle.

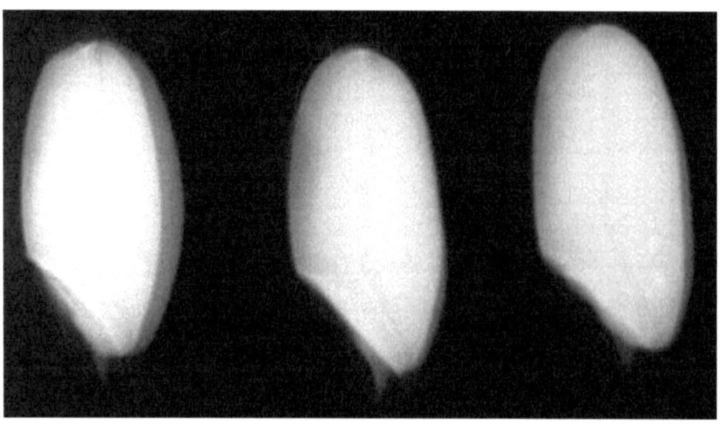

Figure 6. Var. Alenkaya. Side view: Ventral side of the projection is clearly rounded, but in the middle third is nearly straight, that is, the middle third of the kernel can stably lie on a plane. Profile section of the embryo - almost straight or slightly concave line. The back contour is slightly convex. Line at the bottom of the grooves can be seen going in the middle of kernels, with a bend repeating the bend of the back contour.

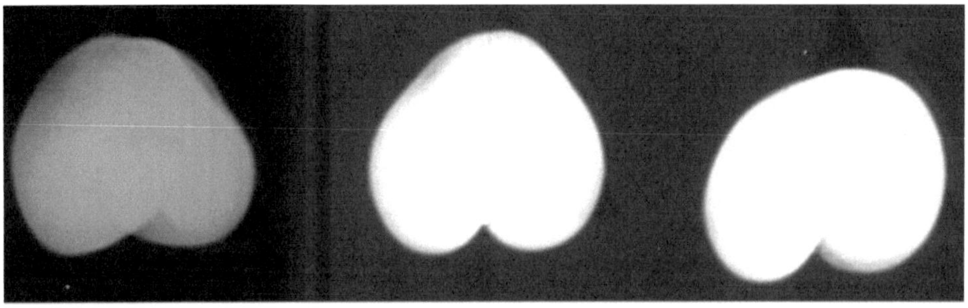

Figure 7. Var. Alenkaya. Up-down view: The up-down projection most often resembles the card colour "diamonds", but there are asymmetries usually caused by a thickening in a random location.

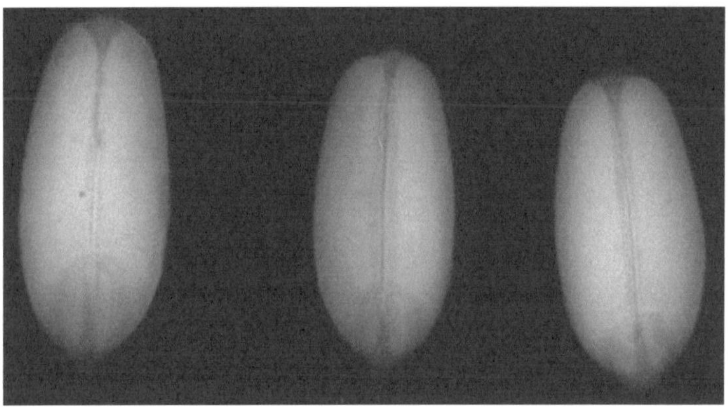

Figure 8. Var. Balaganka. Front view: Elongated projection. The predominant form - with longitudinal and lateral symmetry. The projection of the groove is narrow, but with weak darkening along it. At the top it turns to black in the form of an isosceles triangle with the sharp end down.

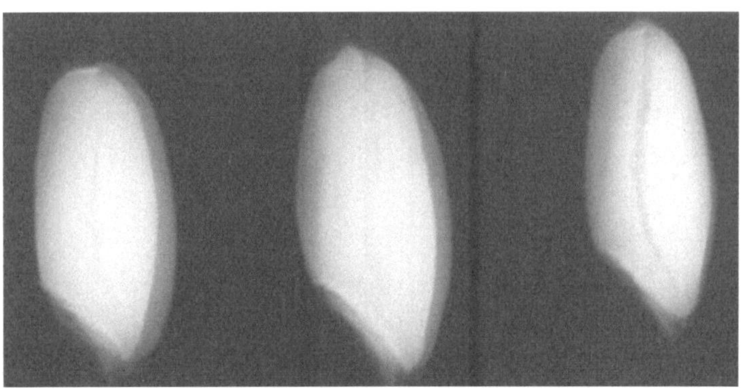

Figure 9. Var. Balaganka. Side view: Ventral and dorsal sides are convex and have an approximately constant radius over the entire length. The embryo contour is straight or slightly concave. The projection of the bottom of the groove is rather broad, indicating an expansion of the groove at its very bottom in the kernel.

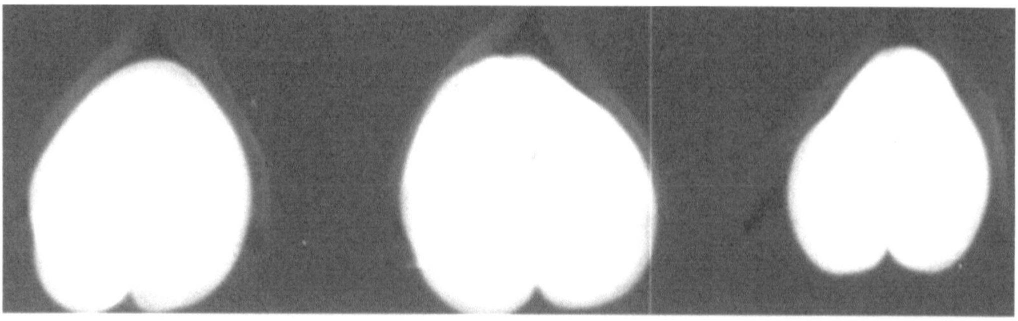

Figure 10. Var. Balaganka. Up-down view: Often asymmetrical, the shape is close to pearform.

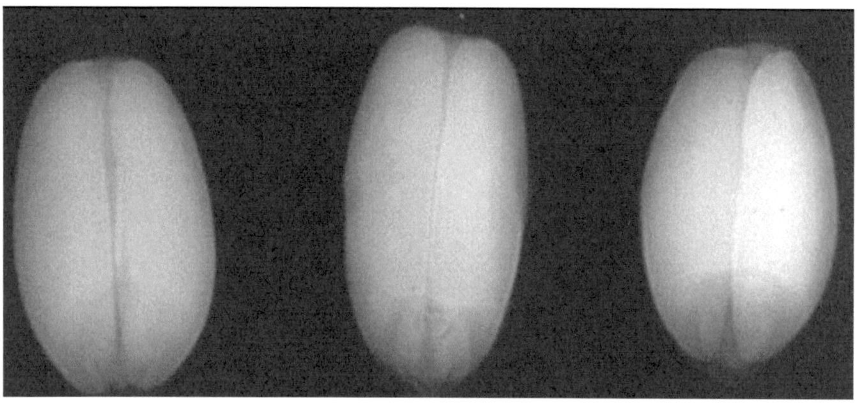

Figure 11. Var. Amurskaya 77. Front view: The projection is mainly barrel-shaped, at least - with a thickening of the end of the embryo. The groove is thin, and at both ends turns into acute angled shadows, almost identical in shape, size and optical density. In some kernels - along the lateral edges of the shadow line - a sign of enzyme-mycosis infection of moderate severity.

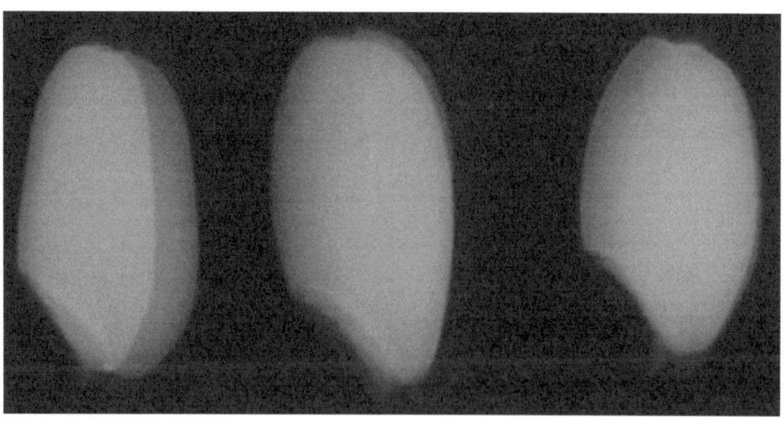

Figure 12. Var. Amurskaya 77. Side view also wide, indicating considerable height of kernel.

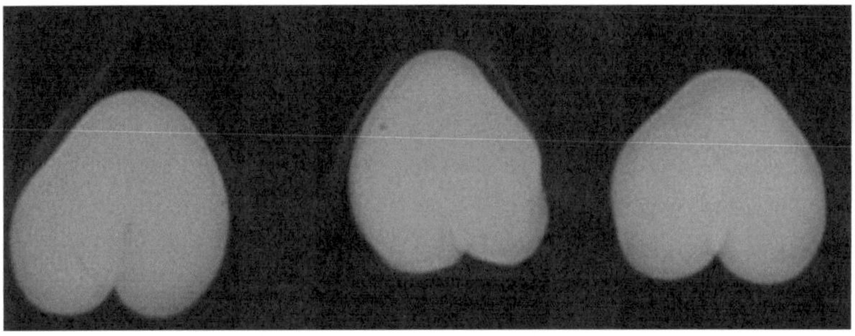

Figure 13. Var. Amurskaya 77. Up-down view. Contours are partially asymmetric, and in a half of the kernels the top is sharpened, that is, on the back there is a narrow ledge. In some kernels a deep-going groove can be observed.

4.2. The second morphophysiological type

This type includes medium-early and medium-late varieties of spring wheat (Henkel, 1969). They are local and selected varieties of temperate latitudes, developing mainly thanks to the winter and early spring precipitation in areas with moisture deficit in the second half of the summer (winter varieties resistant to drought). The length of their growing season is 80-105 days. Under the conditions of increasing length of the day, they pass very quickly through stages I-II, which results in plants with sparse foliage and early transition to stage III. Moisture deficit and low relative air humidity during stages IV-VI of organogenesis inhibit the growth of leaf blades (short and narrow) and contribute to the development of predominantly columnar parenchyma. The first phase is short (stages I-II), the second (stages III-IV) is relatively long. Therefore, the formation of the bottom of the embryonic ear manages to go through thanks to winter and early spring precipitation. The top of the ear, due to moisture deficit in the spring, is often undeveloped. As a result, there is a pronounced spindle in the structure of the ear or even a complete reduction of the upper spikelets. In cultivars of this type the passage through stages III-IV of organogenesis gets accelerated by two or three days (in conditions of 16-20-hour photoperiod), as well as the passage through stages V-VI (with prevalence of red and orange rays in the light spectrum). The cultivars of this type are relatively resistant to high temperatures and to moisture deficit in stages VII-VIII and X-XII of organogenesis. Plant height is 75 - 80cm, but depending on the availability of moisture in stages VI-VIII it varies greatly - from 30 to 110 cm. The ears are of medium size (7-9cm) and medium density.

The specific features of the physiology of development and the high drought tolerance permit the cultivation of many varieties of the second morphophysiological type both in the steppe regions of south-eastern European part of Russia and in many parts of Western Siberia.

The second morphophysiological type, subtype "a", is represented by var. 'Saratovskaya 29' (Fig. 14-16.). Subtype "b" of the second morphophysiological type is represented by variety 'Artemovka' (Fig. 17-19.).

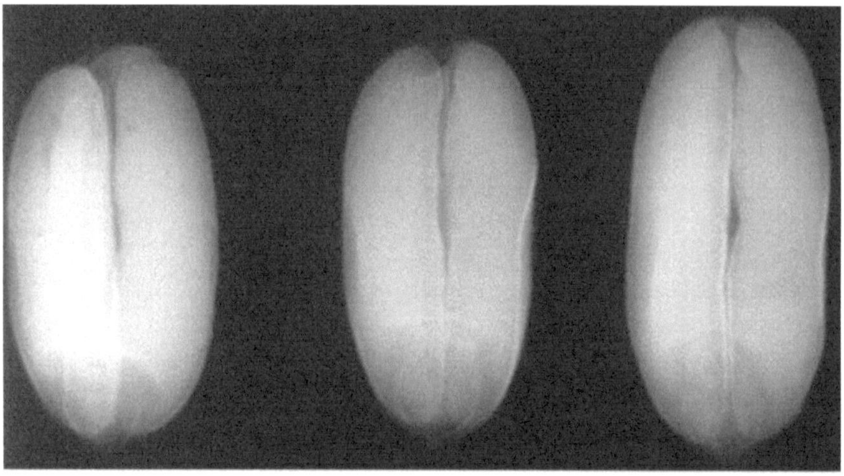

Figure 14. Var. Saratovskaya 29. Front view: The lateral edges of the projection for the most part parallel to each other. The groove is thin and only the upper end has a distinct acute shadow. The shadows along the side edges of the projections under the shell - a consequence of enzymatic-mycosis infection.

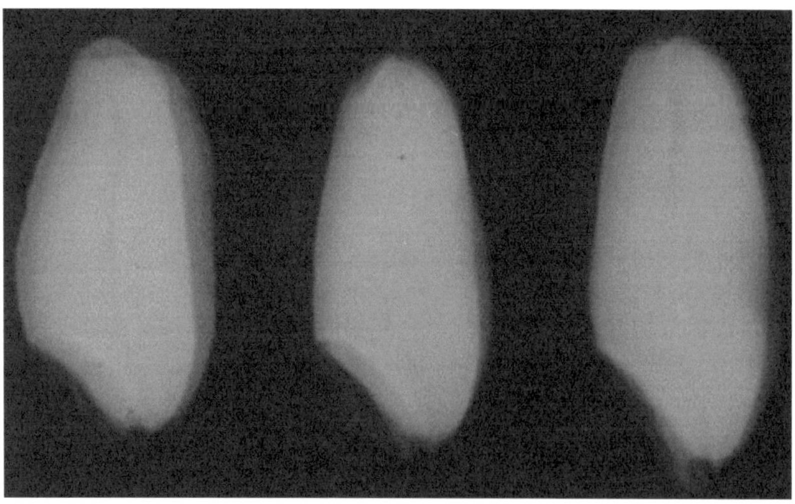

Figure 15. Var. Saratovskaya 29. Side view: Top of the kernel somewhat sharp. Along the shell - shadows, traces of enzymatic-mycosis infection. Embryo section slightly concave.

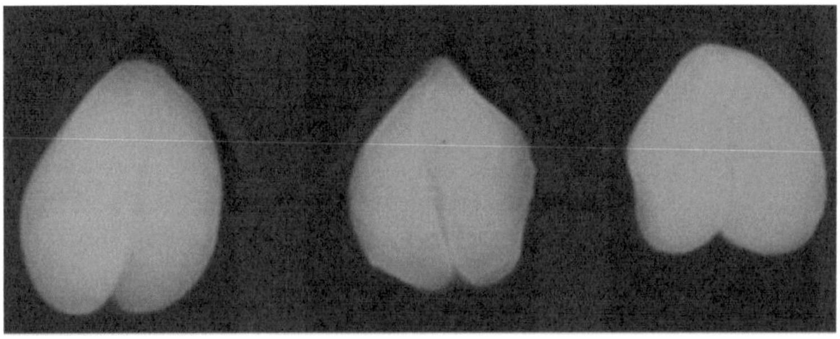

Figure 16. Var. Saratovskaya 29. Up-down view: A characteristic feature - longitudinal grooves and wide lateral recess before the proper groove on the end make the ends of folds sharp and seemingly distant.

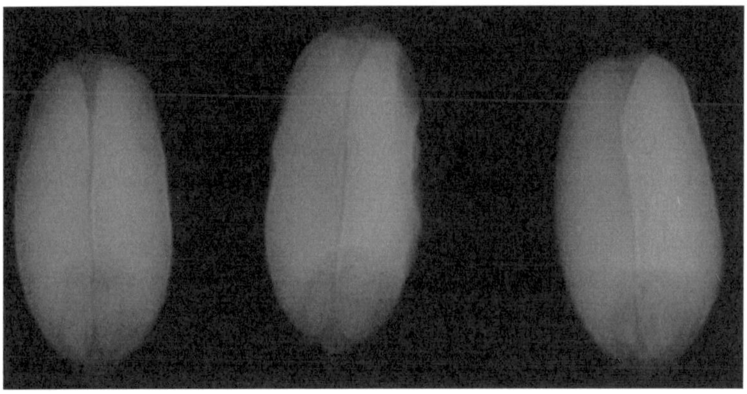

Figure 17. Var. Artemovka. Front view: Kernel elongated with a slight bulge at the bottom. The groove is narrow, with a subtle extension of the middle part, ending with a clear wedge shadow only on the upper end of the kernel. A faint shadow along the groove.

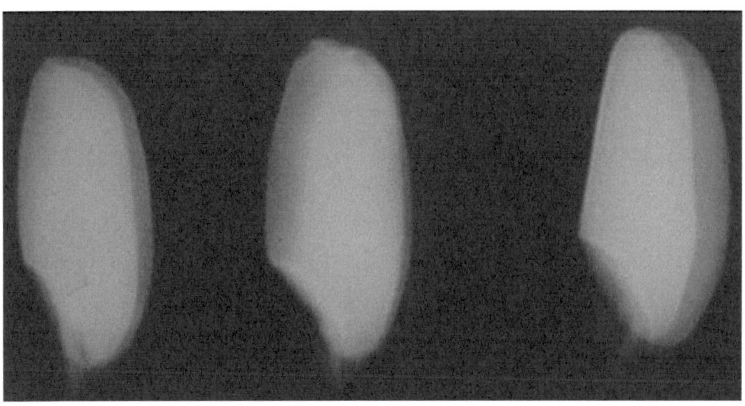

Figure 18. Var. Artemovka. Side view: Line of the ventral side of the kernel is a curve with a single radius, with a slight flattening in the middle. The line of the back of the kernel is a straight line. Concave section of the embryo.

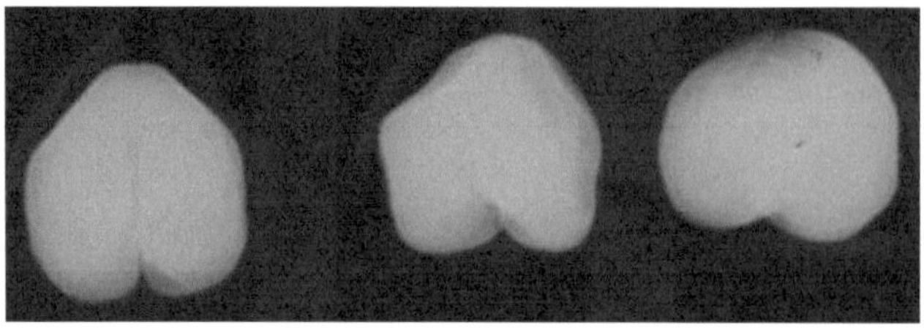

Figure 19. Var. Artemovka. Up-down view: The whole image of up-down view is like a "house with a gable roof". A shallow and narrow deepening in front of the groove, in the form of a small equilateral triangle. Thin groove is visible without any express extension at the end, that is, in the middle of the groove.

4.3. The third morphophysiological type

This type includes medium-early and medium-late maturing cultivars of soft spring wheat from the Siberian-Ural environmental group (subtype "a") as well as from the Northwest Environmental Group (subtype "b") (Henkel, 1969). They are characterised by a relatively long duration of stages I-II of organogenesis and medium duration of stages III-IV. They can develop normally with day length of 16-17 h and predominance of light flux in the red and orange range of the spectrum. Their development is inhibited at 13-14-hour day length. Duration of vegetative period is 85-100 days. Lower temperatures in spring led to a delay of stage II of organogenesis and the formation of 7-9 leaves, and to the growth of mechanical tissues of the lower and middle internodes of the stem. Good moisture availability in stages V-VI of organogenesis is conducive to synchronous formation of spikes and to the formation of a cylindrical and slightly club-shaped ear, as well as to increased growth of the leaves in length and width. Favourable moisture conditions in stages X-XII of organogenesis cause the formation of large kernels, but low temperatures in stage X of organogenesis inhibit the growth of kernels in length. In varieties of this group the kernels are relatively short and often have a low weight of 1000 grains (28 - 30g).

Subtype "a" of the third morphophysiological type is represented by a variety from the forest-steppe (Siberian-Ural) group – 'Viesna' (Fig. 20-22.). Subtype "b" is represented by variety 'Gorkovskaya 20' (Fig. 23-25.).

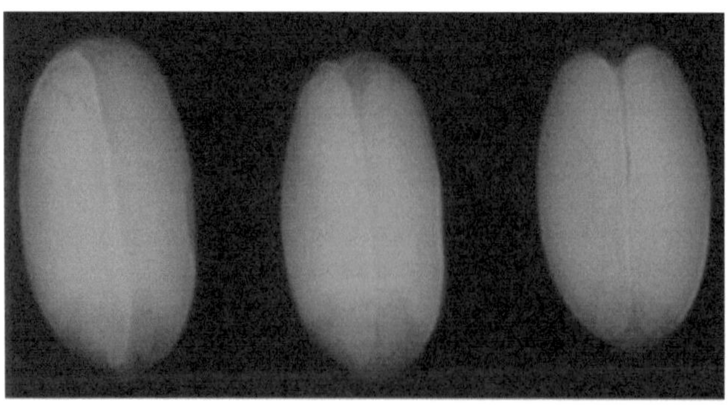

Figure 20. Var. Viesna. Front view: Basically a regular ellipse. The groove is thin, no shadows to be seen around it, wedge-shaped extension visible only at the top of kernel.

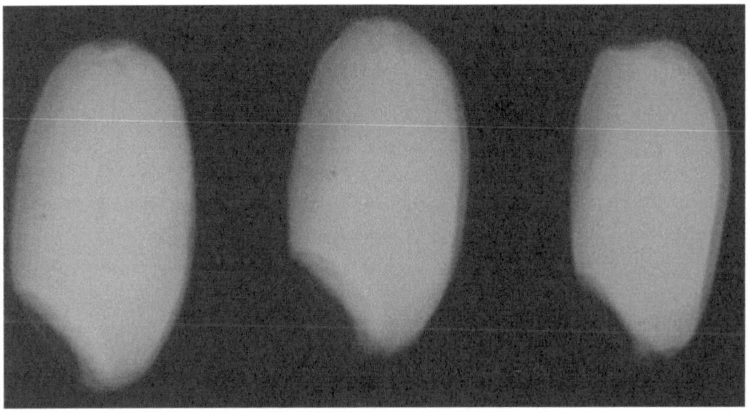

Figure 21. Var. Viesna. Side view: Basically also a regular ellipse, interrupted by the slightly concave section of the embryo. In some kernels the ellipse is slightly distorted by a small bulge in the bottom half. The groove is not visible.

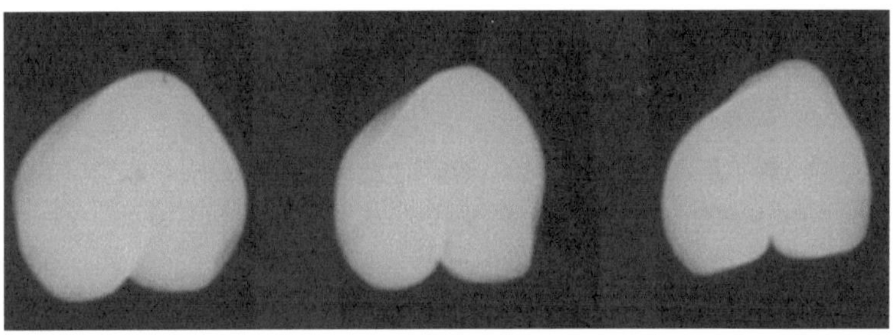

Figure 22. Var. Viesna. Up-down view: General view like a house. The lower edges of the folds are flattened. The groove is thin, with no extension on the end.

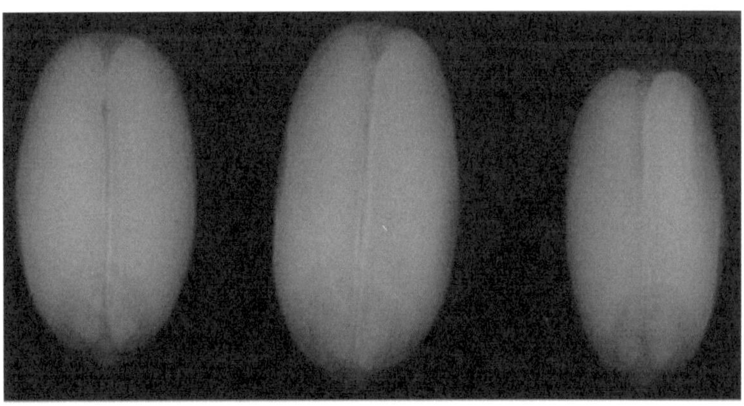

Figure 23. Var. Gorkovskaya 20. Front view. The projection is usually elliptical, with blunt ends. The groove is thin, with a slight expansion in the upper third and a triangular shadow visible at the upper end. Along the groove there is a faint and narrow shadow.

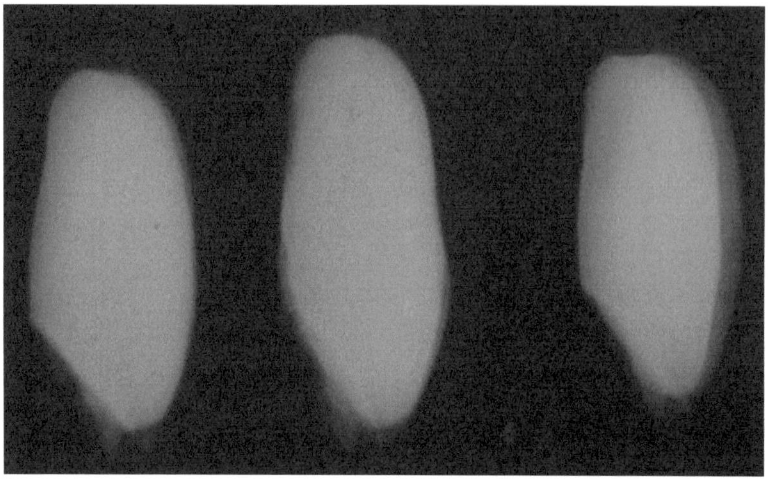

Figure 24. Var. Gorkovskaya 20. Side view: Line of the edge of the ventral part of the kernel - a curve with a single radius. Line of the edge of the back - straight. Visible line the bottom of the groove, running parallel to the back.

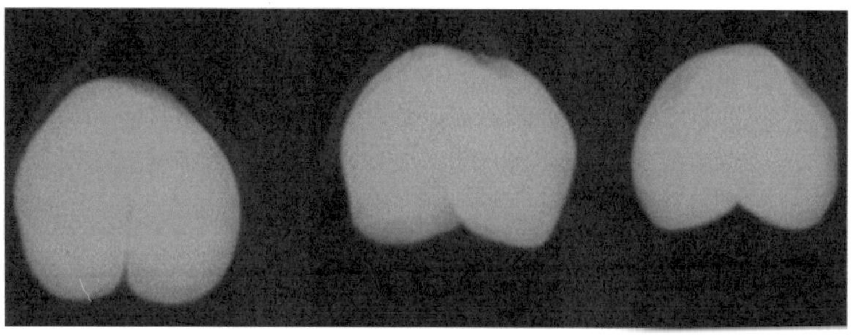

Figure 25. Var. Gorkovskaya 20. Up-down view: Projection of the "worm" type, slightly flattened, so that the width is a bit larger than the height. Before entering the groove - a narrow and low depression (a little groove along the kernel).

4.4. The fourth morphophysiological type

This type includes late-maturing varieties of spring wheat of Western European breeds. The length of growing season is 120-130 days (Henkel, 1969), with a prolonged first stage of development. Under conditions of long day and high light intensity, plants of the fourth type respond with accelerated transition through the second phase (stages III-IV of organogenesis). However, they may be slow to develop in conditions of low light intensity at considerable cloudiness and 14-15-hour day, and in the early stages – also under conditions of a shorter photoperiod. Slow development in such conditions, with good moisture availability and high rates of fertilisation in stages V-VII, causes the formation of large leaves, and of square-headed or even club-shaped forms of ear in the Western European group of varieties. Slow development at a sufficient water supply for plants in stages X-XI leads to the formation of large kernels with a high weight of 1000 kernels (38 - 45g or more).

Spring wheat varieties of the fourth type belong mostly to the forest-steppe and partially to the forest environmental groups. They are mostly prevalent in Germany, the Czech Republic, Slovakia, Denmark, Belgium, Finland and other countries, but compared to winter wheat they occupy small areas. This group includes spring wheat variety 'Peka' (Fig. 26-28.).

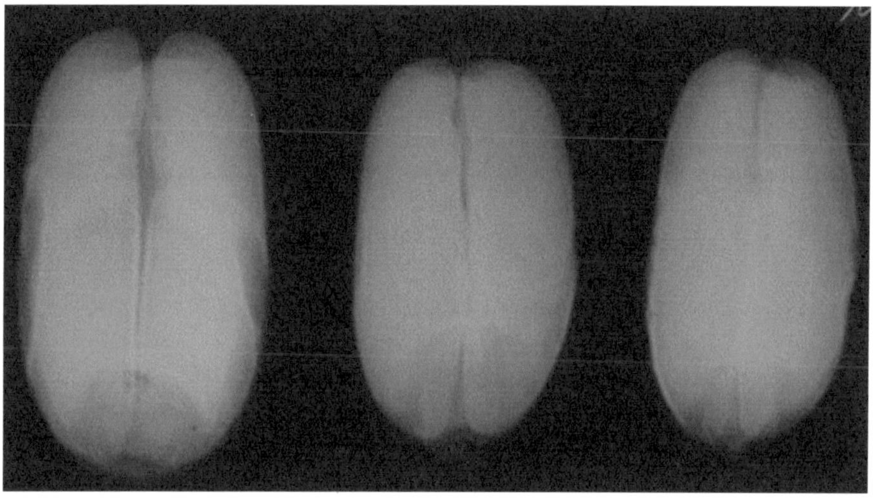

Figure 26. Var. Peck. Front view: Form close to rectangular. The edges are typical of small indentations, as if holes. Shade of the groove is well marked.

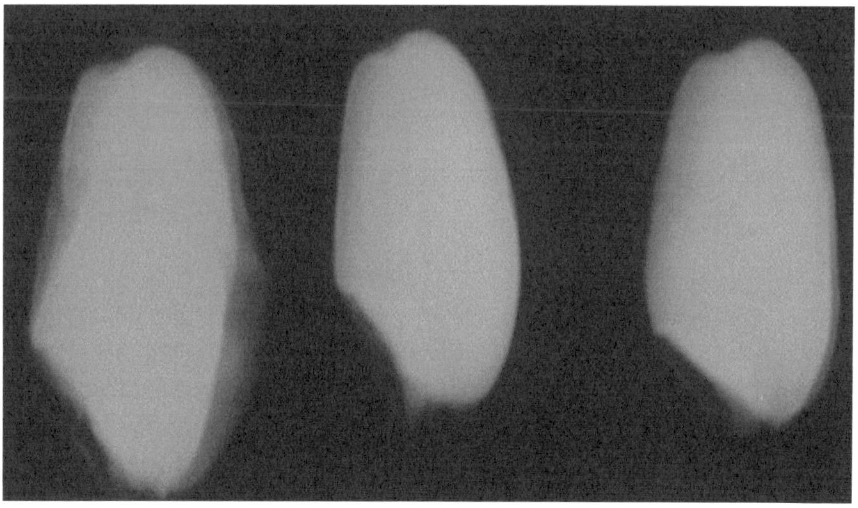

Figure 27. Var. Peck. Side view: As a rule, the contour of dorsal part of kernel is straight. Line of the ventral side is convex. The shadow of the groove can be seen, not in all kernels.

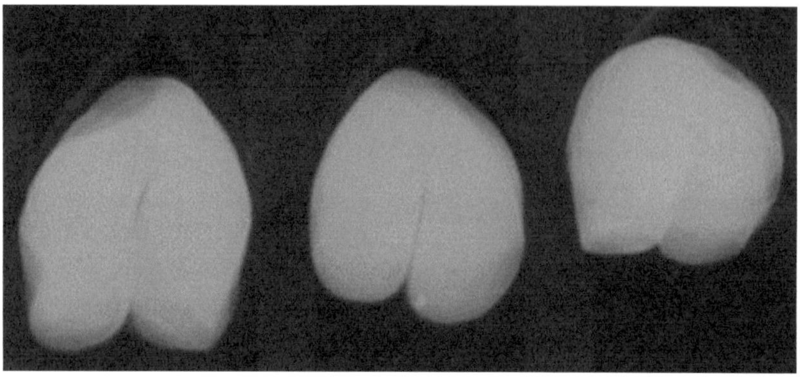

Figure 28. Var. Peck. Up-down view: A depression in front of the groove - either broad or a narrow one. Depending on this the folds are either rounded or sharp. Kernels with rounded folds are seemingly flattened (width greater than height); those with sharp folds have height greater than width, and angular.

4.5. The fifth morphophysiological type

This type comprises mid- and late-ripening varieties of the West Siberian breeding (Henkel, 1969). The specific features of climatic conditions - cold and dry April, May and first half of June, relatively high rainfall in late summer (July), low temperature in August, formed a special type of Siberian forest-steppe ecological forms of wheat.

The length of the growing season is 95-110 days. The slow development and long passage through stage II of organogenesis in the presence of favourable conditions for plant growth lead to increased tillering of plants of this morphological type. Delay in development at stages III-IV of organogenesis causes the possibility of forming an increased number of rudimentary spikelets. Delay at stage II of organogenesis permits significantly more efficient use of late summer rainfall for the formation of large ears and many-flowered spikelets. The ability of going through stages X-XII of organogenesis even at relatively low temperatures ensures the ripening of wheat in late August and in September. The long duration of stage II contributes to the formation of high foliage of plants. The leaves are large, dark green, and with medium- and strong pubescence. No grain of cultivars representing this type was obtained for analysis.

4.6. The sixth morphophysiological type

This type includes early-maturing varieties developed in Central Asia, mainly in the conditions of both spring and autumn sowing under periodic watering, less often in rain-fed crops (Henkel, 1969). They are distinguished by their resistance to soil and air drought, especially in stages II-V and X-XII of organogenesis. They include winter and spring forms, as well as transient forms (spring and winter). In Central Asia, Iran and Afghanistan, high temperatures and direct solar radiation during stages VI-VII of organogenesis often lead to complete closure of the glumes, complicating threshing, and to the development of mechanical tissues in glumes and awns.

The rapid passage through stage II of organogenesis at high temperatures and intense direct solar radiation leads to a sharp decrease in the number of leaves and stem internodes formed at this stage, and to a reduction in the process of tillering.

In stages IV-VI of organogenesis, the plants need watering to ensure normal moisture content of the generative organs. Stages XI-XII proceed normally at high temperatures.

No grain of cultivars representing this type was obtained for analysis.

4.7. The seventh morphophysiological type

This type incorporates spring wheat cultivars with spring and autumn sowing times (?) and intermediate (semi-winter) forms of wheat (Henkel, 1969). These are local and breeding varieties of Georgia, Azerbaijan, Tajikistan, the Mediterranean countries, Ethiopia, etc. They are characterised by a relatively short first phase of development (stages I-II of organogenesis) and medium duration of the second (light) phase (stages III-IV of organogenesis), which plants can normally go through at 14-15-hour day length. In the case of vernalisation of sowing material they require from 15 to 30 days of low temperatures; in the vegetative state the first phase of development of many of them will be completed within 10 to 15 days. Early warm spring and rare cases of the return of cold weather permit the transition to stages III-V of organogenesis under short-day conditions. Delay in development caused by short day in the period of differentiation of the embryonic ear promotes the formation of multi-flower spikelets, and subsequent delay in stage V under conditions of still a relatively short day and high moisture content due to precipitation of spring and the first half of the summer ensures synchronous development of multi-flower spikelets.

Favourable conditions for photosynthesis, large number of leaves (8 - 10 leaves) and high moisture content in stages X-XII of organogenesis ensure the growth and ripening of grains which, in this group of wheat cultivars, reach extreme sizes and weight. No grain of cultivars representing this type was obtained for analysis.

4.8. The eighth morphophysiological type

This type includes wheat varieties formed under the conditions of the Crimea, southern Ukraine, Moldova, Georgia, Armenia, Azerbaijan, South Yugoslavia, Bulgaria, India and other areas with a relatively mild winter (Henkel, 1969). In the vegetative state (in the phase of emergence and tillering), they can pass the first phase of development (stages I-II of organogenesis) at temperatures of +7, +12° C. Therefore, they can quickly move on to stages III- IV of organogenesis in early spring at 12-13-hour day length. As in the beginning of the formation of rudimentary spike there is a certain delay in stages IV-V of organogenesis, so they can form multi-flower spikelets, with simultaneous development of flowers. In conditions of late spring and rapid rise of heat, in many varieties the flowers in upper spikelets may be underdeveloped. In such years the form of the ear is close to the spindle, in the middle part of the ear 3-5 fertile flowers develop, in the top part - 1-2 flowers. In this connection, depending on spring conditions, spike length and number of kernels in ear vary dramatically. The ability of wheat varieties of this type to develop at relatively high temperatures in the first phase and to accelerate their passage through that phase under the effect of low temperatures, as well as their ability to begin the second phase in short-day conditions and dramatically accelerate the development at lengthening photoperiod, resulted in a high flexibility of those varieties. No grain of cultivars representing this type are available.

4.9. The ninth morphophysiological type

This type includes wheat varieties with seed material vernalisation for periods from 40 to 85 days (Henckel, 1969). Beginning in spring and later, the development of plants of this type is similar to the development of varieties of the second morphophysiological type. Representative of this type is variety Mironovskaya 808 (Fig. 29-31.).

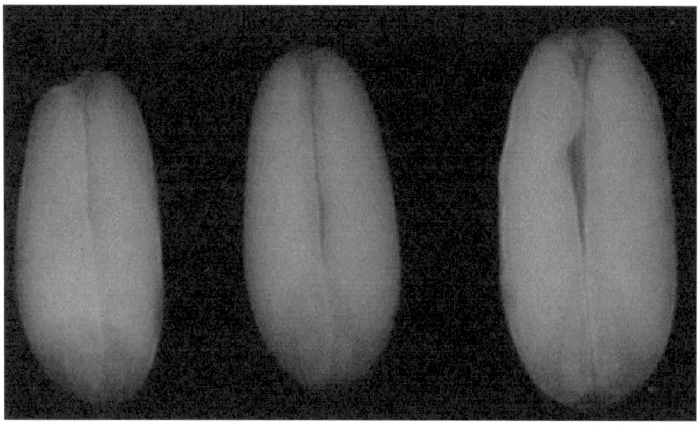

Figure 29. Var. Mironovskaya 808. Front view: Projection elongated, often close to an ellipse, sometimes with a slight thickening towards the embryo end of the kernel. Expansion of groove in the middle due to the particular case of damage by *thysanos*. At the upper part of kernel the groove ends with a well-defined wedge-shaped shadow. Along both sides of the groove rather broad shadows.

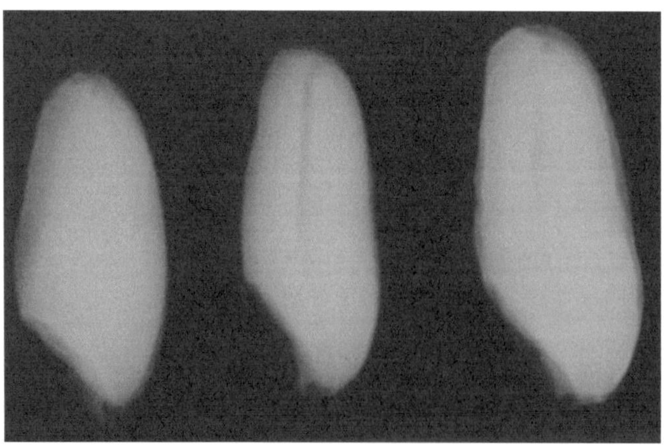

Figure 30. Var. Mironowskaya 808. Side view: The profile of the elipse is narrower with a clear thickening of the bottom. Thye line of the dorsal view is straight line, and at the ventral side – convex, some – with a direct plot in the middle. Section of the embryo – a straight line. Almost always seen the shadow line of the bottom grooves.

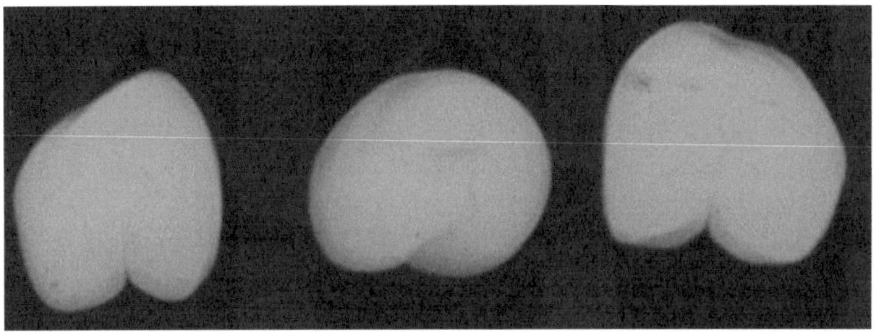

Figure 31. Var. Mironowskaya 808. Up-down view: Asymmetric :round loaf". Distinct broad but not deep ditch in front of the groove itself.

4.10. The tenth morphophysiological type

This group includes winter wheat varieties cultivated in Western Europe and the Baltic states, in the Leningrad region and adjacent areas of the Russian Federation (Henkel, 1969). Since the spring, with the transition to stages III-IV of organogenesis, the development of wheat varieties of this type is similar to the development of varieties of the fourth morphophysiological type. They also form square-headed and club-like ear, late developing and well leaved.

5. Identification of seed appurtenance to variety

The objective of the study was to explore the possibilities of application of X-rays to determine with the morphological features characteristic of wheat varieties. It is proposed to use chosen characteristics for the selection of seeds of regional varieties, which will bring them to the highest sowing condition, prolonging the active "life" of the varieties, as well as to consider those characteristics as a highly efficient tool in plant breeding.

For the study of the varietal morphogenetic specificity we chose the wheat Mironovskaya 808 (in 2004 it was replaced in the registered by the variety 'Volgaskaya 16'), derived from the "spring wheat in winter wheat" cultivars under the guidance of an outstanding breeder of cereals, Academician V.N. Remeslo. The variety is extremely interesting in that under field conditions it allowed, in the central zone and southern regions, to harvest up to 56 quintals per hectare, that is at least twice the average value of the current level of productivity. Mironovskaya 808 is classified as soft winter wheat, morphotype IX, subtype "c", with a long vegetative period of 290-305 days. It was derived through repeated mass selection of morphologically homogeneous plants from initial material obtained through directional modification of the spring wheat 'Artemovka' into a winter form. Group selection of 11 morphologically homogeneous and highly productive plants of the third progeny was the beginning of the variety Mironovskaya 808 (Remeslo, 1977). The leading varieties, in terms of yield and grain quality, arc the winter wheats, Bezostaya 1, and Mironowskaya 808, and the spring wheat Saratowskaya 29 (Grundas & Wrigley, 2004 a).

The variety Mironovskaya 808 is characterised by a broad morphological diversity. The differentiation of its forms bears the character of a separate genotype. These differences are comparable to the differences between individual varieties of wheat. This finding echoes the results of a recent study in the VIP. Among 230 varieties of spring wheat from Asia and Africa, twelve (K-202015 in Egypt; K-43720, K-43730, and K-55728 in Iraq; K-14317, K-14333, K-38598, K-38674, K-38675, and K-60213 in Iran; K-55 733 in Syria, and K-44513, Ethiopia) also proved to be morphologically heterogeneous (Mitrofanova, Wael Al-Youssef, 2008). It was shown that wheat may contain in its composition intravarietal groups of plants whose differences, in terms of their qualitative and quantitative characteristics, are comparable to the intervarietal ones.

Within the scope of the problem, attention was focused on factors affecting the length of the growing season as an approach to the control of subsequent change of form. The morphological varietal specificity of cv. Mironovskaya 808 was studied on combinations of loci Vrn1 - Vrn3, each of which has its registration number in the collection of wheat in the VIP, St. Petersburg, Russia (K-60657, and K-60662). Mironovskaya 808 belongs among the strong varieties of wheat, as a transient variety (Stelmach, 1987).

The variation of ripening time for various Vrn- and Ppd- genotypes in terms of their heading time averaged at a multiple of ±4 days. The distribution by rank of early ripening (acc. to Vrn) was as follows:

1. Vrn1 Vrn2 Vrn3;
2. Vrn1 Vrn3;
3. Vrn1 Vrn2;
4. Vrn1;
5. Vrn2 Vrn3;
6. Vrn3;
7. Vrn2.

The difference in speed of heading between the extreme variants amounts to an average of 19 days. The speed of transition from the second to the third stage of organogenesis was the maximum for Vrn 1 and the minimum for Vrn 2, and in the case of Vrn 3 it was at a medium level for the spring genotypes. This leads to the conclusion that the final stages of organogenesis in plants of various genotypes take place in various periods of vegetation. The duration of those periods and the climatic conditions during those periods determine the number of forming ovules of the reproductive organs.

The differences in the duration of the vegetation period for various genotypes are reflected in the varied form of kernels. This study had the objective of estimation of specific morphological indicators on the basis of variation in the duration of the vegetation period. X-ray images of kernels permit accurate recording of their contours in a form in suitable for identification. Using this method we can register the kernel in three planes, each of which carries its own specific information about its form, suitable for identification.

For morphological classification of cereal grains it is proposed to distinguish two levels of X-ray magnification: the first – image magnification by a factor of 2-10, which allows to study the morphology of kernels as a whole, and the second – magnification of images to x20-x40, to identify variety-specific features of particular elements of the embryo, or signs of kernels damage by mycosis.

The formalisation of the differences for computer identification must be based on the use of a method of recognition of "model of elementary figures" and their combinations, characteristic for wheat kernels.

To detect the presence of fungi in the caryopsis it is necessary to apply image magnification of x27-x30. Fig. 32. presents clearly visible fungal filaments (hyphae).

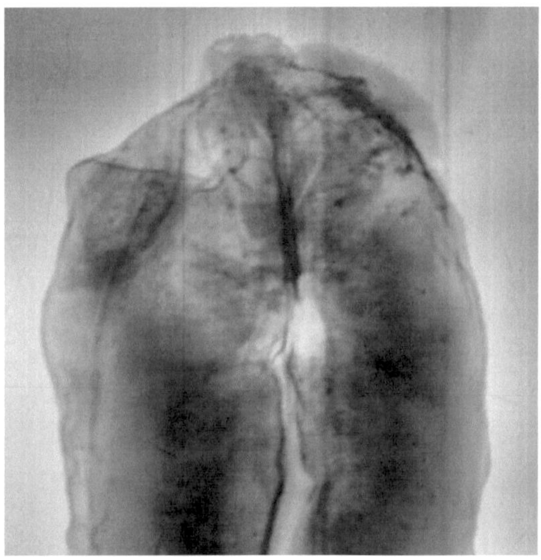

Figure 32. The presence of fungi in the caryopsis of wheat.

6. Features of intravarietal morphological variability

Despite the importance of morphology in breeding work, in estimates of dispersion of the parameters of morphological characteristics of varieties in generations no quantitative studies of the dependence of the shape of kernels on the properties of the parent and daughter plants were previously systematically conducted. The method of X-ray analysis allows us not only to assess mathematically the shapes as a whole, which in itself is important, but also to study the relationships of internal structures of kernels.

Figs. 33-38 present the main and typical forms, identified as characteristic for the variety Mironovskaya 808.

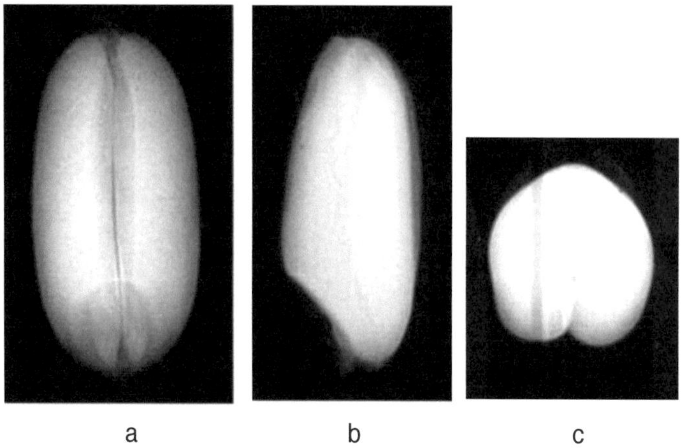

Figure 33. Morphotype 9 (K-60657). Front view (a): The side walls are almost parallel, slightly convex. The ends are approximately the same curvature. Side view (b): The surface from the groove side is flat. Up-down view (c): Close to the shape of a circle ("high round loaf").

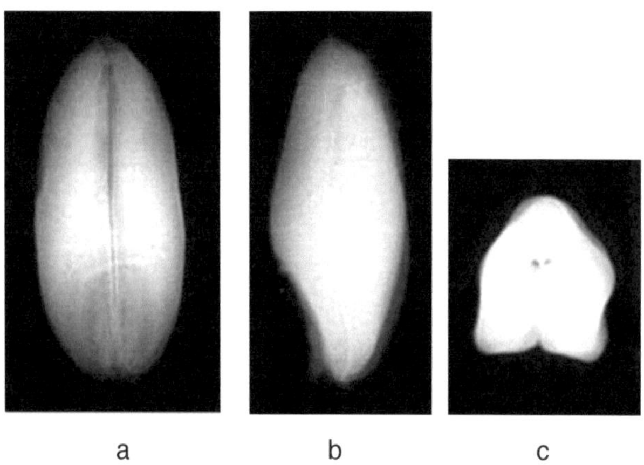

Figure 34. Morphotype 10 (K-60658). Front view (a): Narrowing towards the top of the kernel. Side view (b): The surface of the groove is slightly convex. Up-down view (c): Outline of this shape is rectangular ("house"-type).

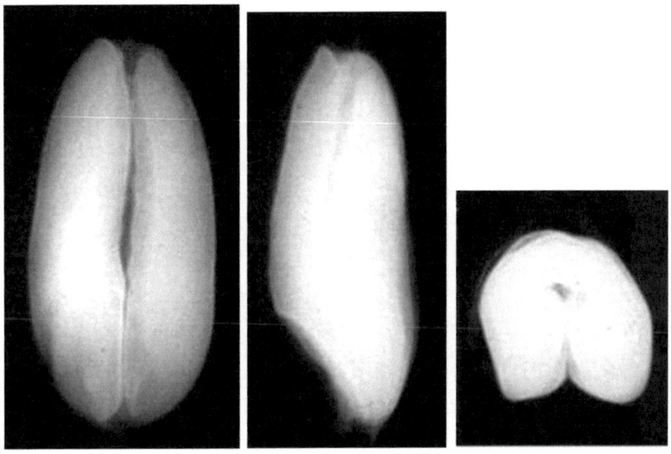

a b c

Figure 35. Morphotype 10 (K-60658). Front view (a): Narrowing towards the top of the kernel. Side view (b): The surface of the groove is slightly convex. Up-down view (c): Outline of this shape is rectangular ("house"-type).

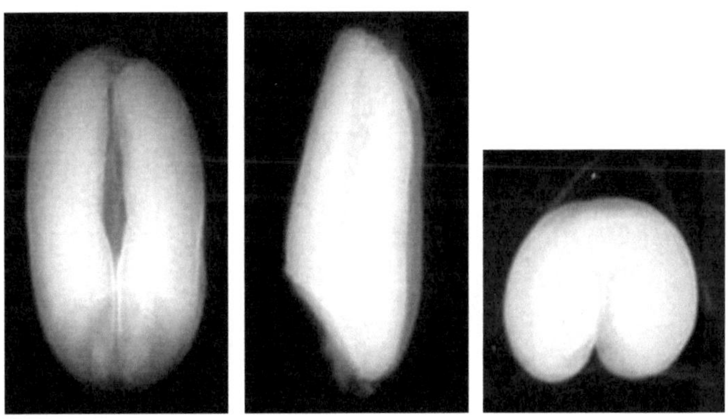

a b c

Figure 36. Morphotype 12 (K 60660). Front view (a): The side lines are parallel, the ratio of length to width is lower than that of morphotype 9 (kernel shorter). Side view (b): The surface of the groove side is slightly concave, from the back side the top is sharpened. Up-down view (c): The ratio of height to width of the projection is lower than in the previous morphotype ("flattened round loaf").

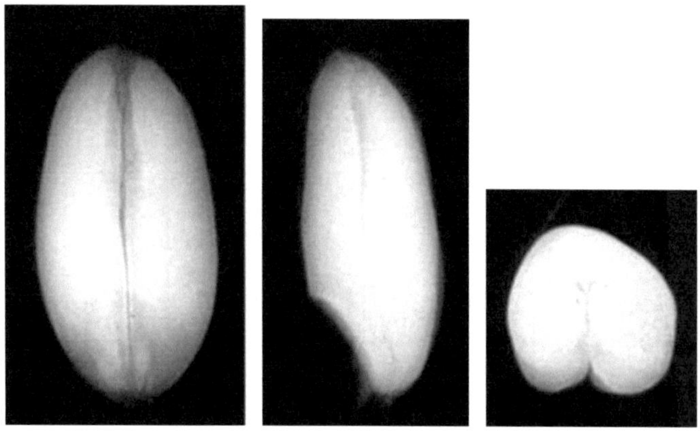

a b c

Figure 37. Morphotype 13, (K-60661). Front view (a): Rounded ends with approximately the same radius, but maximum diameter of the kernel dropped below the middle of the kernel. Side view (b): The surface of the groove side is convex, tapering at the upper end of the centre, on the extension of the projection small grooves, bending towards the back. Up-down view (c): Type "low loaf with a slightly pointed top".

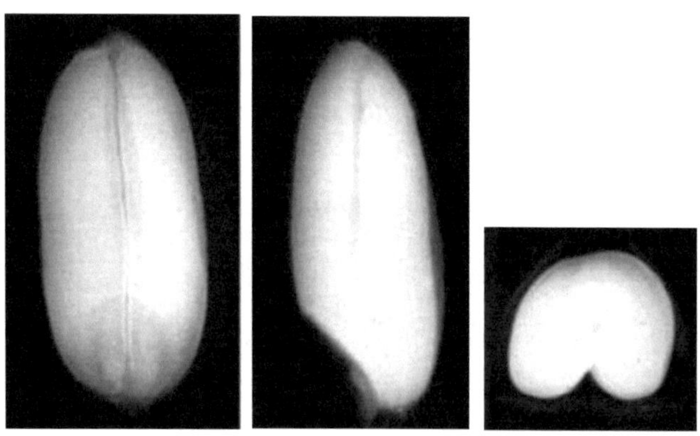

a b c

Figure 38. Morphotype 14 (K-60662). Front view (a): Lateral line projection is slightly concave, in contrast to morphotype 9. Side view (b): The surface of the groove side is flat, slightly concave. The upper end is tapered symmetrically. Up-down view (c): Top contour is rounded with height to width ratio greater than that of morphotype 12.

It was found that in the morphology of kernels and varieties, and in intravarietal variations, there exists a dispersion that is not taken into account in the varietal identification. This variance/dispersion is due to the position of kernels in the ear and to inevitable loss of varietal properties in reproduction of seeds for production sowings. The proposed combination of analytical and X-ray techniques can be regarded as a method of fine separation to ensure varietal purity and to meet industrial requirements for seeds of high productivity.

Analysis of the profiles of typical representatives of kernels allows the following results to be presented for discussion:

1. The forms of kernels from an intravarietal group, with conditionally the same Vrn features and also the same ID number, will be different.

2. Various Vrn groups contain kernels with a similar form, which confirms the need for their purity selection already at the stage of accepting a variety for inclusion in the collection.

3. X-ray images of kernels from a group with an individual number in the collection can be the basis for a description suitable for recognition.

In this study the identification was conducted on the basis of geometric similarity of representatives of morphotypes. Primarily, a differentiation was established with relation to the size of kernels – with the formation of a longer spike the kernels formed are considerably larger. In the selection of seeds, apart from checking for defects, seeds with the highest length/width ratio are chosen. The decision on the selection is based on prior calibration performed on a test batch of kernels.

To determine the form of geometric similarity the following algorithm was applied:

1. Projections of every kernel were set apart and described by surface area corresponding to the area of the projections;

2. The projections of areas in the kernel samples areas were situated in the same orientation of each projection and were scale calibrated to match one of the characteristic linear dimensions;

3. We evaluated the percentage differences on the line of maximum discrepancy of figures on the common surfaces (see Fig. 39-41.).

It was found that the differences in the selected morphological representatives of a variety, at least in one of the projections, amounted to no less than 10%. Assessment of morphological differences was performed using the following formula

$$(100 \times (L_{K.X} - L_{K'.X'}) / L_{K.X}),$$

where: L - the line of maximum divergence of calibrated figures, describing the compared kernels; K and K' - indicate the type of projection and can assume the values of F (Front view), S (Side view) and U (Up-down view); X and X' - indicate the sequence number allocated to identify the morphotype.

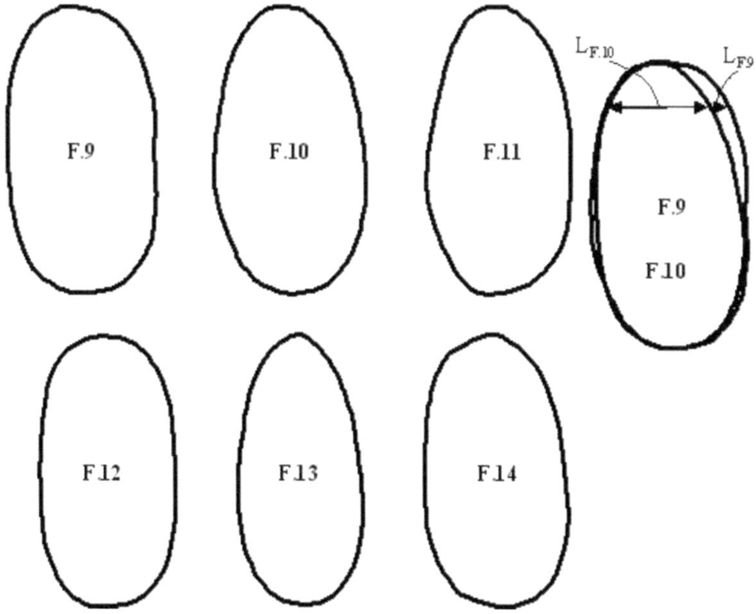

Figure 39. Forms of geometric similarity of wheat var. Mironovskaya 808. Front view (F): To morphotype 9 – F.9, to morphotype 10 - F.10, respectively, etc.

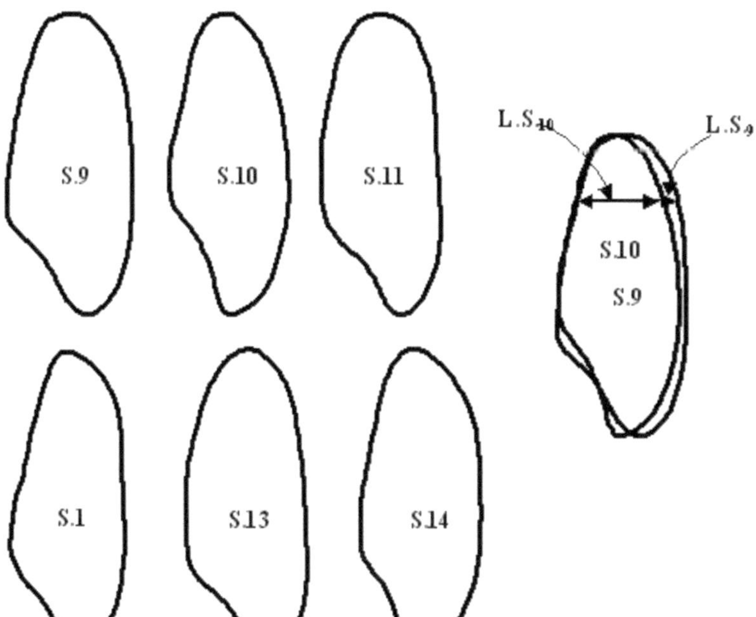

Figure 40. Forms of geometric similarity of wheat var. Mironovskaya 808. Up-down view (U): For morphotype 9 - U.9, to morphotype 10 – U.10, respectively, etc.

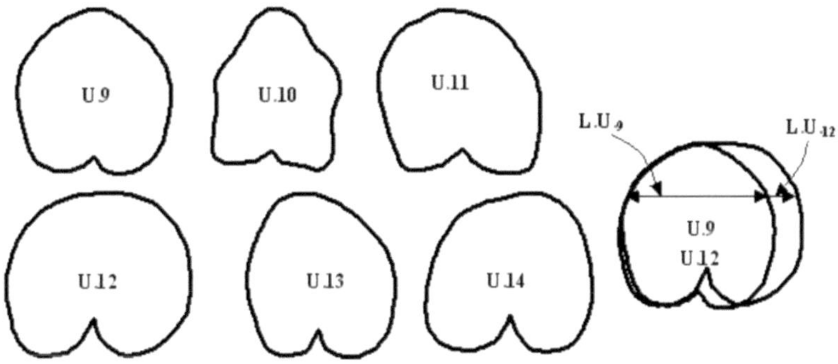

Figure 41. Forms of geometric similarity of wheat var. Mironovskaya 808. Up-down view (U): For morphotype 9 - U.9, to morphotype 10 – U.10, respectively, etc.

7. X-ray inspection of grain

The quality of grain is determined through characterisation of more than 20 of its physical, biochemical and technological properties. Among the approximately three dozen methods for the evaluation of grain quality, the position of the X-ray method (XRM) is more than modest. This method is included in the standard as a method of assessing the rate of infestation and populations of grain insects, including quarantine species (Varshalovich, 1958). XRM allows, without destroying the object, to identify its internal structure and make it observable and accessible for qualitative and quantitative evaluation (Grundas, Velikanov, 1998; Grundas, et al., 1999).

Using XRM we can detect many defects in the internal structure of grains, significantly affecting their quality, such as:

1. Fractures or cracks caused by both natural and technological factors;
2. Damage caused by the sunn pest (*Eurygaster maura*);
3. Occurrence of enzymatic-fungal damage/weakening;
4. Damage to the embryo, of various nature and extent;
5. Infestation with insects, including the earliest stages of larval development;
6. Occurrence of internal sprouting, which had begun in the field or in a pile and was stopped by drying;
7. Infestation and damage by fungi.

All of these defects reduce the quality of grains as seed and as a raw material for processing, and in some cases make the kernels unsuitable for planting or for processing (the presence of insects, embryo broken off).

XRM permits determination of the physical parameters of individual kernels as well as of kernels in bulk or heap (size of kernels, grain nature, the presence of impurities, etc.). In recent period also other desirable capabilities of the XRM are being discovered in its application for analysis of grain quality. Over the past 20 years a lot of attention has been devoted to the XRM by research teams in Russia, Poland, and other East European countries (Velikanov, et al, 1994; Velikanov, et al., 2008; Demyanchuk, et al., 2011; Grundas, et al., 2011).

The results of research carried out by using XRM showed significant differences in grain endosperm cracks between common wheat varieties. Natural wetting of dry grain (below 15% of moisture content) during rainfall when wheat is standing in the field is one of the reasons of its cracking. The susceptibility of wheat grain to mechanical damage is determined by genetic factors (e.g., grain hardness), environmental effects (climatic conditions during preharvest period), and by the conditions of grain storage (especially excessive humidity). The combination of these properties determines the quality of grain material for industrial purposes (Grundas & Wrigley, 2004 b).

Given below are some examples of X-ray images of kernels of basic cereals with internal defects of different nature, with comments:

The kernel can acquire a fracture or a crack in the field during its ripening, due to diurnal changes in temperature and humidity, as well as in the processes of harvesting, drying and transportation of grains (Fig. 42.). Strongly expressed fractures decrease both the sowing quality of seed and the technological properties of grain (oxidation of reserve substances, reduction of allowable storage time, impossibility of obtaining good quality flour).

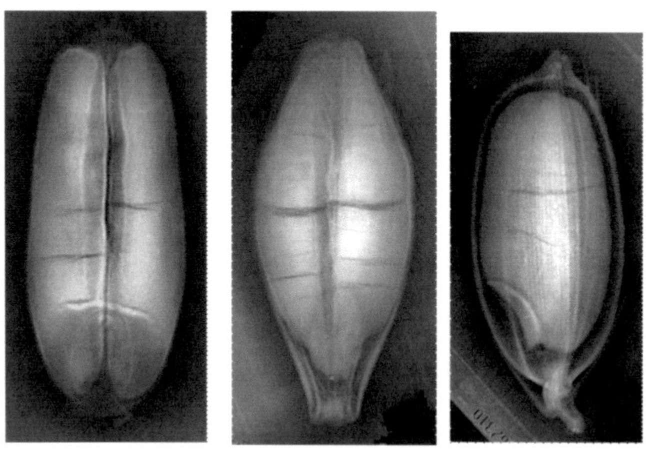

Figure 42. Endosperm cracks in wheat, barley and rice, respectively.

Complete or partial loss of the embryo (for comparison, in the far left picture - wheat kernel with a normal embryo) in kernels of any cereal (in the photo - wheat and rye) leads to non-germination of seeds, and for grain - to damage to seed coat, opening access to air and malicious agents, if that is not done on purpose, within the scope of process conditions directly before milling (Fig. 43.).

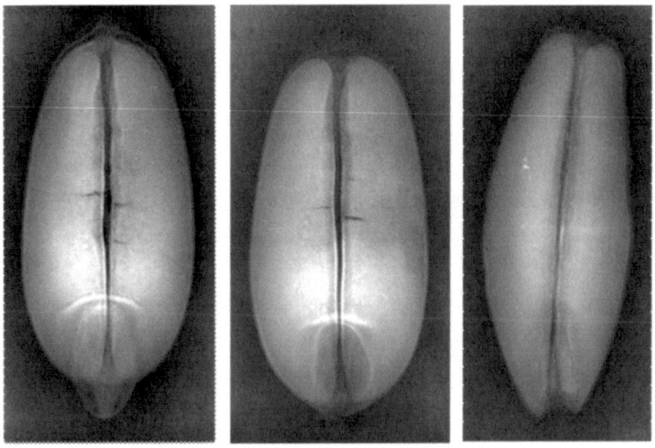

Figure 43. Complete or partial loss of the germ in wheat and rye seeds, respectively.

At high humidity and air temperature, grain can germinate in the field while still standing or in a heap, or in storage (Fig. 44). The initial stages of germination, identified on X-ray images, cannot be detected visually, but the biochemical and morphogenetic changes in kernels reduce their sowing and technological properties (Grundas, 2004; Bechtel, et al 1990). In hulled varieties even advanced internal germination can be visually observed (far right image of two kernels of oats, taken at lower magnification - in the kernel on the left the length of the germ, hidden by the seed coat, almost equal to the length of the endosperm).

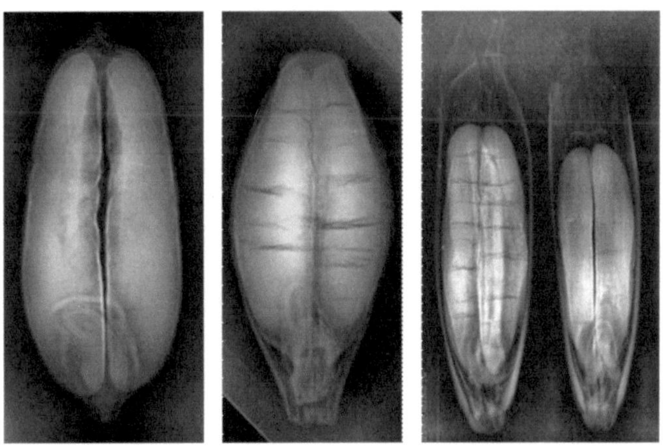

Figure 44. Hidden germination of wheat, barley and oats seeds, respectively.

Cavities, eaten out by insect larvae inside the kernels, are easily detected on the images in the form of characteristic darkened areas (Fig. 45.), regardless of their size (two images of rye kernels on the left). Repeated taking of an image in a few seconds makes it possible to identify live insects in the kernel (two images of wheat kernel on the right). The presence of insects in the grain is not allowed (Varshalovich, 1958; Nawrocka, et al, 2010; Nawrocka, et al, 2012). Selective visual analysis is unreliable. Only X-ray selection can ensure the absence of infected kernels.

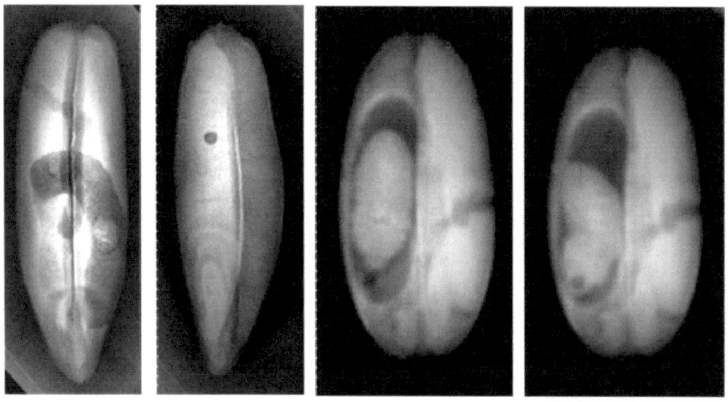

Figure 45. Infestation and populations by granary weevil (*Sitophilus granarius*) in kernels of rye, rice, and wheat, respectively.

Typical moire-effect darkened patterns on kernel images clearly indicate loosening of the endosperm tissue, resulting from the introduction of active hydrolytic enzymes of the sunn pest (Fig. 46.). The presence, in a batch of grain, of 2-3% of kernels with this kind of damage means a change of wheat grain classification from strong to weak one. Visual analysis leads to an underestimation or overestimation. X-ray analysis makes it possible to better quantify the damage.

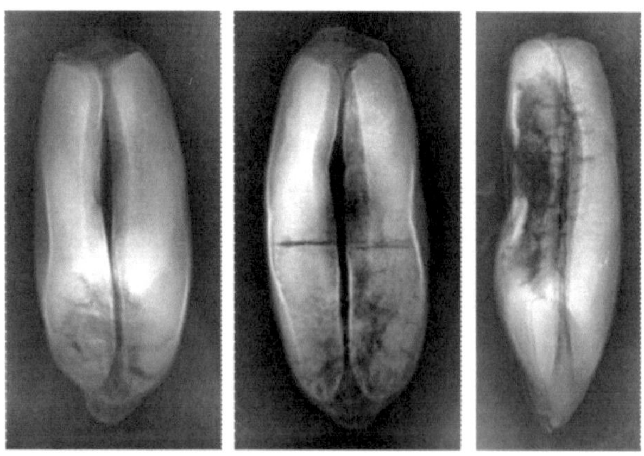

Figure 46. Wheat and rye kernels damage by sunn pest (*Eurygaster integriceps*)

Loss of tissue density in the lateral parts of kernels and along the grooves as a result of the activity of their own enzymes at high humidity in the field, and thus also of enzymes of fungi that have evolved on the surface of kernels, with the resultant hydrolysates (Fig. 47.). Visual identification is difficult, especially in hulled kernels. X-ray analysis is accurate, it is possible to quantify the defect. Kernels with symptoms of enzyme-mycosis damage have lower sowing and technological parameters.

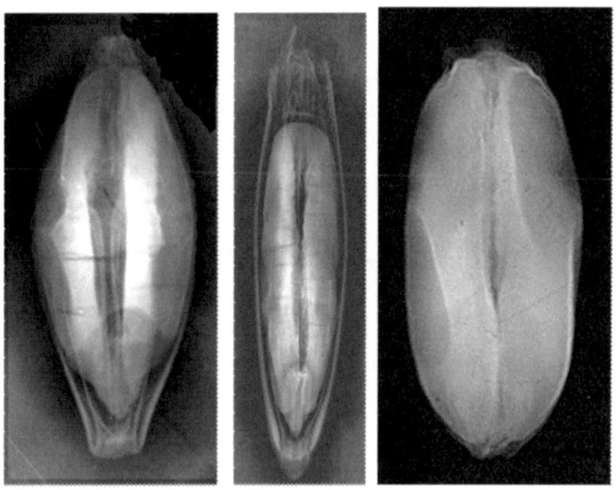

Figure 47. Mycosis-enzyme depletion of barley, oats and wheat kernels.

In spite of the considerable progress in the development of XRM, its application for the diagnosis of the quality of seeds and kernels remains extremely limited. So what is the situation with the whole issue of quality of seeds and kernels?

Great volumes of grain are evaluated for their sowing and technological properties on the basis of extremely small samples. Based on long years of research, the size of these samples (2 kg) is sufficient for a reliable assessment of the material. The risks are great, especially on the grounds of parameters with minimal thresholds of acceptability, such as the presence of insects or the percentage of kernels damaged by the sunn pest. Quantitative evaluation of kernels on the basis of those indicators may be over- or underestimated, and decisions are made on the fate of the entire batch of grain. Getting the diagnosis of the state of a batch of grain in the present conditions has no effect on changing this state in the case of its failure. Such a diagnosis can be the basis for lowering the price, the decision can be taken to treat it with insecticides, the buyer may decide to utilise the grain, using improving additives, but the quality of the original batch has no prospects for improvement. The situation can change radically only through separation based on a critical parameter, that is the presence of insects. However, without the X-ray method such a separation is impossible, because with the current standard diagnosis the source material is destroyed.

The large number of features reliably detected by XRM already today sets this method apart from other methods. In spite of the undeniable advantages, X-ray techniques are limited by two factors. First, identification of the variety and quality assessment of grain often requires assays of proteins and other biochemical parameters of grain. Second, the techniques should allow total control (Demyanchuk, et al., 2007). Today all the prerequisites for the solution of these problems are available.

Thus, as far back as in 1953 Watson and Crick (Watson & Crick, 1953), by comparing the data from X-ray analysis with a cardboard model based on them, determined the structure of the double helix of DNA, for which they were awarded the Nobel Prize. The biochemical diagnostics required for the assessment of quality of kernels and seed can be made using the methods of X-ray structural analysis. This results from the possibility of studying the atomic structure of matter using X-ray diffraction. From the diffraction pattern one can determine the electron density distribution of matter, and based on that the kind of atoms and their arrangement. X-ray structural analysis can determine the structure of crystals, liquids, protein molecules, etc. In order to obtain images of large molecules with atomic resolution rays with shorter wavelengths are applied, i.e., hard X-rays rather than soft.

The solution of the second problem requires rapid analysis of large volumes of grain material. Usually, to ensure the sharpness of images in X-ray filming, the object is translucent only in the period of exposure. For this purpose, the control grid of X-ray tube X-ray machine is fed from the switch current pulses associated with the mechanism of filming apparatus. In this way, by using X-ray tube with cold emission, times of exposure and 10^{-7}s at a frequency of 100 frames per second can be achieved. Currently achievable filming rates are from a few thousand to 100 thousand frames per second, at exposure times of up to 15ns. Adaptation of these techniques of diagnostics to the needs of the selection/separation of kernels will almost completely eliminate poor-quality material. The transition to the industrial application of X-ray separation of grain, from a few kilograms per hour up to several tons per hour, will initiate a review of the whole range of concepts (Demyanchuk, et al., 2007). To address the issues of arbitration it is not enough to say that we see indications of quality, variety or defects. The limit values of parameters should be determined quantitatively and validated by specifically developed regulations. This means that the current problems of X-ray application require a radical revision of existing standards of quality of cereal crops. This will allow the move to full control of parameters of seed variety and kernel quality. This will require the mandatory inclusion of elements of rapid selection of seeds into systems of X-ray diagnostics.

Thus, the X-ray techniques have the potential of becoming a universal method of diagnosing and bringing the quality of original batches of grain to a high status.

http://dx.doi.org/10.5772/52236

8. Conclusions

1. The study showed that using X-ray images with direct X-ray magnification of 2-10 times and computer image analysis the variety-specific morphological parameters of cereal grains can be successfully determined.

2. The algorithm by which the plant is running self-building genome in ontogeny on the principles of complex systems is presented. Its high convergence with the experimental data gives the basis for the application of this approach to the consideration of the morphological characteristics of kernels as a tool for the identification of their morphological type and variety.

3. To identify the kernel, besides three dimensions, it is proposed to determine, by X-ray, the contour of kernels (seeds) in at least two projections. It is advisable to plan a continuation of the research with a collection of cereal cultures to create a database for the identification of kernels and for decision making on the basis of X-ray selection.

4. It is proposed to use the mass-scale X-ray separation of kernels for their selection on the basis of varietal and production characteristics, as reflected in their morphology, and for the elimination of contaminants, including kernels with internal defects, with a view to a permanent renewal and maintenance of resources of industrial high-quality varietal seed.

Author details

Alexander M. Demyanchuk[1], Stanisław Grundas[2] and Leonid P. Velikanov[1]

1 Laboratory of Seed Biophysics, Agrophysical Research Institute, Russian Academy of Agricultural Sciences, St. Petersburg, Russia

2 Bohdan Dobrzanski Institute of Agrophysics, Polish Academy of Sciences, Lublin, Poland

References

[1] Batygin, N., , F., & Demyanchuk, A. M. (1995). Calculation of the ontogeny of wheat (Guidelines). Vavilov Institute of Plants. St. Petersburg, Russia, 36 (in Russian).

[2] Batygina, T. B. (1974). Wheat embryology. Vavilov. Institute of Plants. Leningrad, USSR. Kolos, 204 (in Russian).

[3] Batygina, T. B. (1987). Bread grain. Vavilov Institute of Plants., Leningrad, USSR. Nauka, 203 (in Russian).

[4] Bechtel, D., Zayas, I., Kaleikau, L., & Pomeranz, J. (1990). Size distribution of wheat starch granules during endosperm development. Cereal. Chem. Vol.67, 59-63.

[5] Demyanchuk, A. M. (1997). Algorithms of plants morpho- and ontogenesis. Guidebook Vavilov Institute of Plants. St. Petersburg, Russia, 128 (in Russian).

[6] Demyanchuk, A. M., Velikanov, L. P., Arhipov, M. V., & Grundas, S. (2011). X-ray method to evaluate grain quality. Encyclopedia of Agrophysics. Ed. by Gliński, J., Horabik, J. & J. Lipiec. Publisher by Springer. Pp. 1003-1009. ISBN 978-9-04813-584-4.

[7] Demyanchuk, A. M., Velikanov, L. P., Weitzel, O. O., & Zaycev, A. C. (2007). A perfect result. Effectiveness Solutions in Agriculture. Library TRIZ-PROFI. Moscow. ver. 2.0, 16-17, (in Russian).

[8] Grundas, S. (2004). Characteristics of physical properties of wheat kernels in the ear (Triticum aestivum L.). Acta Agrophysica. Vol. 102 (in Polish).

[9] Grundas, S., Nawrocka, A., & Pecen, J. (2011). Grain physics. Encyclopedia of Agrophysics. Ed. by Gliński, J., Horabik, J. & J. Lipiec. Publisher by Springer. Pp. 323-327. ISBN 978-9-04813-584-4.

[10] Grundas, S., & Velikanov, L. P. (1998). Methodical and technological aspects of X-ray imaging of wheat grain. Book of abstracts of the 16th ICC Conference on "Cereal Science-Its Contribution to Health and Well Being". P. 31. Vienna, Austria. 9-12. 05. 1998.

[11] Grundas, S., Velikanov, L. P., & Archipov, M. M. (1999). Importance of wheat grain orientation for the detection of internal mechanical damage by the X-ray method. International Agrophysics. Vol. 13, 355-361.

[12] Grundas, S. T., & Wrigley, C. (2004a). WHET/Harvesting, Transport, and Storage. Encyclopedia of Grain Science. Ed. by Wrigley, C., Corke, H. & C. Walker. Published by Elsevier Ltd. Vol. 3, 347-355. ISBN 0-12765-490-9.

[13] Grundas, S.T., & Wrigley, C. (2004b). WHET/Ultrastructure of the Grain, Flour, and Dough. Encyclopedia of Grain Science. Ed. by Wrigley, C., Corke, H. & C. Walker. Published by Elsevier Ltd. Vol. 3, 391-400. ISBN 0-12765-490-9.

[14] Henkel, P. A. (1969). Physiology of agricultural plants. Vol. IV. Physiology of wheat. (Ed. by Henkel). Pp. 555. Moscow University.

[15] Kopus, M., Ignatiev, N., Vasyushkina, N. E., Kravchenko, N., & Kopus, E. M. (2009). Genetic polymorphism of amylolytic enzymes and genetics of wheat starch biosynthetic enzymes. Kalynienko All-Russian Scientific-Research Institute of Cereal Plants. Russian Cereal Agriculture. Vol. 4, 23-27. (in Russian).

[16] Larikova, Y. S. (2007). Morphological features of winter wheat grains (Triticum aestivum L.). DrSc Thesis. Pp. 22. (in Russian).

[17] Larikova, Y. S., & Kondratyev, M. N. (2002). Justification of morphological differences of wheat after sorting on the slotted sieve. Abstracts of the 3rd International Iran

and Russia conference "Agriculture and Natural Resources". Moscow, 18-20 September 2002, 89- 90.

[18] Mitrofanova, O., & Wael-Youssef, Al. (2008). Characteristics of the variability of morphological features of local varieties of soft winter wheat. Messages of Sankt Petersburg State Agrarian University, Russia. Vol. 11, 63-65 (in Russian).

[19] Nawrocka, A., Grundas, S., & Grodek, J. (2010). Losses caused by granary weevil larva in wheat grain using digital analysis of X-ray image. International Agrophysics. Vol. 24 (1), 63-68.

[20] Nawrocka, A., Stępień, E., Grundas, S., & Nawrot, J. (2012). Mass loss determination of wheat kernels infested by granary weevil from X-ray images. Journal of Stored Products Research. Vol. 48, 19-24.

[21] Remeslo, V. N. (1977). Breeding, seed production and varietal Agricultural wheat, Moscow, Kolos, 352.

[22] Schmalhausen, I. I. (1968). Integration of biological systems and their self-adjustment. Cybernetics questions of biology. Sibirsk Science. Sibirsk Section of Russian Academy of Sciences, Novosibirsk, , 157-182.

[23] Schrödinger, E. (1944). What is Life? The Physical Aspect of the Living Cell. Cambridge: Cambridge. University Press.

[24] Stelmach, A. F. (1987). Wheat breeding program for earliness. Messages of Agriculture Sciences. Vol. 6 (369), 47-53.

[25] Varshalovich, A. A. (1958). Guide to quarantine entomological expertise seeds by X-ray. Publisher USSR Ministry of Agriculture (in Russian).

[26] Velikanov, L. P., Archipov, M., & Grundas, S. (1994). Signification of kernel orientation related to direction of X-ray beam for evaluation of inner cracks. Abstract of poster session: Testing methods. P.6.8. 14th ICC Congress. The Hague, Netherlands.

[27] Velikanov, L. P., Grundas, S., Archipov, M. M., Demyanchuk, A. M., & Gusakova, L. P. (2008). Agrophysical direction of further development and application of X-ray method. Proceedings of international conference on "New Trends in agrophysics". Lublin, Poland, 147-148.

[28] Watson, J. D., & Crick, F. H. C. (1953). Molecular structure of nucleic acids. Nature. Vol.171, 738-740.

Chapter 11

Physical Properties of Seeds in Technological Processes

B. Dobrzański and A. Stępniewski

Additional information is available at the end of the chapter

http://dx.doi.org/10.5772/56874

1. Introduction

The physical properties of seeds and grain is a wide knowledge that can be useful in the farming, harvesting and storage or in processing such as drying, freezing and other. This knowledge is important in the designing of machinery to harvest and in preparation of processing chain from grain to food. Accurate design of machines and processes in the food chain from harvest to table requires an understanding of physical properties of row material. Considering either bulk or individual units of the agricultural material, it is important to have an accurate estimation of shape, size, volume, density, specific gravity, surface area, and other mechanical characteristics, which may be considered as designing parameters for food production. The measurement techniques allow computation of these parameters, which can than provide information about the effects of processing. Some of characteristics, such as colour, mechanical parameters, rheological properties, thermal and electrical resistance, water content and other physical quantities give excellent description of product quality.

2. Physical properties of agricultural materials

In a Newtonian sense, the physical properties of an object may include many varietals properties. Physical properties are determined for many raw materials and agricultural products, including seeds and grain, however, Barbosa-Cánovas et al. (2004) divided them into the following groups:

- Thermal properties such as specific heat, boiling point, conductivity, temperature, thermal transfer, diffusivity, and boiling point rise, freezing point depression.
- Optical properties, primarily colour, but also gloss and translucency.

- Electrical properties, electric charge, electric field, electric potential, primarily conductivity, resistance, dielectric, conductivity, impedance, inductance, magnetic field, magnetic flux, and permittivity.
- Structural and geometrical properties such as mass, density, particle size, shape, volume, length, location, porosity, surface roughness, and cellularity.
- Mechanical properties (including strength, ductility, tension, compressibility, pressure, deformability moment, and textural properties, malleability,) and rheological properties (such as viscosity, flow rate, fluidity,).
- Others, including mass transfer related properties (diffusivity, permeability), surface tension, cloud stability, gelling ability, and radiation absorbance.

Seeds in general can display large compositional variations, inhomogeneities, and anisotropic structures. Composition can change due to seasonal variations and environmental conditions, or in the case of processed grain, properties can be affected by process conditions and material history.

Early physical property analyses of seeds and grain required constant uniform values and were often oversimplified and inaccurate. Nowadays, computational engineering techniques, such as the finite element method, are much more sophisticated and can be used to evaluate non-uniform properties (for example, thermal properties) that change with time, temperature, and location in grain that are heated or cooled. Improvements measuring the compositions of seeds are now allowing predictions of physical properties that are more accurate than previously.

3. Physical methods for quality evaluation of seeds

It was found that radiation in the near-infrared region of the spectrum can provide information related to many quality factors of agricultural products. Short wave radiations such as X-rays and gamma rays can penetrate through most seeds. The level of transmission of these rays depends mainly on the mass density and mass absorption coefficient of the material, that it allow find internal cracks invisible because of the seed coat (Dobrzański et al, 2003).

Nuclear magnetic resonance (NMR) is a technique that detects the concentration of hydrogen nuclei (protons) and is sensitive to variations in the concentration of liquid in the material. Although NMR imaging (MRI) has been used frequently in the medical field and other quality factors in grain has not been fully explored. Chen (1996) presents an overview of various quality evaluation techniques that are based on one of the following physical property: density, firmness, vibration characteristics, X-ray and gamma ray transmission, optical reflectance and transmission, electrical properties, aromatic volatile emission, and nuclear magnetic response (NMR).

One of the most practical and successful techniques for nondestructive quality evaluation and sorting or separating impurities is the electro-optical technique, based on different optical

http://dx.doi.org/10.5772/56874

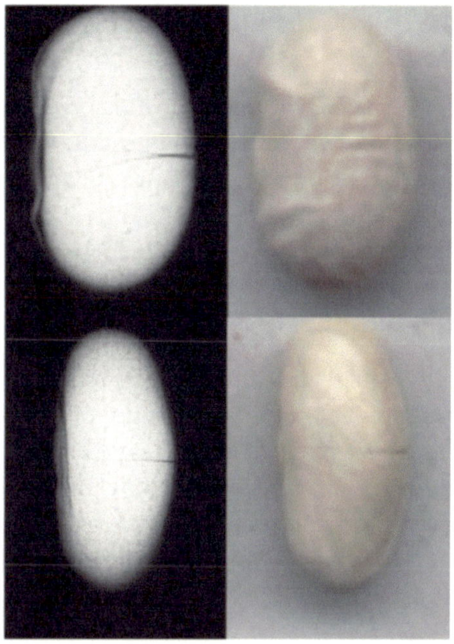

Figure 1. Invisible crack on the seed coat visible in X-ray image

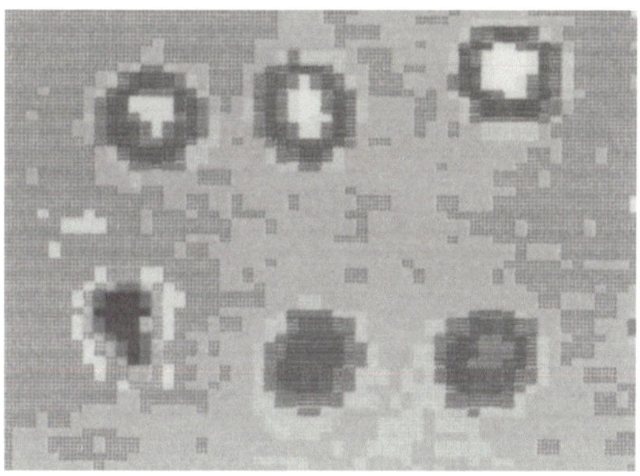

Figure 2. Thermal image of seeds after 3 hours of water soaking and imbibitions

properties of seed and impurities. Thus, determining such optical characteristics of seeds and grain can provide information related to quality factors of the product. Looking for more advanced method of germination capacity estimation, Dobrzański et al (2003) using thermal technique found slightly differences in temperature of dead and alive seeds, being soaked, that thermal properties should indicate live potential of seed in first hours of germination test. According to traditional germination test, that needs two weeks or more for forming germ (depends to species), thermography should be rapid technique for estimation of germination capacity.

4. Instrumental measurement of mechanical properties of seeds and grain

There are many types of mechanical loading: puncture, compression, shearing, twisting, extrusion, crushing, tension, bending, vibration, and impact. And there are four basic values that can be obtained from mechanical properties tests: force (load), deformation (distance, displacement, penetration), slope (ratio of force to deformation), and area under the force/deformation curve (energy). Mechanical characteristics included basic parameters (hardness, cohesiveness, viscosity, elasticity, and adhesiveness) and secondary parameters (brittleness or fracturability, chewiness, and gumminess).

The engineering terms based on these measurements are stress, strain, modulus, and energy, respectively. Stress is force per unit area, either of contact or cross-section, depending on the test. Strain is deformation as a percentage of initial height or length of the portion of sample subject to loading. Modulus of elasticity (tangent, secant, chord, or initial tangent) is a measure of stiffness based on the stress/strain ratio. Force and deformation values are more commonly used in food applications than stress and strain values, that are determined at seed loading.

Figure 3. Universal testing machine Instron 1253 for fatigue test and dynamic load

Generally, determination of mechanical properties based on the high precision measurement, with expensive equipment (Fig. 3 and 4) requires a complete force-deformation curve. From

http://dx.doi.org/10.5772/56874

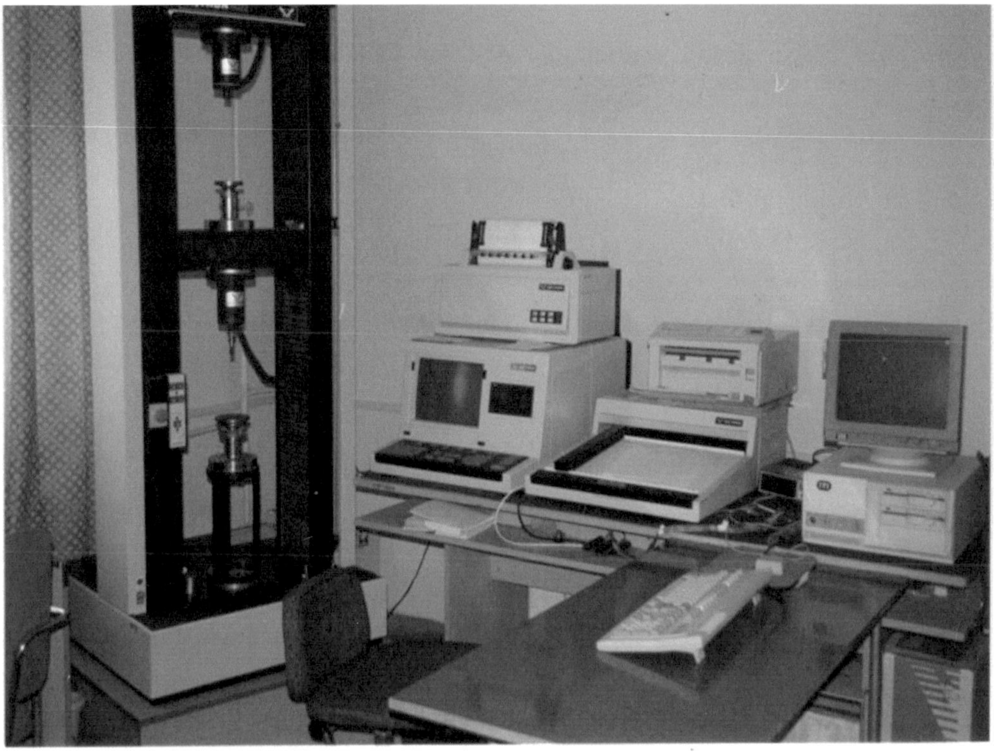

Figure 4. Universal testing machine Instron 6022 for quasi-static tests.

the force-deformation curve, stiffness; modulus of elasticity; modulus of deformability; toughness; force and deformation to point of inflection, to bioyield, and rupture, and maximum normal contact stress or stress index at low levels of deformation can be obtained. The force-deformation curve was frequently used in the study of physical properties of seeds and grain; however this measurement is not widely utilized in practice because of expensive and complicated procedures. A typical force/deformation characteristic for a cylindrical piece of tissue compressed at constant speed gives F/D curve for puncture tests look similar to compression curves. The portion of the initial slope up to point represents nondestructive elastic deformation; point is the inflection point where the curve begins to have a concave-downward shape and is called the elastic limit. The region before this point is where slope or elastic modulus should be measured. Beyond the elastic limit, permanent tissue damage begins. There may be a bioyield point where cells start to rupture or to move with respect to their neighbors, causing a noticeable decrease in slope (Abbott and Harker, 2003). (Dobrzański and Rybczyński, 2011 divided the measurements on direct (contact; compression, shear, impact, rebound) and indirect techniques (non-contact; vibration, sonic).

Seed hardness is related to the quality factors or behaviour of grain in relation to harvest, however, through use of simply test, only the maximum squeezing force has been frequently correlated with numerous quality factors. It is also well known, that including external and

Figure 5. Temperature chamber mounted to 8872 Instron machine (High Wycombe)

internal properties of seed, the hardness should depends to the shape and size of seed; size and contact area with plate; rate of deformation; the way of seed fixing, and other procedures influenced a final accuracy. It is well known, that all mechanical properties and behaviour of agricultural materials are influenced by moisture and temperature, but universal testing machine with temperature chamber (Fig. 5) must be equipped.

5. The cracking mechanisms of grain legume in technological processes

The models describing some geometrical relations for different cracking mechanism were main aim of presented paper. In this study, the verify of cracking mechanism needs to determine the elasticity strain limit of cotyledon and the resistance of the seed coat to tension for various legume seeds. The change of shape as well as the mechanical values of the seed coat and cotyledon at various moisture contents, during swelling and drying were determined. This approach lead to an understanding of the possible causes and mechanisms that affect the cracking of the seed coat. Further, a major problem in the drying of legume seeds with heated air is the splitting of seed coat; affected by the temperature and moisture. The estimation of the effect of drying on the stress and strain development of the seed coat was determined at wide range of moisture.

Legume seed quality is greatly affected by methods of harvesting and handling. In these operations, seeds are subjected repeatedly to many impacts on metal surfaces and against other seeds. Impacts may be result of threshing cylinder or rotor motion, the movement of buckets in vertical elevators, centrifugal discharging, the filling and discharging of screw conveyors, spouting and free-fall dropping (Palusen *et al.*, 1981 a,b). In nearly all handling operations, seeds are accelerated to some velocity and then discharged onto stationary objects or other grains.

Physical properties of agricultural products are needed for adequate design of processing machines but their specific applications should be clearly understood before determining it experimentally (Misra *et al.*, 1981). The modulus of elasticity of seed is a physical property which has been suggested as an evaluation of firmness and hardness (Mac Donald, Jr.,1985). It is also an important property for determination of the stress cracks in seeds. However, in study of stress cracking due to drying of cereal grains from about 25 percent moisture content to about 14 percent moisture content, it is essential to understand the variation of modulus of elasticity according to wide range of moisture (Dobrzański, 1998; 2011).

The moisture content of seed coat and cotyledon influences the cracking mechanism and many researches have already published the results of mechanical properties including a modulus of elasticity for various seeds at a given moisture content and deformation rate. However, in order to clarify the mechanisms of seed cracking, it is necessary to carefully recognise the mechanical properties of cotyledon and seed coat separately.

The major problem in the drying of legume seeds with heated air is the splitting of the seed coat (Liu *et al.*, 1989). Temperature and moisture gradients can cause stresses in the seed coat during drying. Overhults *et al.* (1973) observed the physical damage in soybeans during drying, in the form of indentations and cracks. This condition makes the beans susceptible to microbial attack during storage, and also reduces their germination potential (White *et al.*, 1980). In view of these problems, there is need, specially in the legumineous seed industry, to find the optimum combination of drying parameters which will minimize seedcoat cracking (Mensah *et al.*, 1984).

The following types of cracking mechanisms were observed during mechanical loading of the legume seeds:

- tension of seed coat caused by the shape deformation at axial compression,
- tension of seed coat caused by wet cotyledons sliding,
- shearing of dry seed coat caused by a force that smashed the seed in two cotyledons,
- crack opening displacement of the cotyledon under compression.
- splitting of the seed coat; affected by the high temperature and low moisture (a major problem in drying practice of grain legume). A rapid change of seed moisture during drying caused the shrinkage of seed coat leading to its stress. The splitting of the coat, caused by shrinkage, is frequently the most important mechanism of coat cracking.

6. Tension of seed coat caused by deformation of the seed compressed

With the increase of moisture content seed become very plastic and even slight forces, which do not produce negative biological results, may cause considerable deformations of shape.

The description of the shape deformations of the legume seed compressed between parallel plates accomplishment needs some simplifying assumptions:

- the seed is a sphere surrounded by elastic shell,
- the compressibility of the cotyledon is very low that the volume is steady,
- the seed coat thickness is small that in geometrical calculations could be ignored,
- The seed compressed between parallel plates is circular in initial phase, deforming further in to the barrel of parabolic shape at large deformation.

Using the assumption of constant volume, make it possible to determine the changes of surface area describing the spherical body being compressed between parallel plates. When the deformation is slight the sphere does not lose its shape and hence it was assumed that lateral surface of barrel created by compressing the opposite cups of sphere will also be spherical in its shape.

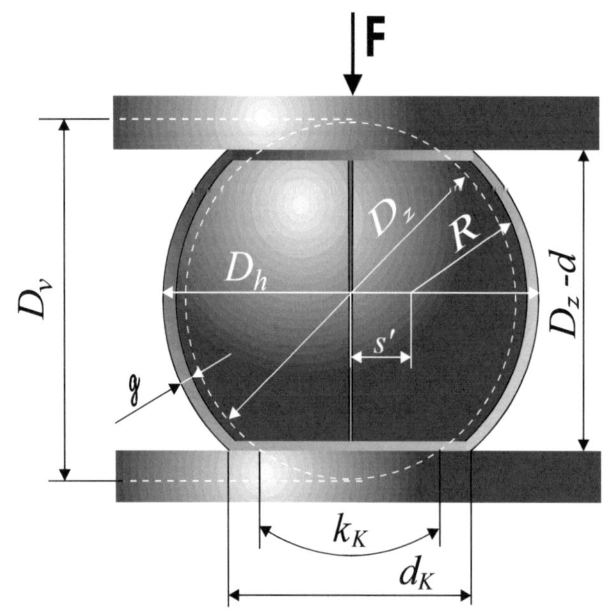

Figure 6. The shape deformation of seed at axial compression.

In initial phase of a shape deformation of viscoelastic seed was observed. Further cotyledons deformation causes seed coat filling, and when this process completed, an increase of the inner pressure leads to the seed coat tension; till its disruption. Assuming that during seed coat tensions the pressure necessary for shape deformation is constant, the increase of pressure Δp_w resulting in the seed coat cracking can be determined from the force-deformation curve.

Substituting characteristic values, taken form the force-deformation curve obtained in compression tests of legume seeds and the diameter of contact area as well as diameter of barrel calculated using geometrical formulas presented in previous papers (Dobrzański, 1998) it is possible to determine the modulus of elasticity of seed coat E_{om} from the following equation:

$$Eom = \frac{D_h \Delta p_w}{\Delta D_h}\left(\frac{(0.5D_h + g)^2 + (0.5D_h)^2}{(0.5D_h + g)^2 - (0.5D_h)^2} + \nu\right) \tag{1}$$

The above method enabled the interpretation of the force-deformation curve for large deformation of non-elastic wet seeds where mechanical strength is mainly connected with the resistance of the seed coat to tension.

7. Tension of wet seed coat caused by the sliding displacement of the cotyledons

The compression of seed leads the shape deformation, however, low increase of lateral surface at seed strain under $\varepsilon<0.3$ proves, that there are any stresses of it. On the other hand, sliding of wet cotyledons during compression caused larger shape deformation of seed leads to tension of the seed coat and stress.

Sliding cotyledons surrounded by the seed coat caused tension strain of the seed coat ε_o, which can be calculated from simple equation based on geometrical relations according to the figure 7:

$$\varepsilon_o = \frac{\pi r_w - k_U + d_U}{\pi r_w} \tag{2}$$

Substituting geometrical relations for above values of formula (2) it is possible to determined strain of the seed coat largest diameter in cross section of body formed by sliding of cotyledons:

$$\varepsilon_o = 1 - \frac{1}{180}\arccos\frac{D_z - 2d}{D_z} + 2\frac{\sqrt{\left(\frac{D_z}{2} + \frac{D_z d}{D_z}\right)^2 - \left(\frac{D_z}{2}\right)^2}}{\pi D_z} \tag{3}$$

The change of seed shape during compression caused increase of the surface area surrounding cotyledon. A sphere surrounded by a shell imitating a seed is deformed on both sides and change the shape into a barrel when compressed between parallel plates. When the deformations are slight the sphere does not lose its shape and hence it was assumed that the lateral surface of the barrel created by compressing opposite cups of sphere will also be spherical.

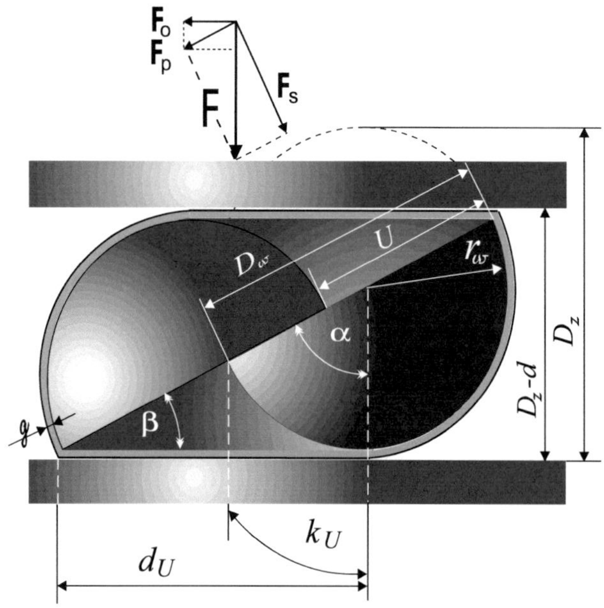

Figure 7. Tension of seed coat caused by the sliding displacement of the cotyledons.

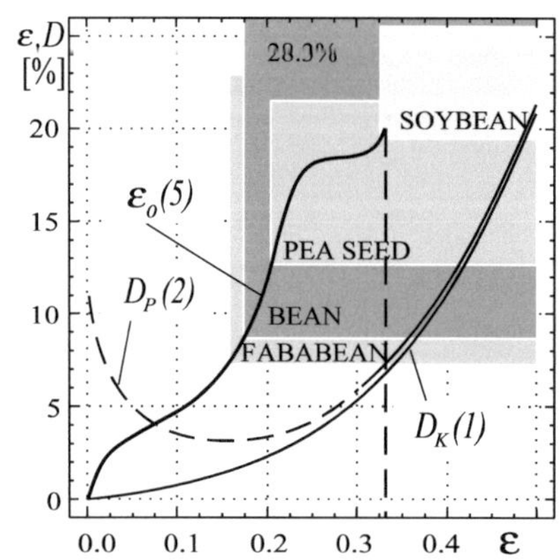

Figure 8. Increase of the seed diameter caused by the shape deformation D_K, D_P and cotyledons sliding ε_o.

Circumferential strain ε_o at tension caused the sliding displacement of the cotyledons determined using formula (3) is represented on the figure 8. The increase of shape deformation is represented by barrel diameter of parabolic D_P and circular shape D_K. Both diameters increases with an increase of axial deformation, however, almost 10 % of the increase was obtained for 0.4 axial strain. In comparison, cotyledon sliding caused coat strain ε_0 up to 15 % for axial strain ε close to 0.2 only. It suggests that sliding may involve rupture easier than shape deformation at compression. The range of legume seed coat elongation at rupture for various moisture are presented on Figure 8. It is easy to conclude, that fababean and bean seed coat is most influential on tension at axial compression and sliding of the cotyledons. At 0.18 of axial strain caused cracking of the seed coat of some fababean and bean.

Fababean *(Vicia faba* L.*)*, soybean *(Glicine hispida* Max.*)*, bean *(Phasoleus* L.*)*, and pea seed *(Pisum sativum* L.*)* cultivars were studied to mechanical resistance of seeds subjected to load at different rates of deformation. The resistance of legume seeds to compression was determined over a wide range of moisture contents. The diameter of seed and shape deformation under compression, as an additionally measurement, were determined with strain gauge. Most resistant species of legume seeds is soybean, for which the seeds could not be damaged in this way. The determination of the cracking mechanism of the cotyledon was carried out at low moisture only with single cotyledons -- a half of bean was compressed between parallel plates at various rates of cross-head movement.

The resistance of seed coat to tension was studied for all seeds at wide range of moisture. The drying test was conducted to evaluate the effect of drying parameters on the stress and strain development of the seed coat. To evaluate the effect of drying conditions on the splitting of seed coat, the samples were dried with air. The time-strain curves at different drying conditions were obtained with 2630-107 Instron strain gauge. Three air velocities: 4.3; 10.5; 15.6 m/s and four temperatures: 22; 42; 52; 62°C were used as parameters of the drying tests.

The results obtained from compression tests of cotyledons of fababean cultivars as an example was presented in previous paper (Dobrzański, 1998). Different values of cracking force for each cultivar indicate that the strength of cotyledons differ. However, the modulus of elasticity is almost the same and there is no significance difference among ($p=0.05$) the cultivars. This was also noted for other legume species.

8. Shearing of the seedcoat at threshing of dry seed

The shearing of dry seedcoat most frequently observed mechanism of bean cracking at trashing, is caused in dry seed only, by a force which is parallel to the plane of cotyledons contact. This mechanism is characteristic also for rape seed and other legume seeds such as pea seed, french bean, soybean and fababean, which covers under the seed coat two cotyledons. Some of dry seed during processing are separated two parts. It may be result of threshing cylinder or rotor motion, centrifugal discharging, the filling and discharging of screw conveyors, sprouting and free-fall dropping. Each part consists the cotyledon and half of the seed coat, which is shearing by a force parallel to the plane of contact of the cotyledons (Fig. 9).

Approximately cross section area of the seed coat, for oblong seeds such as: soybean, bean and french bean, was determined according to the figure 4 from the formula:

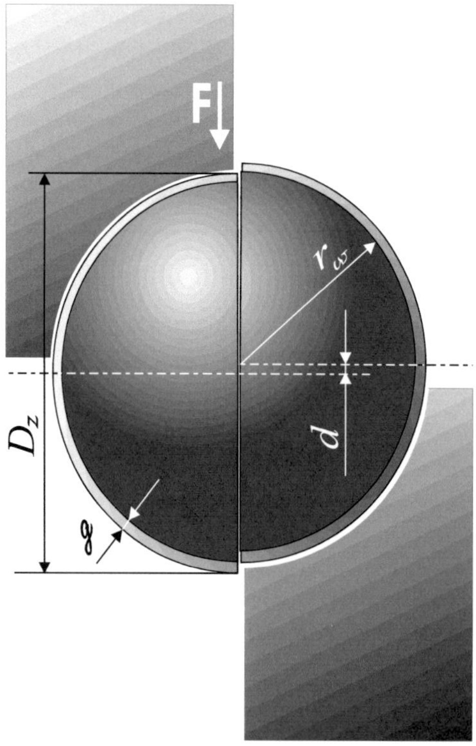

Figure 9. Shearing of the seed coat.

In this mechanism of dry seed cracking the force F made work L at deformation d, leads to the seed coat shearing, which for elastic body may by calculate using simple formula:

$$L = \frac{Fd}{2} \tag{4}$$

Both of these values are determined in experimental test presented on Figure 9. According to the energy conservation law it is possible to determined permissible velocity limit V_p, for which the seed of known mass m did not cause the seed coat cracking. Equating both energies: kinetic energy of spouting seed and work of cracking at seed coat shearing following formula was obtained:

$$V_p = \sqrt{\frac{Fd}{m}} \tag{5}$$

This method of velocity limit V_p calculation covers some simplifying assumptions, however, it allows to estimate the seed cracking of dynamic processes frequently meets in practice, by simply way at quasi-static test.

http://dx.doi.org/10.5772/56874

9. Shrinkage stress of the seed coat at drying

The effect of moisture content on expansion volume of soybean was shown in the experimental results presented in previous paper (Dobrzański, 1996). Seed dimensions increased during swelling for all cultivar. More distinctly effects of moisture content on dimensions of soybean were observed in the initial phase of swelling. Further seed's filling, leads to the seed coat tensioning. Similar behavior and significant decrease of volume in other grain legume was observed (Fig. 10).

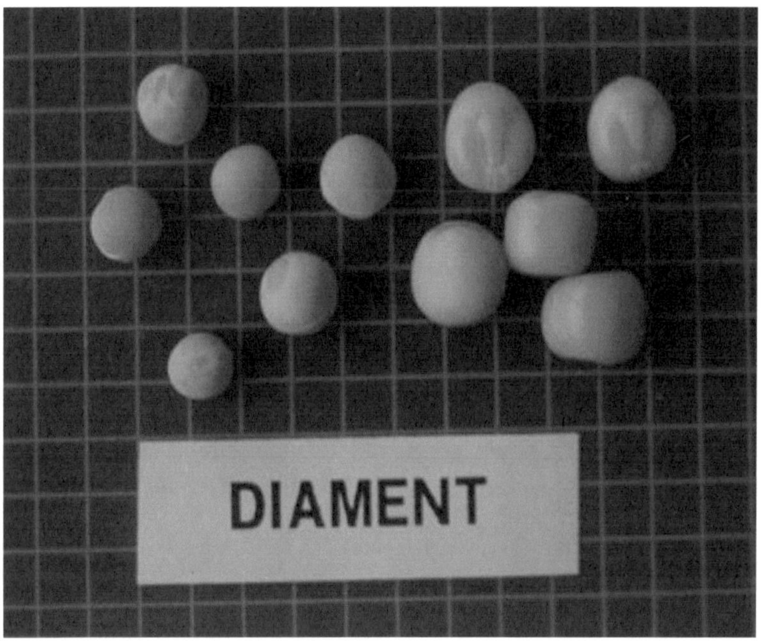

Figure 10. Volume expansion of pea seed (v. Diament) caused by water imbibitions during soaking.

The water vaporizing from the seed is limited by coat and the results presented in previous paper (Dobrzański, 1996) proved that this process was not rapid. On the other hand, all dimensions of seed decreased at this process and exponential relationship was observed.

Rapid change of moisture content of seed coat during drying caused the shrinkage of coat leading its stress. Highest shrinkage for rudest condition of drying was observed. When the drying conditions were milder, the shrinkage of the seed coat was lower. The Figure 11 presents a model of spherical seed in the stream of air T_{tV}. The wet seed of volume V_{Kw} is surrounded by the seed coat of diameter D_W. During drying the seed coat try to reduce its volume to V_{Ks}, following its surface and diameter D_S.

The shrinkage of the seed coat on wet seed of the large volume leads the stress, which involve increase of the inner pressure p_w of incompressibility wet cotyledons.

In order to determine strength criteria for tension seed coat at drying, Dobrzański (1998) found that, it is sufficient to estimate circumferential stress σ_o in cross section of seed. He determined the circumferential stress in the seed coat using Lamé equation concerning radial strain and stress of a pipe subjected to the inner pressure p_w:

$$\sigma_o = \frac{p_w(0.5D_W^2 + D_W g + g^2)}{D_W g + g^2} \tag{6}$$

During drying, the circumferential shrinkage stress σ_o must meet the strength criteria:

$$\sigma_o \le k_r \tag{7}$$

Taking into account, that stress of the seed coat σ_s at drying depends to shrinkage strain ε_s and modulus of elasticity of the seed coat E_o it must be verify experimentally.

The single seed presented on figure 11 is in the stream of hot air, however, most seeds at real condition are in the layer being in contact to other. It limits inflow of air to some parts of seed in contact to other (Fig. 11).

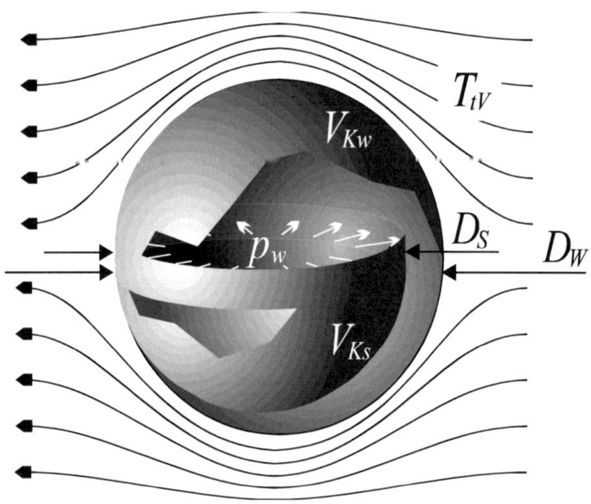

Figure 11. The shrinkage of the seed coat in the stream of air.

Hot air caused the shrinkage of part available for the stream of air. Residual part of seed coat being in the contact to other seed keeps higher moisture content. Dobrzański and Szot, 1997; Dobrzański, 2011) found that mechanical strength of seed coat decreases with the increase of moisture content. Finally, shrinkage of dry seed coat caused tension of wet part of it. Figure 12 presents the cross section of two seed being in the contact.

http://dx.doi.org/10.5772/56874

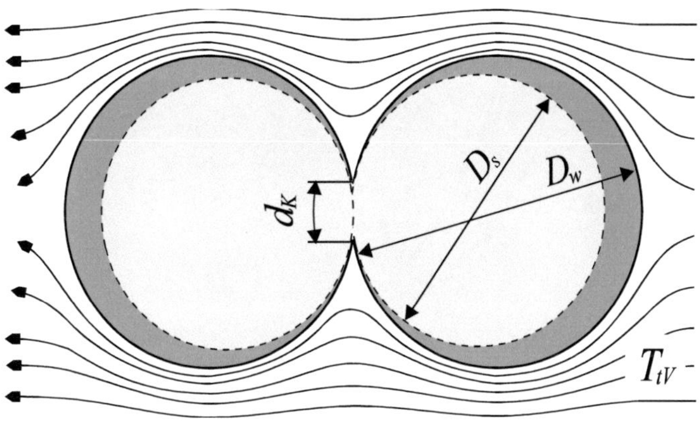

Figure 12. The shrinkage of part of seed coat in the stream of air.

The circle diameter D_W decrease to D_S, however only the part (D_W - d_K) of the seed coat is being subjected to the shrinkage. It allows determination of the shrinkage strain ε_s of dried part of the seed coat:

$$\varepsilon_s = \frac{\pi(D_W - D_S)}{\pi D_W - d_K} \tag{8}$$

The shrinkage leads to the stress σ_s, which depend to the modulus of elasticity of dry seed coat E_{oS} :

$$\sigma_s = E_{oS}\varepsilon_s \tag{9}$$

The shrinkage of dry seed coat caused tension σ_o of wet part d_K. Determining strain ε_o of wet part of the seed coat subjected to extension and its modulus of elasticity E_{oW}, it is possible to estimate the tension strength limit σ_u:

$$\sigma_u = E_{oW}\varepsilon_o \tag{10}$$

For safe condition of drying must be fulfil an inequality:

$$\sigma_u > \sigma_s, \tag{11}$$

For this purpose, the shrinkage stress σ_s and strength R_m, (Fig. 14) strain ε_s, ultimate elongation at rupture ε_o and the modulus of elasticity E_o (Fig. 13) must be determine experimentally in wide range of moisture.

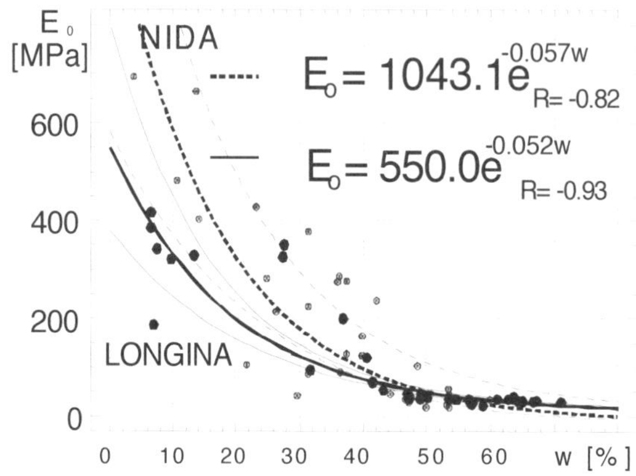

Figure 13. Modulus of elasticity of the seedcoat for Nida and Longina varieties in the range of moisture from 6 to 70 %.

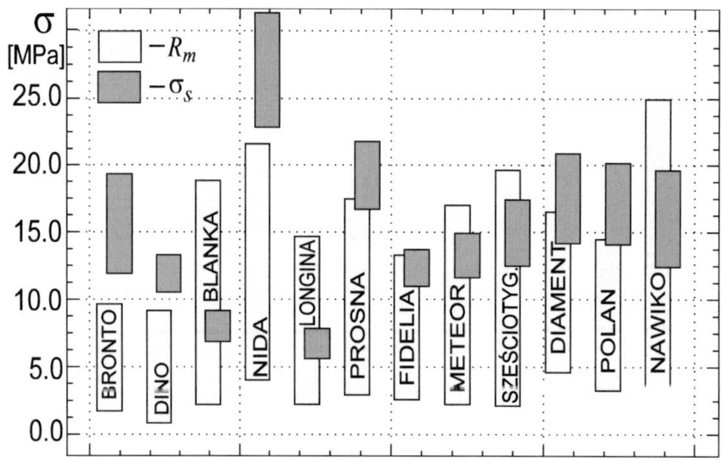

Figure 14. The range of seed coat strength R_m to tension and shrinkage coat stress σ_s at drying for fababean (Bronto, Dino) bean (Blanka, Nida, Igolomska, Prosna), pea seed (Fidelia, Szesciotyg, Diament, Meteor) and soybean (Polan, Nawiko).

A preliminary study was conducted to establish the time required for strain of seed coat after being dried to final moisture content. The second series of drying tests was conducted to evaluate the effect of drying parameters on the stress and strain development of seed coat. To evaluate the effect of drying conditions on the splitting of seed coat, the samples were dried with air. The seed coat ring samples were prepared according to the method elaborated by Dobrzański (1998).

The results obtained at tension of the seed coat shoved that, the increase of moisture content involves the decrease of force at breakage. Cultivar's differentiation of seed coat strength in tension test was observed. The results obtained in all experiments concerning with strength of seedcoat proved that frequently the stress was higher than strength of seed coat (Fig. 14).

Probably, it is one of the most important reason of the seedcoat cracking during drying. The rapid decrease of air temperature during drying causing additional shrinkage of seed coat, clarifies the origin of the coat cracking after disconnection of the heating set in dryers, observed in practice. The highest shrinkage for all cultivar at most severe drying condition was observed. The value of force induced by the shrinkage depended also on drying conditions. Drying of the seed coat at temperature of 62 °C gave highest shrinkage stress. Further drying with the cold air caused the increase of seed coat shrinkage for 25 % more. This phenomenon was observed for other cultivars. Increase of moisture content leads to a decrease of force at breakage. The shrinkage of the seed coat during drying was more than 26 % for all seeds.

The theoretical considerations of cracking mechanisms allows to verify some models describing geometrical relations during shape deformation and drying. The modulus of elasticity of the seed coat was determined at compression and shape deformation. However, this method enabled the interpretation of the force-deformation curve for seed covering moisture over 12%. On the other hand, the seed strain (under 30%) leads shape deformation without stress in the coat. For large deformation of wet seeds mechanical strength is mainly connected with the resistance of seed coat to tension. Sliding of the cotyledons is one of most important reason of seed coat cracking for seed over 20% of moisture content.

Shrinkage of the seed coat at drying leads to its stress and splitting. The results obtained in all experiments concerning with strength of seed coat proved that frequently the stress σ_s was higher than strength R_m of seed coat. Probably, it is one of the most important reason of the seed coat cracking during drying. The additional shrinkage of the seed coat clarifies the origin of cracking, after disconnection of the heating set, frequently observed in practice.

10. Postharvest rapeseed quality according to seeds physical properties

Physics of plant material concentrates on estimation of physical features important for material quality during vegetation, harvest, transport, storage and processing. It also studies microstructure of plant materials and it influence on differentiation of properties. Physics of plant materials develops methods of estimation of damage in agricultural products and the mechanisms of its appearance taking into account external factors. The variations between the varieties are also under investigation. All the data collected from the above mentioned measurements allows elaborating technological improvements for preventing the unfavorable effects of damage. This in consequence increases the quality of raw materials for food production.

Agrophysical metrology allows widen and deeper analysis of processes undergo during each stage of agricultural production with a special emphasis on harvesting and postharvest operations, deals with adaptation, improvement and designing measuring methods, which can be used to explain quantitative and qualitative changes, which take place in agriculture. It develops methods as well as equipment for measuring, monitoring and analyzing the time and space variability of physical parameters of agricultural materials subjected to agrotechnical and technological processing.

The investigations were performed in field conditions and some simulations were also a subject of investigations in laboratory. Samples were collected at regular storage rooms and driers. Physical parameters of rapeseed were obtained in compression tests of single seeds. Compression curves were registered and analyzed with numeric methods to calculate apparent modulus of elasticity and strength of seeds.

Monitoring of quality of storied seeds was made in specially constructed storage silos which allowed measurement of pressure and temperature. The influence of seed moisture content on changes of quality within the storage time was described.

The investigations showed considerably differentiated quality features (percentage of broken seeds) of rapeseed within the whole typical process of their postharvest processing. The amount of broken seeds increased even two times comparing samples taken from harvester (threshed seeds) and samples taken from storage room of dry seeds.

OPERATIONS	BROKEN SEEDS Min [%]	BROKEN SEEDS Max [%]
Unloading of threshed seeds	1,6	7,5
Storage room of wet seeds	1,6	8,0
Before drying	1,5	8,9
After drying	1,7	9,1
Storage room of dry seed	3,2	8,2

Table 1. The changes of the amount of broken seeds during rapeseed postharvest processing

Investigations showed considerable decrease of quality features of rapeseed during harvesting and postharvest operations. The amount of broken seeds considerably depended on their moisture content. The dependence between moisture content of seeds and their resistance to external loads resulted in quality losses was proofed. The most resistant seeds were observed at 6 % of moisture content and decreased both with increased and decreased of seed moisture.

Quality features of stored seeds depended on their moisture content, physical state of seed cover (broken or untouched), pressure and temperature inside silo. Safe time of storage was longer for dryer seeds. In that case of unbroken seeds biological activity was lower as well as biochemical processes underwent slower. The methods could be used to evaluate physical condition of rapeseed. All the methods allowed to monitor changes of quality features of rapeseed as raw material for oil production.

According to the results obtained it can be stated that the most negative influence on quality features of rapeseed (amount of broken seeds) had harvest operation. The amount of broken seeds increase during successive postharvest operations i.e.: transport, cleaning, drying, however the initial quality parameters had considerable significance for final quality of raw material for oil production. The methods used allowed a precise monitoring of changes of rapeseed quality parameters.

http://dx.doi.org/10.5772/56874

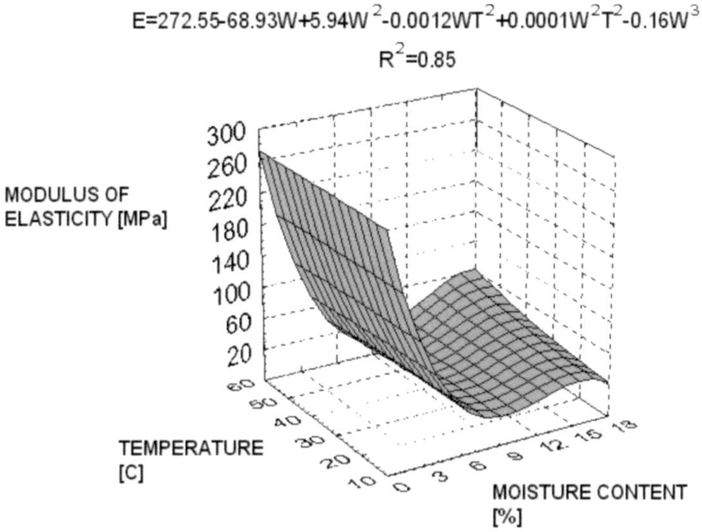

Figure 15. Influence of moisture content and temperature on rapeseed apparent modulus of elasticity

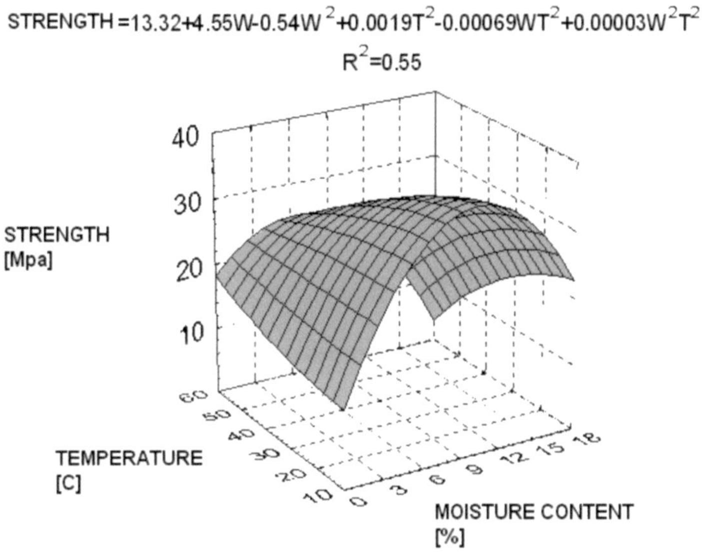

Figure 16. Influence of moisture content and temperature on rapeseed strength

Laboratory studies of mechanical strength of rapeseed allowed indicating range of "safety" moisture content for handling. Seed of 6 % of moisture content had the highest strength. Both the increase and the decrease of moisture content negatively influence strength features of seed. The influence of temperature was much lower than the influence of moisture content in the case of modulus of elasticity as well as in the case of strength. Some methods as well as results could be utilized also for other oilseeds i.e. soybean.

11. Technological parameters of lentil seeds

Lentil is not a significant plant from the point of view of the economy and agronomy (crop rotation), however have unique nutritional properties and most of all is regarded as so called healthy food. Therefore lentil sees become more and more popular and are sought after and desired by consumers as an additive or raw material for the food, pharmaceutical and chemical industry. Physical parameters of such "exotic" plant are under investigation as they are important from the technological point of view.

Air-dry lentils has a mass of 1000 seeds 49.36 g for the variety Tina and 48.05 g for the variety Anita. This parameter increases with increase of seeds moisture content and at 24% seeds achieve mass of 57 g for Tina and 58 g for Anita respectively. The high correlation coefficient was noticed in case of both varieties i.e. 0.987 for Anita and 0,975 for Tina.

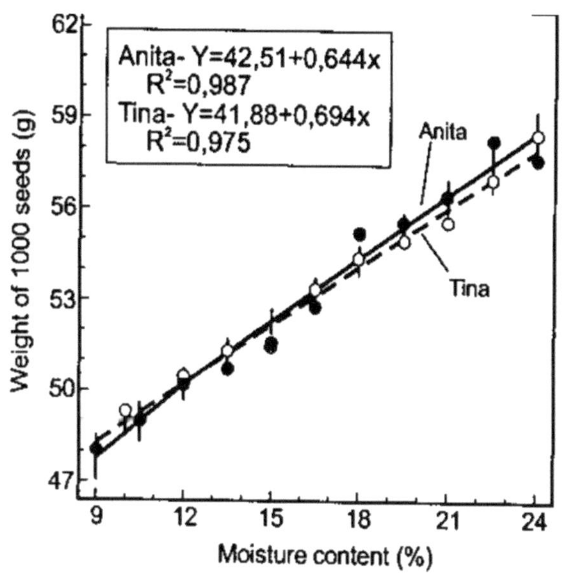

Figure 17. The influence of moisture content of seeds on their weight

Similarly, with an increase of moisture content of the lentils seed its porosity increase. This characteristic was observed for both tested varieties and porosity parameter ranged from 44 to 47% within moisture content range of 8 – 24%. The high correlation coefficient was noticed in case of both varieties i.e. 0.931 for Anita and 0,927 for Tina. The differences between varieties were statistically significant and higher values of this parameter was observed for Tina variety.

Density of lentil seed decreased with an increase in their moisture content. The maximum value of this parameter was observed at 8% moisture content for the variety Tina was 806 kg/m^3. The lowest recorded value is 747 kg/m^3 and this value was similar for both varieties at the humidity of 24%. Also in this case the high correlation coefficient was noticed for both varieties i.e. 0.959 for Anita and 0,913 for Tina.

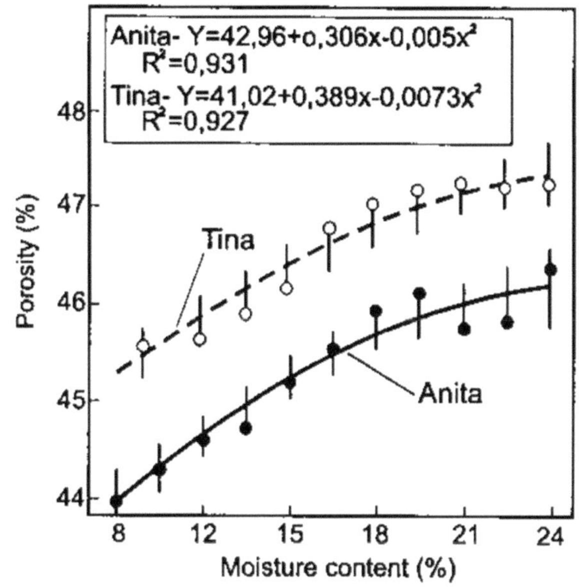

Figure 18. The influence of moisture content of seeds on their porosity

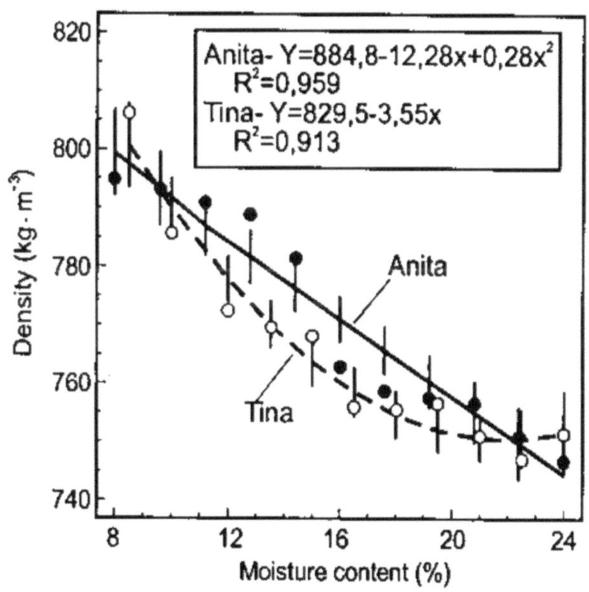

Figure 19. The influence of moisture content of seeds on their density

With the increase of moisture content both angle of repose, and slippery increase. These values of the angle of repose was measured in the range of 27.5 ° for air-dry seeds of Tina variety to 34 º for the same variety at moisture content of 24%.

12. Sweet corn kernels mechanical properties and the effect of harvest time, storage and blanching

Sweet corn attains its consumption and processing ripeness within 24 to 28 days from blooming that is from the appearance of browning that manifests ripeness. In the phase of late milk ripeness, kernels assume yellow-gold colouring. The time of harvest not only determines the physical properties of sweet corn cobs and kernels, but also generally affects the mechanical strength of the kernels. Although the time of harvest mainly affects the mechanical properties of kernels, one should also expect chemical changes related to the ripeness stage. The differences between the mean values of the cutting force were statistically significant and varied within the range from 34.2 for harvest term I to 10.3 N for term III (Fig. 20).

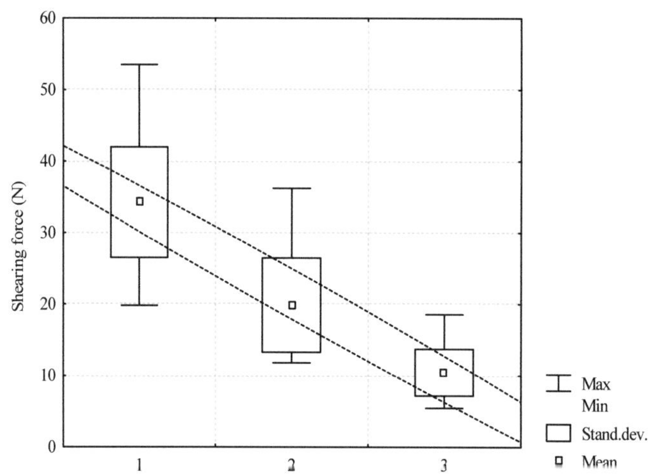

Figure 20. Cutting energy for different harvest terms: 1 – term I, 2 – term II, 3 – term III

The kernels are yellow-gold in the late milk ripeness. The period of corncobs suitability for consumption is relatively short – cobs get over-ripe rapidly, especially at high temperature. The process of over-ripening connected with transformation of sugars into starch, is delayed in with leaves and in cool conditions (Szymanek et al. 2005).

In order to estimate the effect of storage conditions on the possibility of extending the period of corn suitability for processing, tests were performed, consisting in storing corncobs for 7 days under the following conditions:

- conventional storage - temperature approx. 20⁰C (barn, umbrella roof),
- refrigerated storage - cold storage room, temperature approx. 2⁰C,
- soaking in water - temperature approx. 20⁰C.

Immediately after harvest the mechanical properties of cobs at late milk ripeness phase were compared, as well as those of cobs after blanching. Corncob blanching was made in a 4-minute

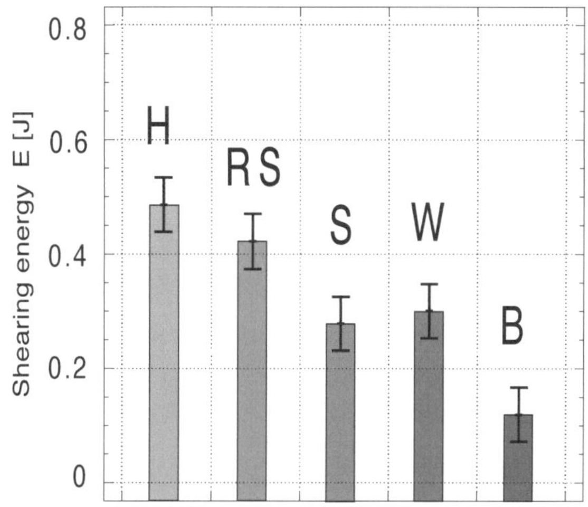

Figure 21. Energy of kernel cutting off from cob core. H– after harvest, W – wetted cobs, S – cobs from regular storage, RS – cobs from refrigerated storage and B – after blanching

water bath with a temperature of 80-90ºC. After the blanching, the cobs were cooled to a temperature of about 20ºC and dried. As follows from Figure 21, kernels from cobs subjected to the process of blanching need lower energy expenditure for the kernel cutting process (0.27J), which provided an inducement for a broader study of the process of cutting of blanched kernels.

Author details

B. Dobrzański and A. Stępniewski

Institute of Agrophysics Polish Academy of Sciences, Lublin, Poland

References

[1] Abbott, J. A, & Harker, R. (2003). Texture, http://usna.usda.gov/hb66/021texture.pdf) , 1-19.

[2] Barbosa-Cánovas, G. V, Juliano, P, & Peleg, M. (2004). Rev.2006), Engineering Properties of Foods, in Food Engineering. In: Barbosa-Cánovas, G.V., (ed.) Encyclopedia of Life Support Systems (EOLSS), Eolss Publishers, Oxford,UK, (http://www.eolss.net)

[3] Bilanski, W, Szot, B, Kushwaha, R. L, & Stepniewski, A. Comparition of strength features of rape siliques and seeds for varieties cultivated in various countries. International Agrophysics, (1994). , 1994(8), 177-184.

[4] Chen, P, & Sun, Z. (1991). A review of non-destructive methods for quality evaluation and sorting of agricultural products. Journal of Agricultural Engineering Research, 49: 85--98.

[5] Chen, P. (1996). Quality evaluation technology of agricultural products. Proc. of ICAME'96, Seul, Korea, Vol. I, , 171-190.

[6] Davison, E, Meiering, A. G, & Middendorf, F. J. (1979). A theoretical stress model of rapeseed. Canadian Agricultural Engineering, , 21(1)

[7] Dobrzański, jr. B., (1993). Geometrical dependences of seed shell of compressed pea. Int. Agrophysics, , 7(4), 259-264.

[8] Dobrzański, jr. B., (1996). The influence of moisture content on the shape deformation of soybean. AgEng'96, Madrid, Paper 96F- , 082, 1-7.

[9] Dobrzański, jr. B., (2000). The analysis of seed cracking at grain legume processing. CD-ROM of AgEng'2000 papers, Paper: 00-PH- , 050, 1-10.

[10] Dobrzański, jr. B., (2002). The cracking mechanisms of grain legume. rozdział- str.: in "Physical Methods in Agriculture", J. Blahovec, M. Kutilek (eds): Kluwer Academic Publishers., 167-194.

[11] Dobrzański, jr., B. (2011). Shrinkage and swelling phenomena in agricultural products. in Encyclopedia of Agrophysics. J. Gliński, J. Horabik, J. Lipiec (eds.) Springer, 730-733.

[12] Dobrzański, jr. B., Szot B., (1997). Strength of pea seed cover. Int. Agrophysics, , 11(4), 301-306.

[13] Dobrzański, jr. B., (1998). Mechanizmy powstawania uszkodzeń nasion roślin strączkowych. Acta Agrophysicca, in polish), 13, 1-96.

[14] Dobrzański, jr. B., Banak E., Grundas S., Sosnowski S., Pecen J., (2003). Metoda rentgenograficzna w identyfikacji uszkodzeń wewnętrznych nasion fasoli szparagowej. Acta Agrophysica, 95, in polish), 2

[15] Dobrzański, jr. B., Mazurek W., Rybczyński R., Geodecki M., Baranowski P., Walczak R., (2003). A new method of the seed viability estimation. New Methods, Means and Technologies for Application of Agricultural Products. Agricultural Engineering LUA, 1(06), 105-110.

[16] Dobrzanski, B, & Stepniewski, A. (1991). Modulus of elasticity of rapeseed shell. Proceedings of th. International Rapeseed Congress, Saskatoon, Canada, 1256-1260, 8.

[17] Dobrzanski, B, Szot, B, & Stepniewski, A. Stress in the shell during large deformation of compressed rapeseed. Proceedings of International Conference on Agricultural Engineering, Beijing, China, (1992).

[18] Dorell, D. G, & Adams, M. W. (1969). Effect of some seed characteristic on mechanically included seedcoat damage in navy beans. Agron. J., , 61(5), 672-673.

[19] Fornal, J, Sadowska, J, & Ostaszyk, A. (1995). Industrial drying of rapeseeds vs. oil extractivity. Proceedings of 9th. International Rapeseed Congress, Cambridge, England IV, , 1406-1408.

[20] Fornal, J, Sadowska, J, Jaroch, R, Kaczynska, B, & Winnicki, T. (1994). Effect of drying of rapeseeds on their mechanical properties and technological usability. International Agrophysics, , 8

[21] Haman, J, Dobrzanski, B, Szot, B, & Stepniewski, A. (1994). Strength of shell in compression test of rapeseed. International Agrophysics, , 8, 245-250.

[22] Ling ChY., Whitney L., (1975). Analysis of spherical shell acted on by two concentric forces in opposite directions. Theory, determination and control of physical properties of food materials., D. Reidel Publ. Co., , 189-196.

[23] Liu, M, Haghighi, K, & Stroshine, R. L. (1989). Viscoelastic characterization of the soybean seedcoat. Trans. ASAE , 32(3), 946-952.

[24] Mac Donald, jr. M.B., (1985). Physical seed quality of soybean. Seed Sci. Technol. , 13, 601-628.

[25] Mensah, J. K, Nelson, G. L, Herum, F. L, & Richard, T. G. (1984). Mechanical properties related to soybean seedcoat cracking during drying. Trans. ASAE, , 27(2), 550-560.

[26] Misra, R, Young, J. H, & Haman, D. D. (1981). Finite element procedures for estimating shrinkage stresses during soybean drying. Trans. ASAE , 24(3), 751-753.

[27] Mohsenin, N. N. Physical properties of plant and animal materials. Gordon and Breach Science Publishers, New York, London, Paris (1978).

[28] Otten, L, Brown, R, & Reid, W. S. (1984). Drying of white beans-effect of temperature and relative humidity on seed coat damage. Canad. Agric. Eng., , 26, 101-104.

[29] Overhults, D. G, White, H. E, Hamilton, H. E, & Ross, I. J. (1973). Drying soybeans with heated air. Trans. ASAE , 16(1), 112-113.

[30] Palusen, M. R, Nave, W. R, & Gray, L. E. (1981). Soybean seed quality as affected by impact damage. Trans. ASAE , 24(6), 1577-1582.

[31] Stepniewski, A, & Dobrzanski, B. Higher temperature of rapeseeds as a factor affecting their mechanical resistance. Proceedings of 8-th. International Rapeseed Congress, Saskatoon, Kanada, (1991). , 1991, 1246-1250.

[32] Stepniewski, A, & Szot, B. Postharvest operations and quality of rapeseed. Proceedings of 9th. International Rapeseed Congress, Cambridge, England (1995). I, , 232-234.

[33] Stepniewski, A, & Szot, B. Quality of rapeseed in postharvest handling. Preprints of workshop on Control Applications in Post-Harvest and Processing Technology, CAPPT'95, Ostend, Belgium (1995).

[34] Stepniewski, A, Szot, B, Fornal, J, & Sadowska, J. Drying conditions and mechanical properties of rapeseed. J. Food Physics, (1994). , 1994, 86-89.

[35] Szot, B, & Stepniewski, A. Studies on mechanical properties of rape in the aspect of its quantity and quality losses. Zemedelska Technika, (1995). , 1995(41), 133-136.

[36] Szot, B, Stepniewski, A, Dobrzanski, B, & Rybczynski, R. The influence of higher temperature of rapeseeds on their mechanical resistance. Proceedings of th. International Conference Physical Properties of Agricultural Materials and Their Influence on Technological Processes, Rostock, Germany, (1989). s. 815- 818, 4.

[37] White, G. M, Bridges, T. C, Loewer, O. J, & Ross, I. J. (1980). Seed coat damage in thin-layer drying of soybeans. Trans. ASAE , 23(1), 224-227.

Chapter 12

Criteria of Determination of Safe Grain Storage Time – A Review

Agnieszka Kaleta and Krzysztof Górnicki

Additional information is available at the end of the chapter

http://dx.doi.org/10.5772/52235

1. Introduction

Cereals, before being consumed as food, go through the process of cultivation, harvesting, drying, preparation and marketing (including storage) all under natural conditions, and therefore, often involve microbiological contamination and infection (Abdullah et al., 2000).

Therefore it can be stated that grain starts deteriorating from the time of harvest, due to interactions between the physical, chemical and biological variables within the environment (Mason et al., 1997). Cereal grains just after being harvested contain microbial contamination coming from several sources, such as dust, water, ill plants, insects, solid, fertilisers and animal feces. Bacteria found in grains mainly belong to the families *Pseudomonadaceae, Micrococcaceae, Lactobacillaceae* and *Bacillaceae,* and moulds are mostly *Alternaria, Fusarium, Helminthosorium* and *Cladosporium*, although other genus can also be present. The microbial composition of the cereals is of great importance for the storage of grains, since at high moisture levels the microorganisms could grow and alter the properties of product (Laca et al., 2006). Grain deterioration is also related to respiration of the grain itself and of the accompanying microorganisms. The evolution of carbon dioxide, water and heat is associated with this respiration or deterioration (Steele et al., 1969).

A 13 % moisture content is considered to be the maximum value for the storage of wheat, corn, barley and rice during short periods, though temperature and oxygen concentration also play an important role (Laca et al., 2006).

Harvesting high moisture grain such as wheat, corn or rice has become, however, common practice to protect the grain from wet weather conditions which can cause weathering and mould infection of grain in the field. High moisture grain is susceptible to deterioration by microorganisms and hence should be dried before unacceptable quality loss occurs. A knowledge of deterioration rates of high moisture grain under various storage conditions would help farmers and grain managers to know how quickly to dry the grain or adjust the storage conditions to prevent further quality loss (Kakunakaran et al., 2001). It is generally accepted that 5-15 % of the total weight of all cereals, oilseeds, and pulses is lost after harvest (Padin et al., 2002). Improved storage conditions would allow a 10 – 20 % increase in the supply of food available to people (Christiansen and Kaufmann, 1969).

Grain quality is critical in today's grain trade because of more stringent food-safety demands and an increase in market competition, therefore to avoid spoilage of grain during storage it is necessary to determine the safe grain storage time.

Safe storage time is the period of exposure of a product at a particular moisture content to a particular relative humidity and temperature below which crop deterioration may occur and beyond which the crop may be impaired. To keep losses low, crops must be dried to the safe storage moisture content (i. e. moisture content required for long term storage) within the safe storage time (Ekechukwu, 1999). Determination of safe grain storage time is an answer to the following question: how long can grains of particular moisture content and temperature be stored without the risk of the quality deterioration (Ryniecki, 2006).

Knows in the bibliography of the subject are tables and graphs of the storage times. Sometimes, however, mathematical formulas are more useful. They can be easily incorporated into the mathematical models of grain drying or aeration and expert systems which are aids for storage-grain management (Arinze et al., 1993; Courtois, 1995; Fleurat-Lessard, 2002; Kaleta, 1996). Such formulas known in the bibliography of the subject and own formulas developed by the authors of the chapter are presented in the paper.

To test the effect of grain parameters on the safe storage period, three criteria have been applied: carbon dioxide (CO_2) production and dry matter loss, appearance of visible moulds, and germination capacity.

2. Carbon dioxide production and dry matter loss

Grain deterioration is related to respiration of the grain itself and of the accompanying microorganisms. Respiration is the process of oxidizing (combusting) carbohydrates and yielding carbon dioxide, water vapour and energy. Therefore, respiration consumes dry matter.

The complete combustion (aerobic respiration) of a typical carbohydrate such as starch is represented by the following equation:

$$C_6H_{12}O_6 + 6O_2 \rightarrow 6CO_2 + 6H_2O + \text{heat} \quad (1)$$

According to this equation during the breakdown of 1 g of dry matter by aerobic respiration using 1.07 g of oxygen, 1.47 g of carbon dioxide, 0.6 g of water, and 15.4 kJ of heat energy are released. It means that a 1 % loss in grain dry matter carbohydrate is accompanied by the evolution of 14.7 g of carbon dioxide per kg of grain matter. Therefore respiration rate is closely related to grain dry matter loss and, consequently, global quality loss. Modelling CO_2 production can be used to simplify the prediction of rate of quality loss, assuming predominantly aerobic respiration.

Contamination of harvested grain by microorganisms is natural and permanent. In temperate climates with medium wet or moist grain at harvest, the genera *Fusarium, Alternaria* and *Helmintosporium* (called "field flora") are predominant. During long term storage, xerophilic fungi of the genera *Aspergillus* and *Penicillium* (called "storage flora") progressively replace the "field flora" over a period of several months of storage. At 15-19 % moisture content, most species of the field flora are inhibited or die whereas storage flora species slowly grow (Fleurat-Lessard, 2002; Frisvad, 1995; Pelhate, 1988). Since the respiratory processes of microorganisms or of hidden insect infestation are similar to those of the grain itself, the combustion of carbohydrates is a representation of grain, microorganisms and insect respiration (Fleurat-Lessard, 2002; Sinha et al., 1986; Steele et al., 1969).

The following mathematical formulas for predicting carbon dioxide production and dry matter loss can be found in the bibliography of the subject.

White et al. (1982) carried out numerous experiments on the carbon dioxide release rates of wheat and developed the following equation for general prediction of the instant rate of CO_2 release from grain:

$$X = a + bT + ct + dt^2 + eM_w \tag{2}$$

where X is the rate of CO_2 production in mg kg^{-1}d. m. per 24-h period, T is the grain temperature in °C, t is the time in h, M_w is the grain moisture content in % w. b., and a, b, c, d, and e are empirical constants.

The following equation was developed by Srour (1988):

$$X = \exp\left(aM_w + bT + c\right) \tag{3}$$

where X is the rate of CO_2 production in mg (100 gd. m.)$^{-1}$ per 24-h period, M_w is the grain moisture content in % w. b., T is the grain temperature in °C, and a, b, and c are empirical constants.

Karunakaran et al. (2001) determined the deterioration rate of wheat stored at 25°C by measuring the respiration rate of grain and microorganisms. The measured rates of CO_2 production during storage at 17, 18, and 19 % m.c. wet basis were combined and fitted to the following equation:

$$\ln X = -15.56 + 0.21t - 0.004t^2 + 1.08M_w \tag{4}$$

where$X$ is the rate of CO_2 production in mg d^{-1} kg^{-1}d. m., t is the storage time in d, and M_w is the moisture content in % w. b.

There are however some problems in using equations (2) - (4) to describe quality changes. They are based on the measurement of CO_2 release rate, either from a grain sample or directly in a grain bin. Such measurements can be done using sophisticated equipment and in laboratory conditions. Additionally, when grain moisture is below 14 % (w. b.) the release rate is very low and therefore it is very difficult to measure it. However, the above formulas are not useful in prediction the storage life.

Another option for the prediction of safe storage life is the calculation of dry matter loss (DML) as a function of grain temperature, grain moisture content, and storage time.

Seib et al. (1980) stated that the amount of dry matter loss from respiration is an indication of grain quality. They also stated that rough rice stored at 15 % and 18 % w. b. moisture content fell below U. S. Grade Nos. 1 and 2 if DML exceed 0.75 %. Some authors assumed that an acceptable level of dry matter loss is 0.5 %. In high moisture maize (corn, 25 % m.c.) a loss of 0.5 % dry matter can occur in 7 days, sometime without any visible moulding. However, this way found to be sufficient to render maize grain unfit for use, and also to produce aflatoxins (Marin et al., 1999). Kreyger (1972) considered grain to be fit for animal feed with DML of up to <2 %. However, Hall and Dean (1978) suggested 1 % DML was acceptable in grain for food use and that this could be applied to both wheat and maize. White et al. (1982) stated that 0.1 % was unacceptable for wheat of premium grade and proposed the absolute limit of DML at 0.04 %. Therefore the problem of what is the limit for an acceptable level of dry matter loss is still controversial.

Seib et al. (1980) developed the following expression to determine DML of long-grain rough rice as a function of grain temperature, grain moisture content, and storage time:

$$\mathrm{DML} = 1 - \exp\left\{-At^{C}\exp\left[D\left(1.8T - 28\right)\right]\exp\left[E\left(M_w - 0.14\right)\right]\right\} \tag{5}$$

where DML is the dry matter loss in decimal form, t is the storage time in h 10^{-3}, T is the grain temperature in °C, M_w is the grain moisture content in decimal w. b., and A, C, D, and E are empirical constants.

Equation (5) was developed for rice with constant airflow being forced through the grain and for the average grain temperature and the average grain moisture content over the storage time in question. The aerobic conditions were moreover assumed. When rice is stored in airtight units a shortage of O_2 would decrease the respiration rate as well as decrease the rate of DML. Therefore, for bunker conditions, equation (5) would be expected to overestimate the actual DML since it was based on the premise of having adequate O_2 to be used by the respiration process (Freer et al., 1990; Hu et al., 2003).

Weinberg et al. (2008) examined the effect of various moisture contents on the quality of maize (corn) grains in self-regulated modified atmospheres during hermetic storage. The experiments were conducted *in vitro*. Maize at 14, 16, 18, 20 and 22 % m.c. was initially conditioned for 28 days in tightly wrapped plastic bags and then stored in sealed containers at 30°C for up to 75 days. Carbon dioxide produced within the containers replaced the oxygen. As the moisture content increased the time for O_2 depletion shortened, from 600 h at 14 % m.c. to 12 h at 22 %. The maize at 20 and 22 % m.c. exhibited the highest DML (0.59 % and 0.74 % respectively after 75 days) and the maize at 14 and 16 % m.c. the lowest (0.02 % and 0.15 %). Adhikarinayake et al. (2006) found out that during airtight storage of 14 % m.c. paddy in a ferro-cement bin, oxygen concentration dropped to 2.7 % within 30 days and carbon dioxide rose to 9.1 %. After 6 months storage, DML was 0.4 %. Varnova et al. (1995) noted that a sealed bulk of barley declined to 4 % O_2 after 50 days.

When grain temperature and moisture content cannot be assumed constant for the entire storage time used, the method of rates was used to calculate DML (Freer et al., 1990):

$$\mathrm{d}(\mathrm{DML})/\mathrm{d}t = ACt^{(C-1)}\mathrm{e}^{y}\mathrm{e}^{z}\mathrm{e}^{-x\mathrm{e}^{y}\mathrm{e}^{z}} \tag{6}$$

where $x=At^{C}$, $y=D(1.8\,T\text{-}28)$, $z=E(M_w\text{-}0.14)$

The values of the constants for long-grain rough rice used in the equations (5) and (6) were found to be (Seib et al., 1980): A=0.001889, C=0.7101, D=0.02740, E=31.63.

Thompson (1972) took into account negative influence of mechanical damages on dry matter loss and obtained the following expression to determine the DML (in %) of shelled corn:

$$\mathrm{DML} = 0.0883\left(\exp 0.006t_r - 1\right) + 0.00102t_r \tag{7}$$

where:

$$t_r = \frac{t}{M_M \cdot M_T \cdot M_D} \tag{8}$$

$$M_M = 0.103\left\{\exp\left[455(100M)^{-1.53}\right] - 0.845M + 1.558\right\} \text{ for } 0.149 \le M \le 0.538\,\mathrm{kg\ H_2O{\cdot}kg^{-1}d.m.} \tag{9}$$

$$M_T = 32.3\exp\left[-3.48(0.03T + 0.53)\right] \text{ for } T \le 15.6°\mathrm{C} \text{ or } M_w \le 19\% \tag{10}$$

$$M_T = 32.3\exp\left[-3.48(0.03T + 0.53)\right] + 0.01(M_w - 19)\exp\left[0.61(0.03T - 0.47)\right] \tag{11}$$

for T>15.6°C and 19<M_w≤28%

$$M_T = 32.3\exp\left[-3.48\left(0.03T + 0.53\right)\right] + 0.09\exp\left[0.61\left(0.03T - 0.47\right)\right] \quad (12)$$

for T>15.6°C and M_w>28%

$$M_D = 0.001\left(\mathrm{MD}\right)^2 - 0.1101\left(\mathrm{MD}\right) + 3.426 \text{ for } 2\ \% \le \mathrm{MD} \le 40\ \% \quad (13)$$

(equation (13) is developed by authors of the chapter on the basis of Steele et al. (1969) data)

where T is the grain temperature in °C, M_w is the grain moisture content in % w. b., t is the storage time in h, M_M is the moisture multiplier, M_T is the temperature multiplier, M_D is the mechanical damage multiplier, and MD is the mechanical damage in %.

Scherer et al. (1980) investigated the dry matter loss of corn. Based on their data we developed the following relationship between monthly DML, grain temperature and grain moisture content:

$$\mathrm{DML} = 6.479 - 0.339T - 0.498M_w + 0.002T^2 + 0.015TM_w + 0.009{M_w}^2 \left(R^2 = 0.9530\right) \quad (14)$$

where DML is the monthly dry matter loss in %, T is the grain temperature in °C, M_w is the grain moisture content in % w. b., and 5°C≤T≤20°C, 14 %w. b. ≤M_w≤35 % w. b.

From equation (5) and from Scherer (1980) data increase in the dry matter loss with the increase of both grain temperature and moisture content can be observed. In such conditions the respiration of grains is more intensive. DML increase with the duration of the grain storage.

Scherer's et al. (1980) investigations on damaged grain confirmed the negative influence of mechanical damages on dry matter loss shown by Thompson (1972). Scherer et al. (1980) stated that increase in amount of damaged corn caused the decrease in safe storage time. They accepted the limit of DML at 0.5 % and observed that 1 % of damaged grain together with 1 % of chaff and fines reduced the safe storage time in almost 6 %, however 20 % of damaged grain and 5 % of chaff and fines reduced the time in almost 38 %. They explained obtained results by more intensive respiration of chaff and fines, and damaged grain comparing with undamaged grain.

Brooker et al. (1974) assumed for stored corn the limit of DML at 1 %. Based on their data the following relationship between safe storage time, grain temperature and grain moisture content can be presented:

$$t = 3774.98 - 88.12T - 252.55M_w + 0.587T^2 + 2.686TM_w + 4.223{M_w}^2 \left(R^2 = 0.861\right) \quad (15)$$

where t is the storage time in d, T is the grain temperature in °C, M_w is the grain moisture content in % w. b., and $1°C \le T \le 24°C$, $15\ \%w.\ b. \le M_w \le 30\ \%\ w.\ b.$

Al-Yahya (2001) examined the conditions of safe storage of wheat. Based on these data the following relationship between storage time, grain temperature, grain moisture content and DML can be presented:

$$t = \exp\left(6.490336 - 0.024165T - 0.163337M_w + 1.292568(\mathrm{DML})\right)\ \left(R^2 = 0.9393\right) \tag{16}$$

where t is the storage time in d, T is the grain temperature in °C, M_w is the grain moisture content in % w. b., DML is the dry matter loss in %, and $4°C \le T \le 40°C$, $15\ \%\ w.\ b. \le M_w \le 24\ \%\ w.\ b.$, $0.25\ \% \le \mathrm{DML} \le 1\ \%$.

From Brooker et al. (1974) data and from Al-Yahya's (2001) data increase in the safe storage time of grains with the decrease of both grain temperature and moisture content can be observed. In such conditions the respiration of grains is less intensive.

According to equation (1) heat energy is released during the respiratory process of grain, microorganisms and insects. The heat produced within the pockets of wet grain is especially harmful. It is not dissipated rapidly because of the low thermal conductivity of the grain (Kaleta, 1999; Kaleta and Górnicki, 2011) and the slow free convection currents in the granular bulk. The elevated grain temperature and moisture content of the pocked provide a favourable environment for further growth of microorganisms, thereby making the heating process self-accelerating. Heat production in stored grain ecosystems was investigated by e. g. Cofie-Agblor et al. (1997), Karunakaran et al. (2001), Scherer et al. (1980), and Zhang et al. (1992). Wilson (1999) proposed a mathematical model for predicting mould growth and subsequent heat generation in bulk stored grain. Unlike previous models, it was intended to be applicable in conditions that change with time. Starting from a model for mould growth in varying conditions the work of a number of authors was combined to produce a model to predict the heat production at all parts in a grain bulk. The effect of temperature and relative humidity on the mould growth rate was decoupled, so that the resulting equation for mould growth was a product of one-parameter terms. The heat generation rate was then written as a specific function of the mould population and mould grow rate. The model's current predictions for very wet grains was good, but for dried grain model performs less well.

3. Appearance of visible moulds

Spoilage of grains is the result of microorganisms (bacteria, yeast, fungi, and moulds) utilizing the nutrients present in the grain for growth and reproductive processes, spoilage may result in a loss of nutrients from the grain since microorganisms use these nutrients in much the same way as livestock. Also, microorganisms produce heat and moisture during growth which can cause a temperature rise in stored grain. Heating initiated by microbial growth can cause "heat damage" and can sometimes render grain unfit for feed. Such conditions have been known to cause fires and dust explosions in storage structures (Ross et al., 1979).

Certain microorganisms, when allowed to grow under the proper environmental conditions, can produce toxins or other products which, if consumed by either livestock or humans, can cause serious illness and even death. A number of these toxins and the microorganisms which produce them have been identified.

Toxigenic fungi infect agricultural crops both in the field and in storage. Converse et al. (1973) found the following variety of fungi in the corn at harvest and after 28 days of aeration in bins: *Fusarium, Cephalosporium, Alternaria, Cladosporium, Mucor, Nigrospora, Penicillium, Aspergillus flavus, A. glaucus, A. niger*, and *A. ochraceus*. Pronyk et al. (2006) noted that initial fungal counts showed that canola seeds were infected with high levels of pre-harvest fungi *Alternaria alternata*(Fr.) Keissl. and *Cladosporium* spp. and low levels of storage fungi *Eurotium* spp., *Aspergillus candidus* Link, and *Penicillium* spp.

Fungal infections can be discolour grain, change its chemical and nutritional characteristics, reduce germination and, most importantly, contaminate it with mycotoxins, the poisonous metabolites produced by certain fungal genera.

Ergot is a disease of cereal crops caused by the fungus *Claviceps purpurea*. It causes reduced yield and quality of grain. The effect of the ergot's alkaloid toxins on man and animals is, however, of much greater significance (Moreda and Ruiz-Altisent, 2011).

Aflatoxins are secondary metabolites produced by *Aspergillus flavus* Link and *A. parasiticus* Speare. These compounds are only few of over 120 mycotoxins produced by fungi. The aflatoxin's dietary effect upon poultry can result in poor growth, increased mortality, poor feed conversion, and increased condemnations. A number of other animal species are also subject to alfatoxicosis. Aflatoxin has been know to act as a potent toxin, a carcinogen, a teratogen, and a mutagen (Brekke et al., 1977; Liu et al., 2006; Wieman et al., 1986).

The fumonisins are secondary metabolites produced by *Fusariummonili forme* Sheldon and *F. proliferatum* (Matsushima) Niremberg. They show a worldwide distribution and can be isolated from maize and maize-based food and foodstuffs naturally contamined with *Fusarium*. The fumonisins have been associated with leukoencephalomalacia (ELEM) in equines, porcine pulmonary edema (PPE), diarrhea and reduced body weight in broiler chicks, carcinogenicity in rats and leukoencephalomalacia and hemorrhage in the brain of rabbits. In addition, epidemiological evidence suggest a correlation between the consumption of *F. moniliforme* contaminated maize and a high incidence of human esophageal carcinoma (Marin et al., 1999; Orsi et al., 2000).

Mites also infect stored cereals. These arthropods contaminate grains and are a matter of great concern in the medical and veterinary fields, since they may act as carriers of bacteria and toxigenic fungi. Grains contaminated by mites may cause acute enteritis when ingested, and severe dermatitis and allergy in cereal handlers. Furthermore, mites can feed on the germ of kernels, thereby reducing the content of iron and vitamins of the B complex and germination ability. Stored – product mites can survive and multiply by feeding on several species of seed – borne fungi. Fungal spores and mycelia contain small amounts of essential nutrients (e. g. vitamins of the B complex and steroids), and moisture levels adequate for the metabolic demands of mites. The constant migration of mite populations within a granary ecosystem efficiently contributes to the dispersal of viable fungal spores of several species, including *Aspergillus* spp. and *Penicillium* spp., carried on the vector's body surface or deposited with its feces (Franzolin et al., 1999).

Conditions favouring the development of mycotoxins in cereals before and after harvest are important to grain – exporting countries concerned with marketing high – quality products. In post-harvest situation, crop spoilage, fungal growth, and mycotoxin formulation result from the interaction of several factors in the storage environment. These factors include: moisture, temperature, time, insect vectors, damage to the seed, oxygen levels, composition of substrate, fungal infection level, prevalence of toxigenic strains of fungi, and microbiological interactions. An understanding of the interactions involved would facilitate prediction and prevention of mycotoxin development in grains (Abramson et al., 2005).

Investigations of conditions favouring the development of mycotoxins in grains before and after harvest were carried out by many researches for the following grains: barley (Abramson et al., 1999; Gawrysiak-Witulska et al., 2008), canola (Pronyk et al., 2006), maize (corn) (Franzolin et al., 1999; Liu et al., 2006; Marin et al., 1999; Orsi et al., 2000; Reed et al., 2007; Wicklow et al., 1998), rice (Liu et al., 2006; Abdullah et al., 2000), wheat (Abramson et al., 2005; Padin et al., 2002).

Abramson et al. (1999) stated that ochratoxin A, citrinin and sterigmatocystin reached mean levels of 24.38 and 411 ppb by 20 weeks in the 19 % moisture content barley, but were absent in the 15 % m.c. barley, and no other mycotoxins were detected. *Penicillium* species and *Aspergillusversicolor* (Vuill.) Tiraboschi comprised the predominant microflora. The effect of storage time was apparent at both 15 and 19 % moisture content for grain temperature, *Alternaria alternata* (Fr.) Keissler, *Penicillium* species and *Aspergillus versicolor*. At 19 % moisture content, storage time also affected moisture content, CO_2 level, ergosterol content, seed germination, and mycotoxin production. At 19 % m.c., elevated ergosterol levels at weeks 4 and 8 appears to offer early warning of the appearance of sterigmatocystin at week 12, and of ochratoxin A and citrinin at week 20.

Pronyk et al. (2006) found that total ergosterol (fungal – specific membrane lipid used as an indicator of fungal infection in grain) levels in stored canola increased with storage time, temperature, and seed moisture content.

Liu et al. (2006) noted that no significant linear relationship existed in whole grain rice and brown rice between the amount of aflatoxins and the length of storage. The amount of aflatoxins in whole grain rice samples from 1 to 10 yr ranged from 2.79 to 2.93 μg kg^{-1} and peaked in the samples that were storage for 7-8 yr (6.23 μg kg^{-1}). With increasing storage length, the aflatoxin content in brown rice was consistently low ranging from 0.74 to 1.19 μg kg^{-1}. However, in maize samples, the amount of aflatoxins significantly increased with storage length. The average amount of aflatoxins in 1-yr maize was only 0.84 μg kg^{-1}, while in 2-

yr maize it was as high as 1.17 μg kg^{-1}. Practically, no maize grains were kept in storage for more than 3 yr.

Franzolin et al. (1999) examined the ability of mites of the species *Tyrophagus putrescentiae* to spread the toxigenic fungus *Aspergillus flavus* from contaminated maize to sterile grains under controlled conditions. The obtained results confirms that *T. putrescentiae* is a means of dispersal for toxigenic fungi in stored grain kept under warm and moist conditions. The levels of aflatoxin contamination recorded after 90 days of incubation exceeded the safe limits established by Brazilian legislation.

Abdullah et al. (2000) examined the average numbers of days before visible fungal development at 25°C on, among others, ordinary rice and glutinous rice. They found that ordinary rice at 13.0 % moisture content (d. b.) and glutinous rice at 12.9 % m.c. (d. b.) would be safe for about 2 months (57±2 days and 73±1 days respectively). However, ordinary rice at 14.1 % m.c. and glutinous rice at 14.2 % m.c. may spoil in about 20 days. Hence, an error in the moisture content of 1.1 % for rice and 1.3 % for glutinous rice is disastrous. At 21.9 % m.c. ordinary rice and 25.6 % m.c. glutinous rice the data indicated a shell-life of about 7 days.

Abramson et al. (2005) stated that ochratoxin A and citrinin reached mean levels of 6.5 and 11.6 mg kg^{-1}, respectively, by 20 weeks at 20 % m.c., but were absent at 16 % m.c., and no other mycotoxins were found. *Penicillium* species were the predominant microflora. Ergosterol levels remained between 3.9 and 8.4 mg kg^{-1} at 16 % m.c., but increased from 3.9 to 55.5 mg kg^{-1} at 20 % m.c. during 20-week trial period.

There is, however, lack of simple equations predicting the length of safe storage period by a combination of, at least, moisture content of grain and storage temperature.

Bailey and Smith (1982) (cited after Bowden et al. (1983)) developed the following empirical formula predicting the duration of a safe barley storage period without occurrence of visible mould under the good aeration conditions:

$$t = 67 + \exp\left\{5.124 + \left(39.6 - 0.8107T\right)\left[\left(M_w - 12\right)^{-1} - 0.0315\exp 0.0579T\right]\right\} \quad (17)$$

where t is the storage time in h, T is the grain temperature in °C, M_w is the grain moisture content in % w. b., and 5°C≤T≤25°C, 16 % w. b. ≤ M_w ≤26 % w. b.

Kreyger (1972) investigated the safe storage times of several grains. He assumed that the best criterion for safe storage times is the one that is based on the time to the appearance of visible moulds. Based on Kreyger's (1972) data, we developed the following formula:

$$t = \exp(A + BT + CM_w) \quad (18)$$

where t is the storage time in weeks, T is the grain temperature in °C, M_w is the grain moisture content in % w. b., A, B, C are empirical constants given in Table 1, and 10°C≤T≤25°C.

Grain	Coefficients			R^2	Range of application
	A	*B*	*C*		
wheat	50.66928	-0.272909	-2.52755	0.9707	16.1 %w.b.≤M_w≤23.0 %w.b.
barley	27.04320	-0.174362	-1.17856	0.9727	15.6. %w.b.≤M_w≤22.7 %w.b.
oats	31.60300	-0.201594	-1.55997	0.9728	14.8 %w.b.≤M_w≤22.0 %w.b.
rye	34.58371	-0.283607	-1.58288	0.9779	15.4 %w.b.≤M_w≤24.4 %w.b.

Table 1. Values of coefficients in equation (18) and range of application

Equation (17) and (18) confirm that the duration of the safe storage time increases with the decrease of both grain temperature and grain moisture content. Such conditions are not favourable for the mould development.

There are, however, controversies about the criterion of appearance of visible moulds. Several researches (Ryniecki and Nellist, 1991; Nellist, 1998), followed Kreyger (1972), took it as the best criterion for safe storage time. Some of them (Armitage, 1986; Fleurat-Lessard, 2002) mentioned, however, several drawbacks of this criterion. The main drawback of this kind of prediction of safe storage life of stored grain is the subjective determination of visible mould on the kernel. Another drawback is the lack of progressiveness in the prediction. Before the onset of visible spoilage, grain is theoretically sound and its quality is not altered. The day after spoilage is seen, the grain is deteriorated and should be downgraded.

4. Germination capacity

Various factors can reduce the storage life of some premium grade quality cereals. Moisture content of the harvested grains and storage temperature can encourage mould and insect pest damage. The best studied quality parameter is germination capacity, which is only of direct importance for grains. Nevertheless, this is probably the best surrogate measure of cereal grain soundness (Pomeranz, 1982). Cereals retaining a high level of viability in storage are also likely to retain the other main parameters of commercial or technological quality (Fleurat-Lessard, 2002).

Germination is defined as the appearance of the first signs of growth, i. e. the visible protrusion of the radical (Black, 1970). Germination can be affected by many factors such as grain temperature, grain moisture content, grain damages, fungus and insect infection. Much research has been conducted to determine the effect of various factors on germination.

McNeal (1966) found that soybean can be kept for 12 months without an expressive decline in germination if the temperature is kept below 16°C and the moisture content is not higher than 16.2 %, dry basis. Mayeux et al. (1972) noted than the germination of soybean seed is influenced by the percentage of split beans in stored seed, and storage temperature and moisture play an important role in maintaining the soybean seed quality. Kreyger (1972) used percentage germination as an indicator of grain deterioration. He studied the effect of many levels of grain moisture content and grain temperature on the percentage germination. His findings will be discussed below.

Parde et al. (2002) studied the storage behaviour of soybean seed and the loss in quality due to free-fall from different heights (0.5-2 m) on to different surfaces (cement and galvanized iron) were studied. They found that soybean seed is susceptible to mechanical damage. The severity of damage varies with moisture content of seed because the dryer seed is harder. The hight of fall produces significant effects on germination. An average germination loss of 10 % and 31 % was noticed when the seed fell from a height of 1 and 2 m, respectively, on to the cement floor. This drop in germination was 7.5 % and 22 % when dropped from the same heights on to galvanized floor. The seed lots held at 12 % moisture content, dry basis, suffered less damage during free-fall from different heights than the lots held at 10 % and 11 % m.c. Soybean seed lots at 12 % m.c. retained germination ability for a longer period than the seed lots at lower m.c.

Pronyk et al. (2006) stated that germination decreased with storage time, temperature, and moisture content. After 56 days, germination of canola stored at 12 % m.c., wet basis and at 25-30°C dropped till 73 %. The same value of germination stored: at 12 % m.c. and at 30-35°C showed after approximately 27 days, at 14 % m.c. and at 25-30°C showed after 29 days, at 14 % m.c. and at 30-35°C showed after 12 days.

Weinberg et al. (2008) examined the germination percentage of the maize (corn) stored in the self-regulated atmospheres in the sealed containers. They noticed that the germination percentage decreased during the storage period, and decreased as the moisture content increased. With 18 % m.c. and above the germination percentage decreased to zero after 35 days of storage.

Genkawa et al. (2008) tested airtight storage of brown rice with a low moisture content. They stated that the germination rate of brown rice with 16.2 % m.c., wet basis, at 25°C declined from 97 % to 27 % but for rice with less than 12.8 % m.c. at 25°C germination was above 90 %.

There is, however, lack of simple equations predicting the length of safe storage period by combination of, at least, moisture content of grain and storage temperature.

Fraser and Muir (1981) developed a set of two regression equations for predicting allowable storage times for wheat before the germination capacity drops by 5 %:

$$\begin{aligned} \log t &= 6.234 - 0.2118 M_w - 0.0527 T, \text{ for } 12\ \%\ \text{w.b.} \le M_w < 19\ \%\ \text{w.b.} \quad (a) \\ \log t &= 4.129 - 0.0997 M_w - 0.0576 T, \text{ for } 19\ \%\ \text{w.b.} \le M_w \le 24\ \%\ \text{w.b.} \quad (b) \end{aligned} \tag{19}$$

where t is the storage time in d, M_w is the grain moisture content in % w. b., T is the grain temperature in °C, and 10°C≤T≤40°C.

Kaleta (1996) used equation (19) in her computer program developed to simulate wheat drying in silos with radial (horizontal) and vertical airflow, predict grain spoilage under the simulated conditions, and determine the most advantageous conditions of conducting the process of wheat drying in silos.

Muir and Sinha (1986) developed a set of two regression equations for predicting allowable storage times for canola before the germination capacity drops by 5 %.

$$\begin{array}{l} \log t = 6.224 - 0.302\,M_w - 0.069T, \ \text{for}\, M_w < 11\ \%\ \text{w.b.}\ \text{(a)} \\ \log t = 5.278 - 0.206\,M_w - 0.063T, \ \text{for}\, M_w{}^{3}11\ \%\ \text{w.b.}\ \ \text{(b)} \end{array} \tag{20}$$

where t is the storage time in d, M_w is the grain moisture content in % w. b., T is the grain temperature in °C, and 10°C≤T≤40°C.

Arinze et al. (1993) used equation (20) in their computer program developed to simulate in-bin drying of canola grain, predict grain spoilage under the simulated conditions, and obtain an economic analysis of various drying schemes for canola for use as a management tool in the selection of the appropriate drying system. Equation (20) predicts the allowable storage times when canola is stored at constant temperatures and moisture contents. During a drying process, however, both temperature and moisture content vary with time. To predict grain spoilage or deterioration under dynamic or changing conditions, Arinze et al. (1993) used spoilage index (SI). A value of t was computed at each interval Δt from equation (20), and the calculated ratios of $\Delta t/t$ were calculated. Theoretically, grain loses 5 % of its germination when the sum of the computed $\Delta t/t$ values for each layer over the simulated drying period equals unity:

$$\mathrm{SI} = \sum_{i=1}^{n} \left(\frac{\Delta t}{t} \right)_i = 1 \tag{21}$$

where n is the number of simulated time steps. SI is a spoilage or storage index and its instantaneous value represents the progress of grain spoilage. A spoilage index of 1 or greater indicates that the allowable storage time has elapsed and the 5 % loss in germination has occurred to the canola.

Karunakaran et al. (2001) defined the safe storage time of wheat as the storage time for the germination to decrease to 90 % and developed the following correlation equation for 19 % m.c., wet basis, wheat at 10-35°C:

$$\log t = 2.057 - 0.049T \tag{22}$$

where t is the storage time in d, and T is the grain temperature in °C. They stated also, that the safe storage times of 17 % m.c. wheat were 5, 7, and 15 d at 35, 30, and 25°C, respectively.

The germination capacity of wheat at 17-19 % m.c., wet basis, stored at 25°C can be predicted from the measured respiration rate and moisture content by the equation (Karunakaran et al., 2001):

$$Y = 100 - 0.1X + 0.093M_w \tag{23}$$

where Y is the germination capacity of grain in %, X is the rate of CO_2 production in mg d^{-1} kg^{-1}d. m., and M_w is the grain moisture content in % w. b.

Equation (23) is useful to determine the condition of the grains coming to grain-handling facilities, for which the storage conditions (time and temperature) are not known but the moisture content and respiration rate of the grain can be determined in 2 h rather than the 7 d required for germination. For wheat stored for a known length of time at 25°C and moisture levels of 17-19 %, the germination capacity can be predicted from the storage time, moisture content of the stored grain, and CO_2 production (Karunakaran et al., 2001):

$$Y = 54.56 - 1.213t + 2.823M_w - 0.076X \tag{24}$$

where Y is the germination capacity of grain in %, t is storage time in d, M_w is the grain moisture content in % w. b., and X is the rate of CO_2 production in mg d^{-1} kg^{-1}d. m.

Based on the germination data of Kreyger (1972), we developed the following formulas for predicting allowable storage times:

$$t = \exp\left(A + BT + CM_w\right) \tag{25}$$

were t is storage time in weeks, T is the grain temperature in °C, M_w is the grain moisture content in % w. b., A, B, C are empirical constants given in Table 2, and $10°C \le T \le 20°C$.

Al-Yahya (2001) explored some of the factors and conditions, such as grain moisture, grain temperature, and mechanical grain damage, that influence the germination of grain at various levels of dry matter loss during wheat storage. The objective was to determine the changes in percentage germination of stored wheat at different levels of DML under different storage conditions, i. e. different grain moisture contents, temperatures and levels of mechanical damage (MD). Based on Al-Yahya's (2001) data, we developed the following formulas:

the first one

$$Y = A + BM_w + C(\mathrm{MD}) + DM_w^{\;2} + EM_w(\mathrm{MD}) + F(\mathrm{MD})^2 \tag{26}$$

were Y is the germination capacity of grain in %, M_w is the grain moisture content in % w. b., MD is the mechanical damage in %, A, B, C, D, E, and F are empirical constants given in Table 3, and 15 %w. b. $\leq M_w \leq$24 %w. b., 0≤MD≤30 %.

Grain	Coefficients			R^2	Range of application
	A	*B*	*C*		
wheat	12.28039	-0.128973	-0.473026	0.9929	12.0 %w.b.$\leq M_w \leq$23.0 %w.b.
barley	13.12305	-0.174000	-0.452103	0.9965	11.0. %w.b.$\leq M_w \leq$23.0 %w.b.
oats	13.96125	-0.148378	-0.604968	0.9940	11.0 %w.b.$\leq M_w \leq$22.0 %w.b.
rye	10.13185	-0.087999	-0.426973	0.9931	11.5 %w.b.$\leq M_w \leq$24.0 %w.b.

Table 2. Values of coefficients in equation (25) and range of application

T, °C	DML, %	*A*	*B*	*C*	*D*	*E*	*F*	R^2
	0.25	-113.667	22.569	0.032	-0.594	-0.022	0.004	0.940542
4	0.5	-45.074	15.904	0.184	-0.475	-0.069	0.023	0.836703
	1	246.235	-19.763	-0.074	0.507	-0.062	0.017	0.729865
	0.25	-83.9483	19.1422	-1.207	-0.5037	-0.03733	0.048	0.72055
15	0.5	-204.267	31.578	-0.412	-0.852	-0.024	0.008	0.928671
	1	-80.551	15.347	-0.29	-0.42	-0.046	0.012	0.875763
	0.25	-34.73	12.798	0.914	-0.323	-0.091	0.012	0.869765
25	0.5	-37.747	13.559	-0.473	-0.381	-0.037	0.02	0.905906
	1	207.167	-17.628	-0.06	0.481	-0.063	0.016	0.734423
	0.25	-4.483	9.097	0.606	-0.214	-0.067	0.004	0.448631
40	0.5	-303.503	37.932	-2.536	-0.926	0.049	0.031	0.732198
	1	5.479	-2.086	-1.064	0.207	-0.03	0.026	0.917996

Table 3. Values of coefficients in equation (26)

and the second one

$$Y = A + BT + C(\mathrm{MD}) + DT^2 + ET(\mathrm{MD}) + F(\mathrm{MD})^2 \tag{27}$$

where Y is the germination capacity of grain in %, T is the grain temperature in °C, MD is the mechanical damage in %, A, B, C, D, E, and F are empirical constants given in Table 4, and 4°C≤T≤40°C, 0≤MD≤30 %.

Equations presented in this section confirm that the changes in germination capacity of stored grain are lower with lower following parameters: grain temperature, grain moisture content, mechanical damage and storage time. In general, the conclusions are the same as in previous section: longer storage times are possible with lower both grain moisture contents and temperatures and with lower levels of mechanical grain damages.

At the end of the chapter it is worth to mention shortly the other grain quality criteria which can be important to consumer and food manufacturer.

Colour of white rice is an important criterion for judging quality and price. The white colour becomes yellow after a period of storage. Dry matter loss of grain and heat liberated from its respiration and biological activities may accelerate rice yellowing. Parameters affecting the rice yellowing are temperature and relative humidity (water activity) (Soponronnarit et al., 1998; Tirawanichakul et al., 2004).

M_w, % w.b.	DML, %	*A*	*B*	*C*	*D*	*E*	*F*	R^2
	0.25	96.676	-0.72	-0.623	0.008	0.003	0.009	0.874369
15	0.5	87.017	0.262	-0.572	-0.027	-0.015	0.008	0.90452
	1	67.972	0.178	-1.315	-0.033	-0.002	0.022	0.963953
	0.25	100.072	-0.627	-0.144	0.014	-0.004	-0.001	0.8498037
18	0.5	82.486	0.004	-0.352	$1.166 \cdot 10^{-4}$	$-7.455 \cdot 10^{-4}$	$5 \cdot 10^{-4}$	0.92672
	1	46.691	0.168	-0.955	-0.018	0.008	0.007	0.8205147
	0.25	102.912	-1.133	-0.719	0.019	-0.024	0.022	0.95634
21	0.5	78.891	0.084	-1.352	-0.002	-0.019	0.039	0.837519
	1	60.816	-0.15	-0.737	0.006	-0.027	0.006	0.963041
	0.25	95.373	-0.941	-1.736	0.024	-0.005	0.037	0.75977
24	0.5	69.078	-0.288	-2.259	0.012	0.021	0.034	0.971917
	1	70.357	-1.923	-2.238	0.049	0.016	0.03	0.895284

Table 4. Values of coefficients in equation (27)

Corn quality can mean wet-milling quality. It corresponds to the amount of survival thermo-sensitive proteins inside the grains and is very well correlated with the thermal history of the grains (Courtois, 1995).

The rate of quality changing can be represented with a simple zero- or first-order reaction (Labuza, 1980):

$$\pm\frac{dA}{dt}=k_0\exp\left(-\frac{E_A}{RT}\right)A^n \tag{28}$$

where A is amount of a quality factor, $\pm dA/dt$ is the rate loss of a quality factor or production of undesirable effects, k_0 is the pre-exponential factor, E_A is the activation energy in J mol^{-1}, R is the gas constant in J mol^{-1} K^{-1}, T is the temperature in K, and n is the reaction order (1 for first-order, 0 for zero-order).

Somponronnarit et al. (1998) stated that the yellowing rate of paddy can be explained by temperature and water activity and developed the following empirical equations to predict the change in the yellow colour:

$$\frac{db}{dt}=k \tag{29}$$

and

$$\ln k=71.87-25.32(\mathrm{RH})-\frac{25919.13}{T}+\frac{10712.78(\mathrm{RH})}{T} \tag{30}$$

where b is yellowness of rice in Hunter b unit, t is the time in d, k is the constant value for the yellowing rate in Hunter b unit d^{-1}, RH is the relative humidity in decimal, T is the temperature in K, and 308 K$\le T \le$338 K, 0.80$\le$RH$\le$0.95.

Courtois (1995) developed the following empirical equation to predict the change in the wet-milling quality of corn:

$$\frac{dQ}{dt}=-k_0\exp\left(-\frac{E_A}{RT}\right)Q^2 \tag{31}$$

and

$$k_0=-1.9561{\cdot}10^{16}+5.4287{\cdot}10^{17}M+6.8210{\cdot}10^{17}M^2 \tag{32}$$

where Q is the wet-milling quality, t is the time in s, k_0 is the pre-exponential factor in s^{-1}, T is the temperature in K, M is the grain moisture content in decimal d. b., and R is the gas constant in J mol^{-1} K^{-1}, E_A=-133200 J mol^{-1}.

5. Conclusion

Nowadays grain is harvested with a combine harvester. Therefore it is possible to delay the process and to harvest ripe and dry grain, without any bigger losses caused by ridging of grain, yet in certain parts polluted with green parts of plants, straws and seeds of weeds or with unripe caryopses, moisture content can even exceed 30% w. b., and temperature is often above 30°C. This state can cause self-heating processes even when the grain itself is considered as dry. In such grain and even in grain considered as dry, vital functions connected with metabolism still exist, namely grain respiration, growth of moulds and other microorganisms as well as growth of insects. These processes lead to a decline in the quality of grain and even to its entire damage. The intensity of these processes depends mainly on the moisture content of grain and its temperature. For the purpose of safe grain storage one ought to limit its vital functions as soon as possible through lowering moisture content and temperature reduction. It can be realized by drying, and then cooling the grain. Due to economy in thermal energy consumption, grain is often dried with the atmospheric air or slightly heated air, but such a process runs very slowly, and grain has to stay in the drying chamber for quite a long time. During harvest, when granaries accept large quantities of harvested grain, it is not always possible to immediately clean, dry and cool the grain due to the limited capacity of devices. Therefore there is a necessity of periodic storage of the fresh grain mass, so there is a risk that undesirable processes will occur, which can lead to a decline in quality, and even entire damage of grain. It is therefore necessary to determine the time of safe grain storage, i. e. the time in which the growth of undesirable processes does not cause any essential changes in the quality of grain. The basic criteria of determination the length of this period are: CO_2production and connected with it loss of the dry matter of grain, appearance of visible moulds, and germination capacity.

The dependencies for determining the time of safe grain storage were discussed. The general conclusions for all discussed criteria are the same: longer storage times are possible with lower both grain moisture contents and temperatures and with lower levels of mechanical grain damages.

Author details

Agnieszka Kaleta and Krzysztof Górnicki

Faculty of Production Engineering, Warsaw University of Life Sciences, Poland

References

[1] Abdullah, N., Nawawi, A., Othman, I. (2000). Fungal spoilage of starch-based foods in relation to its water activity (aw). *Journal of Stored Products Research*, Vol. 36, No. 1, (January 2000), pp. 47-54 , ISSN 0022-474X.

[2] Abramson, D., Hulasare, R, York, R. K., White, N. D. G., Jayas, D. S. (2005). Mycotoxins, ergosterol, and odor volatiles in durum wheat during granary storage at 16 % and 20 % moisture content. *Journal of Stored Products Research*, Vol. 41, No. 1, (January 2005), pp. 67-76, ISSN 0022-474X.

[3] Abramson, D., Hulasare, White, N. D. G., Jayas, D. S., Marquardt, R. R. (1999). Mycotoxin formation in hulless barley during granary storage at 15 % and 19 % moisture content. *Journal of Stored Products Research*, Vol. 35, No. 3, (July 1999), pp. 297-305, ISSN 0022-474X.

[4] Adhikarinayake, T. B., Palipane, K. B., Müller, J. (2006). Quality change and mass loss of paddy during airtight storage in a ferro-cement bin in Sri Lanka. *Journal of Stored Products Research*, Vol. 42, No. 3, (July 2006), pp. 377-390, ISSN 0022-474X.

[5] Al-Yahya, S. A. (2001). Effect of storage conditions on germination in wheat. *Journal of Agronomy & Crop Science*, Vol. 186, No. 4, (June 2001), pp. 273-279, ISSN 1439-037X.

[6] Arinze, E. A., Sokhansanj, S., Schoenau, G. J. (1993). Development of optimal management schemes for in-bin drying of canola grain (rapeseed). *Computer and Electronics in Agriculture*, Vol. 9, No. 2, (September 1993), pp. 159-187, ISSN 0168-1699.

[7] Armitage, D. M. (1986). Pest control by cooling and ambient air drying. In: *Spoilage and Mycotoxins of Cereals and other Stored Products*, B. Flannigan, (Ed.), 13-20, CAB International, Wallingford, UK.

[8] Bailey, P. H., Smith, E. A. (1982). *Strategies for control of near-ambient grain driers – simulation using 1968 Turnhouse (Edinburg) weather*. Dept. Note SIN/330, Scot. Inst. Agric. Engng., Penicuik (unpublished)

[9] Black, M. (1970). Seed germination and dormancy. *Science Progress*, Vol. 58, ISSN 0036-8504, pp. 379-393.

[10] Bowden, PJ., Lamond, W. J., Smith, E. A. (1983). Simulation of near-ambient grain drying. 1. Comparison of simulations with experimental results. *Journal of Agricultural Engineering Research*, Vol. 28, No. 4, (July 1983), pp. 279-300, ISSN 0021-8634.

[11] Brekke, O. L., Peplinski, A. J., Lancaster, E. B. (1977). Aflatoxin inactivation in corn by aqua ammonia. *Transactions of the ASAE*, Vol. 20, No. 6, (November 1977), pp. 1160-1168, ISSN 0001-2351.

[12] Brooker D. B., Bakker-Arkema, F. W., Hall, C. W. 1974. *Drying cereal grains*. AVI Publ. Comp. Inc., ISBN 0870551612, WestportConn. USA.

[13] Christiansen, C. M., Kaufmann, H. H. (1969). *Grain storage: The role of fungi quality loss*. University of Minnesota Press, ISBN 0816605181, Minneapolis, MN, USA.

[14] Cofie-Agblor, R., Muir, W. E., White, N. D. G., Jayas, D. S. (1997). Microbial heat production in stored wheat. *Canadian Agricultural Engineering*, Vol. 39, No. 4, (October/November/December 1997), pp. 303-307, ISSN 0045-432X.

[15] Converse, H. H., Sauer, D. B., Hondges, T. O. (1973). Aeration of high moisture corn. *Transactions of the ASAE*, Vol. 16, No. 4, (September 1973), pp. 696-699, ISSN 0001-2351.

[16] Courtois, F. (1995). Computer-aided design of corn dryers with quality prediction. *Drying Technology*, Vol. 13, No. 1&2, pp. 147-164, ISSN 1532-2300.

[17] Ekechukwu, O. V. (1999). Review of solar-energy drying systems I: an overview of drying principles and theory. *Energy Conversion & Management*, Vol. 40, No. 3, (April 1999), pp. 593-613, ISSN 0196-8904.

[18] Fleurat-Lessard, F. (2002). Review. Qualitative reasoning and integrated management of the quality of stored grain: a promising new approach. *Journal of Stored Products Research*, Vol. 38, No. 3, pp. 191-218, ISSN 0022-474X.

[19] Franzolin, M. R., Gambale, W., Cuero, R. G., Correa, B. (1999). Interaction between toxigenic *Aspergillus flavus* Link and mites (*Tyrophagus putrescentiae* Schrank) on maize grains: effect on fungal growth and aflatoxin production. *Journal of Stored Products Research*, Vol. 35, No. 3, (July 1999), pp. 215-224, ISSN 0022-474X.

[20] Fraser, B. M., Muir, W. E. (1981). Airflow requirements predicted for drying grain with ambient air and solar-heated air in Canada. *Transactions of the ASAE*, Vol. 24, No. 1, (January 1981), pp. 208-210, ISSN 0001-2351.

[21] Freer, M. W., Siebenmorgen, T. J., Couvillion, R. J., Loewer, O. J. (1990). Modelling temperature and moisture content changes in bunker-stored rice. *Transactions of the ASAE*, Vol. 33, No. 1, (January 1990), pp. 211-220, ISSN 0001-2351.

[22] Frisvad, J. C. (1995). Mycotoxins and mycotoxigenic fungi in storage. In: *Stored-grain Ecosystems*, D. S. Jayas, N. D. G. White, W. E. Muir, (Eds.), 251-288, ISBN 0824789830, M. Dekker, Inc. New York, USA.

[23] Gawrysiak-Witulska, M., Wawrzyniak, J., Ryniecki, A., Perkowski, J. (2008). Relationship of ergosterol content and fungal contamination and assessment of technological quality of malting barley preserved in a metal silo using the near-ambient method. *Journal of Stored Products Research*, Vol. 44, No. 4, pp. 360-365, ISSN 0022-474X.

[24] Genkawa, T., Uchino, T., Inoue, A., Tanaka, F., Hamanaka, D. (2008). Development of a low-moisture-content storage system for brown rice: Stability at decreased moisture contents. *Biosystems Engineering*, Vol. 99, No. 4, (April 2008), pp. 515-522, ISSN 1537-5110.

[25] Hall, C. W., Dean, P. E. (1978). Storage and preservation of cereal grains. In: *Cereals 78: better nutrition for the worlds millions*, Y. Pomeranz, (Ed.), 223-243, American Association of Cereal Chemists, St Paul, MN, USA.

[26] Hu, W., Akimoto, K., Hamanaka, D., Sorour, H., Hori, Y. (2003). Respiration and quality of rough rice under unsteady atmospheric conditions. *Journal of Faculty of Agriculture, Kyushu University*, Vol. 47, No. 2, pp. 427-435, ISSN 0023-6152.

[27] Kaleta, A. (1996). *Modelling of Grain Drying in Silos*. Fundacja Rozwój SGGW, ISBN 83-86980-03-6, Warszawa, Poland.

[28] Kaleta, A. (1999). *Thermal Properties of Plant Materials*. Warsaw Agriculture Press, ISBN 83-7244-039-5, Warsaw, Poland.

[29] Kaleta, A., Górnicki, K. (2011). Databases on physical properties of plant and agricultural products. In: *Encyclopedia of Agrophysics*, J. Gliński, J. Horabik, J. Lipiec, (Eds.), 189-194, Springer, ISBN 978-90-481-3584-4, Dordrecht, The Netherlands.

[30] Karunakaran, C., Muir, W. E., Jayas, D. S., White, N. D. G., Abramson, D. (2001). Safe storage time of high moisture wheat. *Journal of Stored Products Research*, Vol. 37, No. 3, (July 2001), pp. 303-312, ISSN 0022-474X.

[31] Kreyger J. (1972). *Drying and storing grains, seeds and pulses in temperate climates*. Bulletin 205. Institute for Storage and Processing of Agricultural Products, Wageningen, The Netherlands.

[32] Labuza, T. P. (1980). The effect of water activity on reaction kinetics of food deterioration. *Food Technology*, Vol. 34, pp. 36-41, 59, ISSN: 0015-6639.

[33] Laca, A., Mousia, Z., Diaz, M., Webb, C., Pandiella, S. S. (2006). Distribution of microbial contamination within cereal grains. *Journal of Food Engineering*, Vol. 72, No. 4, (February 2006), pp. 332-338, ISSN 0260-8774.

[34] Liu, Z., Gao, J., Yu, J. (2006). Aflatoxins in stored maize and rice grains in Lioaning Province, China. *Journal of Stored Products Research*, Vol. 42, No. 4, pp. 468-479, ISSN 0022-474X.

[35] Martin, S., Homedes, V., Sanchis, V., Ramos, A. J., Magan, N. (1999). Impact of *Fusariummoniliforme* and *F. proliferatum* colonisation of maize on calorific losses and fumonisin production under different environmental conditions. *Journal of Stored Products Research*, Vol. 35, No. 1, (January 1999), pp. 15-26, ISSN 0260-8774.

[36] Mason, L. J., Rulon, R. A., Maier, D. E. (1997). Chilled versus ambient aeration and fumigation of stored popcorn. Part 2: Pest management. *Journal of Stored Products Research*, Vol. 33, No. 1, (January 1997), pp. 51-58, ISSN 0260-8774.

[37] Mayeux, M. M., Esphaphani, M., Kiver Jr., O. J. (1972). *The effect of environmental conditions on germination of soybean*. Paper No. 72-319. ASAE. MI.

[38] McNeal, X. (1966). *Conditioning and storage of soybeans*. Bulletin 714. Arkansas Agricultural Experiment Station, University of Arkansas, Fayetteville, AR.

[39] Moreda, A., G. P., Ruiz-Altisent, M. (2011). Quality of agricultural products in relation to physical conditions. In: *Encyclopedia of Agrophysics*, J. Gliński, J. Horabik, J. Lipiec, (Eds.), 669-678, Springer, ISBN 978-90-481-3584-4, Dordrecht, The Netherlands.

[40] Muir, W. E., Sinha, R. N. (1986). Theoretical rates of flow of air at near-ambient conditions required to dry rapeseed. *Canadian Agricultural Engineering*, Vol. 28, No. 1, (January 1986), pp. 45-49, ISSN 0045-432X.

[41] Nellist, M. E. (1998). Bulk storage drying in theory and practice. *Journal of the Royal Agricultural Society of England*, Vol. 159, pp. 120-135, ISSN: 0080-4134.

[42] Orsi, R. B., Carrêa, B., Possi, C. R., Schammass, E. A., Nogueira, J. R., Dias, S. M.C., Malozzi, M. A. B. (2000). Mycoflora and occurrence of fumonisins in freshly harvested and stored hybrid maize. *Journal of Stored Products Research*, Vol. 36, Vol. 1, (January 2000), pp. 75-87, ISSN 0260-8774.

[43] Padin, S., Dal Bello, G., Fabrizio, M. (2002). Grain loss caused by *Tribolium castaneum*, *Sitophilus oryzae* and *Acanthoscelides obtectus* in stored durum wheat and beans treated with *Beauveri abassiana*. *Journal of Stored Products Research*, Vol. 38, No. 1, (January 2002), pp. 69-74, ISSN 0260-8774.

[44] Parde, S. R., Kausal, R. T., Jayas, D. S., White, N. D. G. (2002). Mechanical damage to soybean seed during processing. *Journal of Stored Products Research*, Vol. 38, No. 4, (October 2002), pp. 385-394, ISSN 0260-8774.

[45] Pelhate, J., (1988). Microbiology of seeds in relation to conditioning: its influence of germinative capacity. In: *Preservation and Storage of Grains, Seeds and their By-Products*, J. L. Multon, (Ed.), 310-327, Lavoisier Publishing, ISBN 2-85 206-4367, New York, USA.

[46] Pomeranz, Y. (1982). Biochemical, functional and nutritive changes during storage. In: *Storage of cereal grains and their products, 3rd edition*, C. M. Christiansen, (Ed.), 56-114, American Association of Cereal Chemists, ISBN 0913250236, St Paul, MN, USA.

[47] Pronyk, C., Abramson, D., Muir, W. E., White, N. D. G. (2006). Correlation of total ergosterol levels in stored canola with fungal deterioration. *Journal of Stored Products Research*, Vol. 42, No. 2, pp. 162-172, ISSN 0260-8774.

[48] Reed, C., Doyungan, S., Ioerger, B., Getchell, A. (2007). Response of storage molds to different initial moisture contents of maize (corn) stored at 25°C, and effect on respiration rate and nutrient composition. *Journal of Stored Products Research*, Vol. 43, No. 4, pp. 443-458, ISSN 0260-8774.

[49] Ross, I. J., Loewer, O. J., White, G. M. (1979). Potential for aflatoxin development in low temperature drying systems. *Transactions of the ASAE*, Vol. 22, No. 6, pp. 1439-1443, ISSN 0001-2351.

[50] Ryniecki, A. (2006). *Drying and Cooling Grain in Bulk. Handbook. Questions and Answers. PartI.* MR INFO, ISBN 8390978415, Poznań, Poland.

[51] Ryniecki, A., Nellist, M. E. (1991). Optimization of control systems for near-ambient grain drying. *Journal of Agricultural Engineering Research*, Vol. 48, (January–April 1991), pp. 1-35, , ISSN 0021-8634.

[52] Scherer, R., Kutzbach, H. D., Thaler, M., Müller, H. -M. (1980). Atmungswärme und Atmungsverluste von Körnermais. *Grundlagen der Landtechnik*, Vol. 30, No. 4, pp. 104-110, ISSN 0017-4920.

[53] Seib, P. A., Pfost, H. B., Sukabdi, A., Rao, V. G., Burroughs, R. (1980). Spoilage of rough rice measured by evolution of carbon dioxide, *Proceedings of the 3rd annual workshop on Grains Post-harvest Technology*, pp. 75-93, Kuala Lumpur, Malaysia, 29-31 January 1980.

[54] Sinha, R. N., Waterer, D., Muir, W. E. (1986). Carbon dioxide concentrations associated with insect infestation of stored grain 1: Natural infestation of corn, barley and wheat in farm granaries. *Science des Aliments*, Vol. 6, pp. 91-98, ISSN 0240-8813.

[55] Soponronnarit, S., Srisubati, N., Yoovidhya, T. (1998). Effect of temperature and relative humidity on yellowing rate of paddy. *Journal of Stored Products Research*, Vol. 34, No. 4, (October 1998), pp. 323-330, ISSN 0260-8774.

[56] Srour, S. (1988). Thermic properties of grains-production of heat and CO2, In: *Preservation and Storage of Grains, Seeds and their By-Products*, J. L. Multon, (Ed.), 189-202. Lavoisier Publishing, ISBN 2-85 206-4367, New York, USA.

[57] Steele, J. L., Saul, R. A., Hukill, W. V. (1969). Deterioration of shelled corn as measured by carbon dioxide production. *Transactions of the ASAE*, Vol. 12, No. 5, pp. 685-689, ISSN 0001-2351.

[58] Thompson, T. L. (1972). Temporary storage of high-moisture shelled corn using continuous aeration. *Transactions of the ASAE*, Vol. 15, No. 2, pp. 333-337, ISSN 0001-2351.

[59] Tirawanichakul, Y., Prachayawarakorn, S., Varanyanond, W., Soponronnarit, S. (2004). Simulation and grain quality for in-store drying of paddy. *Journal of Food Engineering*, Vol. 64, No. 4, (October 2004), pp. 405-415, ISSN 0260-8774.

[60] Varnava, A., Navarro, S., Donahaye, E. J. (1995). Long term hermetic storage of barley in PVC – covered concrete platforms under Mediterranean conditions. *Postharvest Biology and Technology*, Vol. 6, No. 1-2, (June 1995), pp. 177-186, ISSN 0925-5214.

[61] Weinberg, Z. G., Yan, Y., Chen, Y., Finkelman, S., Ashbell, G., Navarro, S. (2008). The effect of moisture level on high-moisture maize (*Zea mays* L.) under hermetic storage conditions – *in vitro* studies. *Journal of Stored Products Research*, Vol. 44, No. 2, pp. 136-144, ISSN 0260-8774.

[62] White, N. D. G., Sinha, R. N., Muir, W. E. (1982). Intergranular carbon dioxide as an indicator of biological activity associated with the spoilage of stored wheat. *Canadian Agricultural Engineering*, Vol. 24, pp. 35-42, ISSN 0045-432X.

[63] Wicklow, D. T., Weaver, D. K., Throne, J. E. (1998). Fungal colonists of maize grain conditioned at constant temperatures and humidities. *Journal of Stored Products Research*, Vol. 34, No. 4, (October 1998), pp. 355-361, ISSN 0260-8774.

[64] Wieman, D. M., White, G. M., Taroba, J. L., Ross, I. J., Hicks, C. L., Langlois, B. E. (1986). Production of aflatoxin in damaged corn under controlled environmental conditions. *Transactions of the ASAE*, Vol. 29, No. 4, pp. 1150-1155, ISSN 0001-2351.

[65] Wilson, M. T. (1999). A model for predicting mould growth and subsequent heat generation in bulk stored grain. *Journal of Stored Products Research*, Vol. 35, No. 1, (January 1999), pp. 1-13, ISSN 0260-8774.

[66] Zhang, Q., Muir, W. E., Sinha, R. N., Cenkowski, S. (1992). Heat production in wet wheat under adiabatic conditions. Canadian Agricultural Engineering, Vol. 34, No. 3, pp. 233-238, ISSN 0045-432X.

Chapter 13

Extrusion-Cooking of Starch

L. Moscicki, M. Mitrus, A. Wojtowicz,
T. Oniszczuk and A. Rejak

Additional information is available at the end of the chapter

http://dx.doi.org/10.5772/52323

1. Introduction

The extrusion technology, well-known in the plastic industry, has recently become widely used in food industry, where it is referred to as extrusion-cooking. It has been employed for the production of so called engineered food and special feed.

Generally speaking, extrusion-cooking of vegetable raw materials consists in the extrusion of grinded material at baro-thermal conditions. With the help of shear energy, exerted by the rotating screw, and additional heating by the barrel, the food material is heated to its melting point, than is conveyed under high pressure through a series of dies and the product expands to its final shape. That results in much different physical and chemical properties of the extrudates in comparison to raw materials used.

Food extruders – processing machines (see fig. 1), belong to the family of HTST (High Temperature Short Time) equipment, with a capability to perform cooking tasks under high pressure. This aspect may be explained for vulnerable food and feed as an advantageous process since small time span exposures to high temperatures will restrict unwanted denaturation effects on e.g. proteins, amino acids, vitamins, starches and enzymes. Physical technological aspects like heat transfer, mass transfer, momentum transfer, residence time and residence time distribution have a strong impact on the food and feed properties during extrusion-cooking and can drastically influence the final product quality (Mościcki et al., 2009, Moscicki, 2011, Mościcki, 2011).

Nowadays, extrusion-cooking as a method is used for the production of different food staff, ranging from the simplest expanded snacks to the highly-processed meat analogues. The most popular extrusion-cooked products are following:

- direct extrusion snacks, RTE (ready-to-eat) cereal flakes and diverse breakfast foods produced from cereal material and differing in shape, colour and taste and easiest to implement in terms of production;
- snack pellets - half products suitable for fried or hot air expanded snacks, pre-cooked pasta;
- baby food, pre-cooked flours, instant concentrates, functional components;
- pet food, aquafeed, feed concentrates and calf-milk replacers;
- textured vegetable protein (mainly from soybeans, though not always) used in the production of meat analogues;
- crispy bread, bread crumbs, emulsions and pastes;
- baro-thermally processed vegetable components for the pharmaceutical, chemical, paper and brewing industry;
- confectionery: different kinds of sweets, chewing gum, and many others.

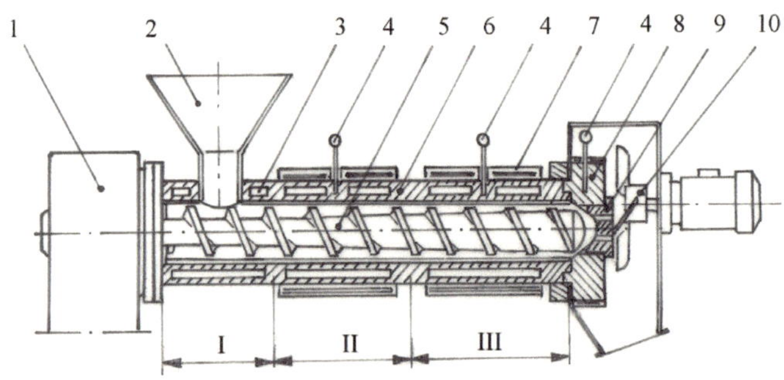

Figure 1. A cross-section of a single-screw extrusion-cooker: 1 - engine, 2 - feeder, 3 - cooling jacket, 4 - thermocouple, 5 – screw, 6 - barrel, 7 - heating jacket, 8 - head, 9 - net, 10 -cutter, I - transport section, II – compression section, III – melting and plasticization section (Mościcki et al., 2009).

Figure 2. Different type of extrusion-cooked food and feed products (Mościcki et al., 2009).

http://dx.doi.org/10.5772/52323

2. Starch and starchy products

Extrusion-cooking is accompanied by the process of starch gelatinization, involving the cleavage of intermolecular hydrogen bonds. It causes a significant increase in water absorption, including the breakage of starch granules. Gelatinized starch increases the dough viscosity, and high protein content in the processed material facilitates higher flexibility and dough aeration. After leaving the die hot material rapidly expands as a result of immediate vaporisation and takes on a porous structure. In the extruded dough protein membranes closing occur creating cell-like spaces, and starch, owing to dehydration, loses its plasticity and fixes the porous nature of the material. Rapid cooling causes the stiffening of the mass, which is typical for carbohydrate complexes embedded in a protein matrix and their total enclosure by the membrane of hydrated protein. The resulting product is structurally similar to a honeycomb shaped by the clusters of molten protein fibres.

Starch occurs primarily in cereal grains and potatoes. It takes the form of granules of different and characteristic shape, depending on the origin as well as on the variety and type of fertilization. As commonly known, two main components of starch are amylose and amylopectin displaying different physical and chemical properties, for example, chemical structure. The technological assessment of extrudates takes two factors into account: the water solubility index (WSI) and the water absorption index (WAI). These properties were studied in many laboratories and the conclusions were that WAI of many starch products increases together with the temperature rising in the extruder's barrel. It has been assumed that the maximum value is obtained around the temperature ranges from 180 to 200 °C. When these temperatures are exceeded, WAI drops and causes the WSI increase. The lower material initial moisture content used in extrusion, the higher extrudate's WSI rate can be obtained. A noticeable influence on the product properties has the percentage of amylose and amylopectin and its ratio in the processed material.

The extrusion processing of starchy materials certainly impacts the changes in product viscosity (pasting characteristic) after dissolving in water. This feature is very important for the technological point of view. Using Brabender viscometer we can see that the characteristic viscosity curve for starch is clearly reduced through extrusion; at the same time, the decrease of viscosity is greater if higher temperatures were applied during the extrusion-cooking. The application of higher pressure during the extrusion (compression changing) does not affect the extrudate viscosity; however, it affects on viscosity stability of products retained at a temperature of 95 °C. In some cases, the properties of extrudate may be arranged by amylose bounding with fatty acids or monoglycerides.

Another factor determining changes in the starch molecules during the extrusion-cooking process is the pressure and the values of existing shearing forces. In order to obtain certain technological properties of extrudates, which are often semi-finished products intended for further processing, it is necessary to set proper parameters of the extrusion process. This is achieved by the use of screws with varying compression degrees, relevant rpm of the working screw, appropriate die size, SME input, etc.

3. Starch transformation by thermo-mechanical treatment

Starch can be modifying by enzymatic, chemical or physical methods depend on processing and application fields of final products. Different types of processing, based on disruption and melting the semi-crystalline structure of starch may be used to transformation of native starch to form a thermoplastic starch (TPS) starch. Thermo-mechanical treatment that combines temperature and shear stress, like extrusion or injection moulding, is useful to transform granular starch into TPS. TPS modify by these methods may be applied as basic raw material or partial replacement of plastics for packaging materials applications in selected areas of food industry, horticulture, agriculture, but also for biomedical and cosmetic industries as gels, foams, films or in the form of a membrane with defined properties or biodegradability (Yimlaz, 2003).

Starch modification with thermomechanical treatment is difficult because of important increase of starch viscosity during heating and shearing, what may be the reason of uncontrolled dextrin's formation and browning reaction, especially when temperature above 100 °C is used. Some kind of plasticizers may improve intensity of starch transformations, i.e. water is most commonly used and the minimum moisture content required for starch gelatinization is around 33%. There are many studies about the different transformations of starchy material to thermoplastic forms with intermediate and high water level. High water content in the mixture also influences the onset temperature, glass transition temperature and rheological properties of molten materials (Della Valle et al., 1995, Igura et al, 2001, Nashed et al., 2003, van Soest et al., 1996a). In many scientific publications also other plasticizers were examined i.e. monoglycerides or glycerol, as flexibility improvers (Schogren, 1993). Addition of glycerol is of influenced on the onset of gelatinization and results in an increase in the activation energy for the melting of the starch crystallites and results in higher glass transition temperatures and higher interactions forces between glycerol and starch polymers (Della Valle et al., 1995, van Soest & Knooren, 1997, van Soest et al., 1996b, You et al., 2003). During the extrusion process high shear stresses and high values of energy input take place and under these conditions the melting process may be enhanced (Della Valle et al., 1995). Specific mechanical energy (SME) values necessary to transformations decrease with increasing water level in raw material. Corn or waxy corn and wheat or barley starch were most often investigated as basic thermoplastic raw materials (Nashed et al., 2003, Barron et al., 2000).

Modifications in the presence of plasticizers can be done by thermomechanical processing techniques like heating, kneading, injection, compression or vacuum moulding or extrusion below 150 °C (Yimlaz, 2003). Depending on the starch origin some specific differences are observed. Wheat, corn or potato starches behave different for different plasticisers or lubricants content and different processing conditions.

Test results achieved with different types of rheometers used to simulate the thermomechanical conditions differ and depend on the equipment used. The influence of the intensity of the treatment can be tested with a batch mixer, a cone and plate rheometer, a Shear Cell device with well defined shear rate, a Couette-type device with variable eccentricity or a compression moulding device. All these devises can simulate to a certain extent different temperature and treating time processes i.e. mixing-kneading, extrusion-cooking or injection moulding (Yimlaz, 2003, Peighambardoust et al., 2004, Peighambardoust et al., 2007, van den Einde et al., 2004, van den Veen, 2005). The shear rate, temperature profile, residence time during treatment influence simultaneously the starch behaviour and properties. Also in these research results the changes in properties in excess of water were observed. Most studies on the influence of water on the properties were carried out at high water content, but it is well known that extrusion-cooking or injection moulding processes are done with a limited (10-30%) water content. Properties of products can be measured by WAXS or intrinsic viscosity (Barron et al., 2000). Product behaviour and properties after extrusion with a high level of water content are not always acceptable because of crystallinity, retrogradation and stress-strain profiles of the materials expressed by the tensile strength or elongation (van Soest & Knooren, 1997). Depending on the amount of glycerol in TPS the product may be in its glassy or rubbery state at ambient temperature (Yimlaz, 2003). In the presence of sufficient water or glycerol under gelatinization conditions native starch becomes gel-like in appearance or properties but during the thermoplastic processing behaves like a polymer melt. The tests with different starch origins can be found in literature (Yimlaz, 2003, Peighambardoust et al., 2004, van den Einde et al., 2004).

Tests of starch melting usually are performed in a two-bladed counter rotating batch mixer Brabender Mixograph simulating mixing-kneading conditions interfaced with a computer and control unit. Wójtowicz (2009) tested various starches with water and glycerol according to following procedure: mixer temperature was set to 85 °C and heating was started directly after chamber closing and rotation of screws was increased from 5 to 100 rpm during 3 minutes. Comparison of starches origins were performed at the same temperature-time but screw rotation increased from 5 to 80 rpm. Samples were treated during 10 minutes in total. During mixing the torque was recorded continuously.

In the study presented by Wójtowicz & van der Goot (2005) investigations results on similar starch–glycerol mixtures with limited water addition (below 30%) subjected on the heating and shearing behaviour are presented. The purpose of the treatment was to use special designed shearing device - Shear Cell for obtains a starchy molten phase under thermo-mechanical processing similar to extrusion. This new equipment is based on the cone and plate rheometer ideology on a pilot scale (Figure 3). There is possibility to isolate singular parameters during processing like temperature, rotation speed or shear stress in this equipment (van den Einde et al., 2003). The details of this shearing device can be found elsewhere (Peighambardoust et al., 2004, van den Einde et al., 2004, van den Veen et al., 2004). The based material was potato starch with addition of glycerol 20-25% and with 5-20% amount of added water (w/w). Treatment temperature was selected on 85 °C for samples with 15 and 20% of water added, 88 °C for samples with 5 and 10% of water added and 115 °C for starch – glycerol mixtures. The rotation speed was 10 rpm during first 2 minutes and increased simultaneously to 100 rpm during 3 minutes. The torque changes during the treatment were recorded.

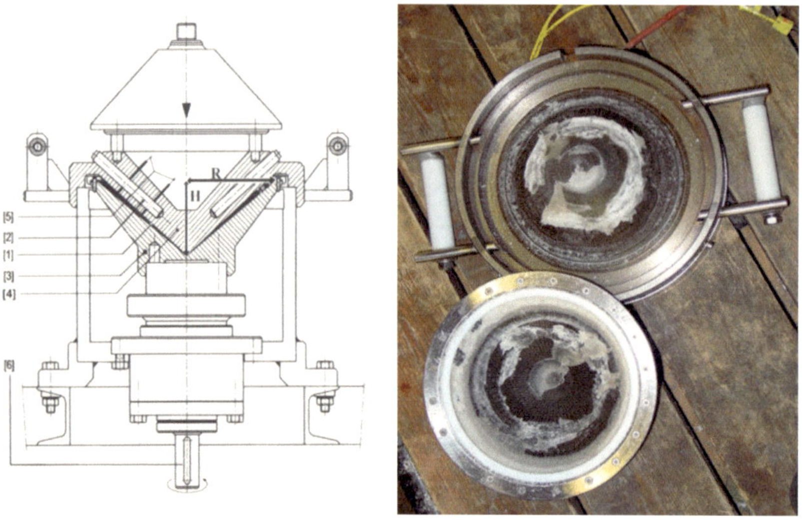

Figure 3. Shear Cell equipment scheme and after starch processing: 1- shearing zone, 2 – heating elements, 3 – rotating plate, 4 – non-rotating cone, 5 – thermocouple, 6 – torque measurement point. Cone angle = 100°, shear zone angle = 2,5°, r = 0,1 m, h = 0,082 m (Wójtowicz & van der Goot, 2005)

Research results showed that the addition of water in amount from 5 to 20% influenced on almost every recorded parameter during treatment in Brabender Mixograph. During treatment starch-glycerol-water mixtures it was observed decreasing start melting temperature from 80 °C for mixtures with limited water addition (5%) to 65-70 °C for samples with 20% of added water. Also the time needed to start melting of samples decreased with increasing of water addition. It seems to be that water becomes a plasticizer for starch and this is in accordance with previous reports (van Soest et al., 1996a). Also the effect of water addition on torque values and decrease of torque with increasing water content in mixtures was observed.

It was also found that potato starch required lower melting temperature than wheat and corn starch with glycerol and water addition but simultaneously higher maximum torque values were reported during the melting tests. The lowest torque values were reached for corn starch samples independent on water addition. The temperature range needed to melting beginning was 80-95 °C for corn starch-glycerol mixtures, 78-91 °C for wheat starch-glycerol mixtures and 78-88 °C for potato starch-glycerol mixtures mixed under 80 rpm used, depend on water content in the treated sample. Higher water level influences on lower temperature reached and lower max torque values during tests (Fig. 4).

During the tests the effect of water addition on torque values and decrease of torque with increasing water content in mixtures was observed. Torque values reported for mixing-kneading were quite low comparing the extrusion process and Shear Cell treatment. On the Figure 5 and 6 there are shown potato starch-glycerol (80-20) mixtures behaviours during treatment with different water addition in similar condition (85 °C heating temperature and 100 rpm) in Brabender Mixograpf and Shear Cell, respectively. Unfortunately mixtures with low amount 5% of water added showed the highest values of torque and increasing the water content influence on lower torque in both types of treatment.

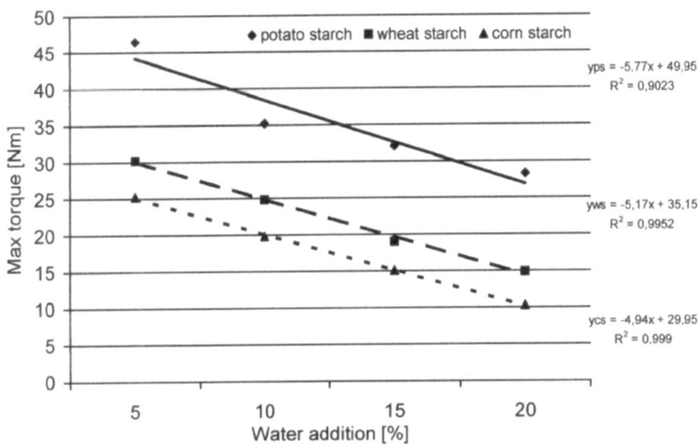

Figure 4. Torque values during mixing-kneading of various starch origins with 20% of glycerol addition and different level of water added (Wójtowicz, 2009)

Differences between curves in Fig. 5 and 6 may be important for definition of shearing influence on samples behaviour and properties. Also time of the beginning of rheological changes is similar except sample with 5% of added water. During intensive thermomechanical treatment in the Shear Cell shorter time required to starting changes inside the structure is observed. It may be explained much higher shear stress during shearing-heating in Shear Cell (Wójtowicz, 2009). It was not possible to start the melting process in potato starch-glycerol mixture without water addition because of to low temperature of heating in Brabender device equipped with water heating system and maximum temperature which can be achieved is 98 °C during heating. After tests in Shear Cell it is known now that start-melting temperature for these recipes is about 115 °C (Wójtowicz& van der Goot, 2005).

It is also important that increasing of total amount of plasticizers (water and glycerol) influenced on lower torque values during measurements, as showed on the Figure 7 and 8. Increased glycerol addition also has a strong effect on torque results. Nashed et al. (2003) reported through DSC that glycerol behaves as an anti-plasticizer because of hindering the gelatinization process and linear increase of onset temperature with increasing glycerol content was observed during treatment of wheat starch-water-glycerol mixtures. During thermo-mechanical treatment of starch-glycerol mixtures it was clear that higher glycerol addition influenced on decreasing melting or gelatinization time and temperature and also torque during treatment decreased (Wójtowicz & van der Goot, 2005).

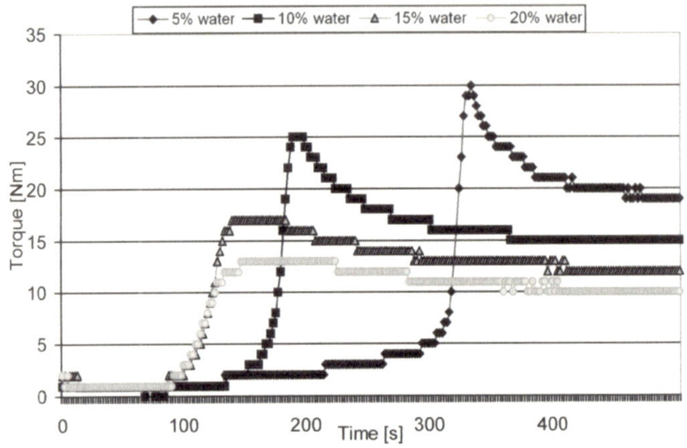

Figure 5. Torque values during treatment in the Brabender Mixograph of potato starch-glycerol (80-20) mixtures with different water addition (Wójtowicz, 2009)

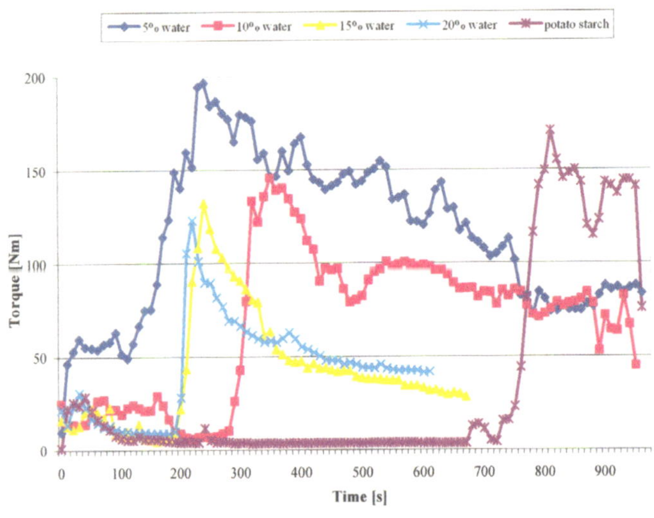

Figure 6. Torque values during treatment in the Shear Cell of potato starch-glycerol (80-20) mixtures with different water addition (Wójtowicz & van der Goot, 2005)

Wójtowicz and van der Goot (2005) noted that there were also visible differences in transparency and flexibility of achieved samples after different treatment type (Fig. 9). After mixing-kneading in Brabender equipment samples became elastic, with foam consistency and non-transparent, milky white colour. Only for samples of potato starch with small amount of water added material became a little brittle and partially transparent with 5% added water and completely transparent, glassy look like with higher level of water added. But these last samples were much stickier after treatment then the others. When they were warm behaves flexible and easy undergo elongation and formation different shapes. After cooling at room temperature material became hard to formulation, and no more flexible. All the samples with addition of glycerol and heated-sheared in the Shear Cell had visibly transparent glassy-like appearance and smooth surface, directly after processing in worm stage they were easily to elongation, showed rubbery properties and they were longer elastic after cooling to room temperature (Wójtowicz & van der Goot, 2005). This phenomenon may be the result of pressure differences between both types of equipment during processing, in Shear Cell the pressure was much higher than in mixer and air bubbles or steam formed in transformed material disappeared pushed out through silicone seal by pressure inside the apparatus. Mixing-kneading equipment had lower pressure in testing chamber and it was not possible venting processed material, it may be the reasons of foamy structure of modify starch.

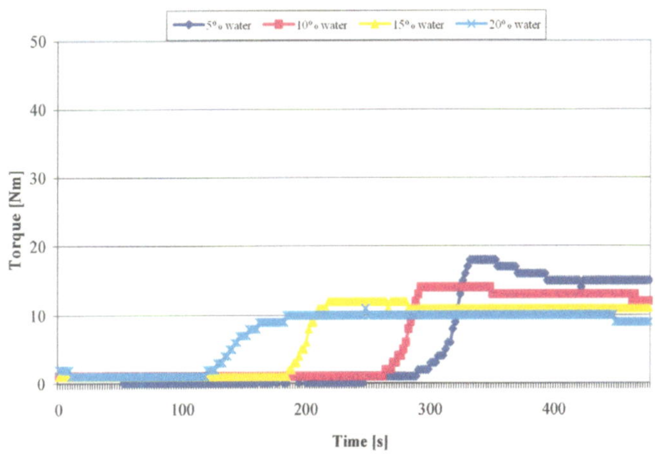

Figure 7. Torque values during treatment in the Brabender Mixograph of potato starch-glycerol (75-25) mixtures with different water addition

The intrinsic viscosity of starch (η) is very sensitive for thermomechanical treatment and the degradation of molecular weight compounds in starch, therefore it can be used as a method to molecular weight measurement. The authors (Wójtowicz & van der Goot, 2005) report an intrinsic viscosity of native starch and sheared-heated samples measured according the Ubbelohde viscometer method (Cunningham, 1996) at 25 °C.

Intrinsic viscosity of native potato starch was 369,8 ml/g. Intrinsic viscosity of samples treated in Brabender Mixograph varied from 174 ml/g for starch-glycerol (80:20) mixtures with 5% water added to 202 ml/g for starch-glycerol (75:25) with 20% water added. These values compare to native starch were much lower and it may suggest breakdown of starch granules during Brabender treatment. But differences between samples were smaller than observed after shearing-heating process in Shear Cell. Results for samples processed in Shear Cell also were lower than for native starch and values variation was between 108 for samples of starch-glycerol (80:20) mixtures with 10% water added and 215 ml/g for starch-glycerol

(75:25) without water added (Wójtowicz & van der Goot, 2005). It means that during shearing-heating treatment in Shear Cell high macromolecular degradation takes place. Intrinsic viscosity values were slightly dependent on glycerol content in mixtures and with increasing glycerol content decrease of intrinsic viscosity were observed in potato starch-glycerol-water mixtures treated in the Shear Cell under similar conditions.

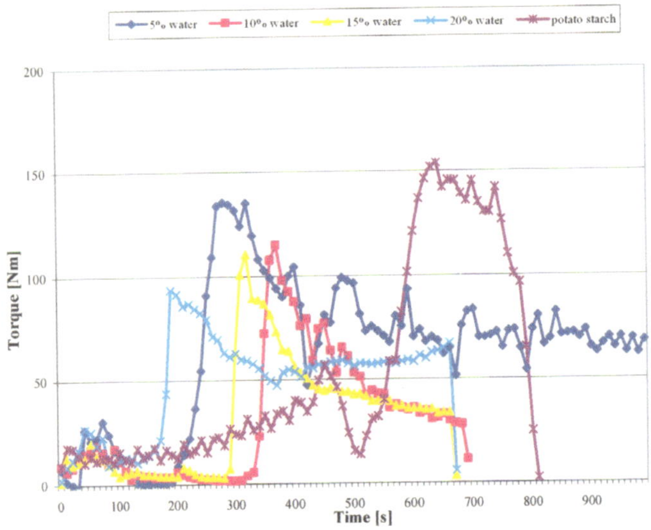

Figure 8. Torque values during treatment in the Shear Cell of potato starch-glycerol (75-25) mixtures with different water addition

Figure 9. Overview of samples after treatment in: a) BrabenderMixograph, b) Shear Cell

These results are in accordance with Fujio et al. (1995) for potato, corn and wheat starches, van den Einde et al. (2003) for cornstarch and also Rushing & Hester (2003) for polymers. Intrinsic viscosity is very sensitive especially on thermomechanical degradation. Low values of shear stress influenced on lower intrinsic viscosity of treated starch-glycerol mixtures. But in all cases variations between intrinsic viscosity's values are quite small and it can be concluded that almost the same degree of starch molecules degradation was noted. In comparing with native starch intrinsic viscosity values reduced over 50%, so residue starch molecules was broken by thermomechanical treatment. By comparing those data with data generated by van den Einde et al. (2003), it can be concluded that potato starch was less thermostable than corn starch.

4. Starch modification by extrusion-cooking

Native starch is not always suitable for practical use. Therefore, various starch modification techniques have been developed for food and non-food applications. Generally there are chemical methods. In many cases, especially in the food sector, chemically modified starch can be replaced by that extrusion-cooked. During the extrusion-cooking physical and chemical transformation of starch takes place and no chemicals are needed. Baro-thermal treatment causes gelatinisation of starch, accompanied by rupture of intermolecular bonds, resulting in rupture of starch grains and significantly increase of water absorption.

The degree of changes in starch depends on properly selected process parameters and the residence time of raw material in the extruder. That allows us to create the expected properties of the obtained modified starches, including the degree of gelatinization and viscosity of the gels. These products may find wide application in food industry as food additives, very often by replacing chemically modified starchy products. Extrusion-cooked starch may find its use as a component of food products in the manufacture of instant products, different kinds of fillings in the confectionery industry, as a gelling agent, structure stabilizer and water- or fat-absorbent fillers. That may be very attractive from the consumer point of view. Application extrusion-cooking is a relatively cheap alternative in the production of modified starches.

4.1. Materials and methods

Potato starch Superior type was purchased from PEPEES Company (Lomza, Poland). Its moisture content was 17%. During the extrusion-cooking process the 4 levels of moisture content of raw material (17, 20, 25 and 30%) were used.

Extrusion-cooking of potato starch was carried out using a modified single screw extrusion-cooker TS-45 (Polish design) with L / D = 16, and the die with one opening with a diameter of 3 mm was uses. During the study three temperature of extrusion process (100, 120 and 140 °C) and a variable screw's speed (1.00, 1.33, 1.66 and 2.00 s^{-1}) were used. The process energy consumption was measured with a wattmeter connected to the extruder and the specific mechanical energy (SME) input was calculated (Janssen et al, 2002, Mitrus, 2005a, Mitrus & Moscicki, 2009, Wolf, 2010).

Degree of starch gelatinization was measured by enzymatic method in accordance with Polish standard PN-A-79011-11:1998.

Cross-sectional expansion index was determined as the diameter of extrudates divided by the diameter of the matrix opening (Moscicki, 2011). Measurements were done in 10 repetitions.

Water absorption index was determined according to the method of Anderson et al. (1970) with own modification. The extrudates were crushed using a laboratory mill to particles with a diameter less than 0.3 mm. A 0.7 g ground sample was suspended in 7 ml of distilled water at 20 °C in a tared centrifuge tube, stirred intermittently over a 10 min period. The resulting suspension was centrifuged at speeds 250 s^{-1} for 10 minutes in T24D type centrifuge. The supernatant liquid was poured into a tared evaporating dish. The remaining gel was weighted and the WAI was calculated as WAI = w_g/w_s(%), where w_g is a weight of gel and w_s is the weight of dry sample. Measurements were performed in 6 replications.

Water solubility index was determined from the amount of dried solids recovered during evaporation of supernatant obtained from the WAI analysis according to the method of Harper (1981). Results were calculated from formula WSI = w_{ds}/w_s (%), where w_{ds} is the weight of dry solids of supernatant and w_s is the weight of dry sample. Measurements were performed in 6 replications.

The data reported was subjected to analysis of variance (ANOVA) by Duncan's test ($P<0.05$) using SAS 9.1 software.

4.2. Results and discussion

Specific mechanical energy during potato starch extrusion-cooking depends on the process parameters. The research revealed that the values of SME were within a range 298.8-990 $kJkg^{-1}$ (0.083-0.275 $kWhkg^{-1}$). These values were lower than the values obtained by Della Valle et al. (1995). The lowest energy consumption was observed during the extrusion-cooking process of potato starch at 140 °C. The highest energy consumption was recorded during the extrusion-cooking process at a temperature of 100 °C (at moisture content of 17 and 20%) and at 120 °C (moisture content 25 and 30%).

Studies have shown the energy consumption dependence from extruder screw speed (Fig. 10). Effect of moisture content on the SME was inconclusive. When carrying out the process at 100 °C, little change was observed (decrease) in specific mechanical energy. At higher temperature of extrusion it was observed that an increase in moisture content of starch increases the rate of the SME. It was most likely caused by increasing viscosity of processed slurry. Due to the presence of water the starch melts and underwent liquefaction, resulting in lower glass transition temperature.

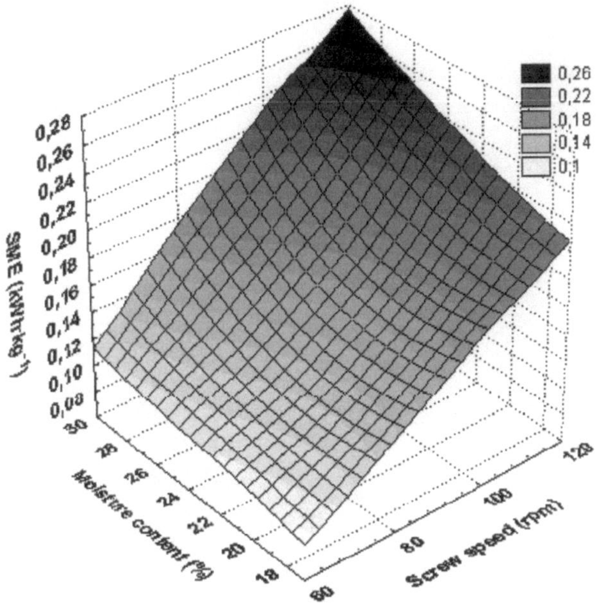

Figure 10. SME changes during potato starch extrusion-cooking at 120 °C

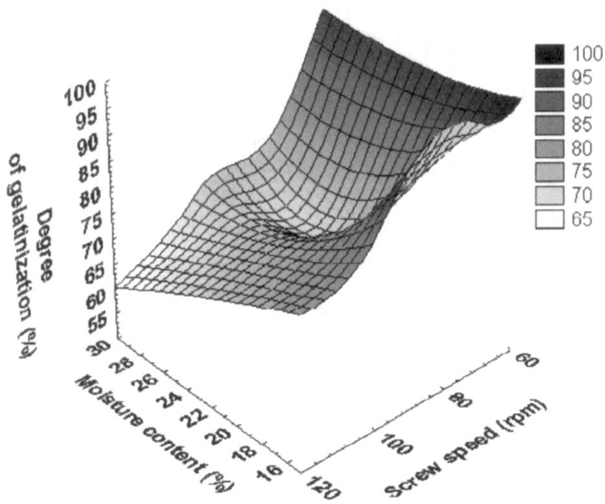

Figure 11. Degree of gelatinization of the potato starch extrusion-cooked at 100 °C

During our research has been noticed that the degree of potato starch gelatinization decreased with increase of extrusion-cooking temperature and when processed starch contained more water (Fig. 11). That was evident especially when high rpm of the screw was used. The highest degree of gelatinization (96.5%) was recorded for starch extruded at 100 and 120 °C at the moisture content of 17%. The lowest degree of gelatinization (41%) was recorded for starch extruded at 140 °C at the moisture content of 30%. Such relatively low degree of gelatinization may happen due to relative short residence time of the dough in the extruder, which could stimulate the process. Nevertheless that has to be proved in more detailed measurements. Extrudates obtained under these conditions had properties of thermoplastic starch.

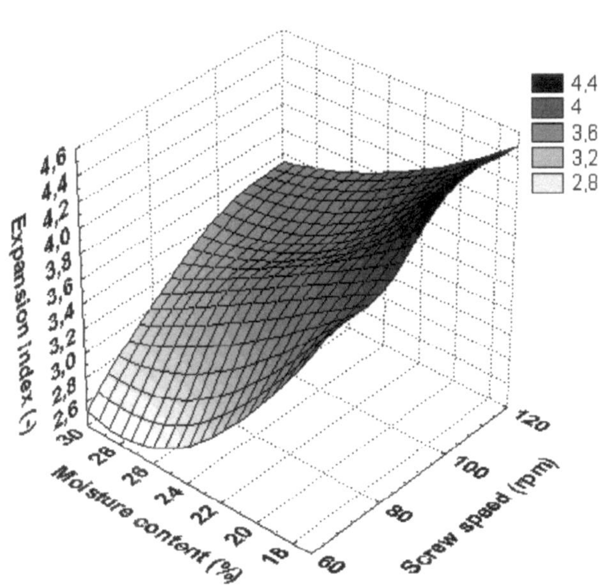

Figure 12. Expansion index of the potato starch extrusion-cooked at 100 °C

Measurements of the expansion index of extruded potato starch showed that its value decreases with moisture content increase (Fig. 12). Extruder screw speed increase caused increase of expansion index. This is a common phenomenon for the most of the extrudates. Extrudates were characterized by a typical structure, resembling a honeycomb structure. At high moisture contents of raw material (25 and 30%) the formation of "glassy" extrudates with a low degree of expansion was observed. This effect was particularly visible for extrusion temperatures of 120 and 140 °C. Extrudates obtained under these conditions had homogeneous, amorphous structure without pores and steam bubbles. Such behavior of the processed material is related to the glass transition temperature (T_g) of starch and temperature of steam bubbles formation (T_p) (Della Valle et al., 1997, van Soest et al., 1996c). When the product temperature is higher than T_g and close to T_p, bubbles growth stops and the extrudate obtained its structure. At high moisture contents of raw material T_g and T_p may be lower than the temperature at which extrudate shrinkage begins as a result of condensation (about 100 °C). At low moisture contents T_p is much higher than 100 °C and, therefore, an increase of steam bubbles before they end up collapsing, resulting in a high degree of expansion.

Native potato starch has WAI approximately 97% and WSI approximately 0.25%. The study showed that the baro-thermal modification of starch significantly affects on its water absorption and solubility in cold water.

The research revealed that the value of WAI initially increased with increasing screw speed, and decreased at high speeds of extruder screw. With the increase in moisture content of processed potato starch the increase of water absorption was observed (Fig. 13). WAI values of the extruded potato starch ranged from 282 to 569% and generally did not deviate from the values obtained for a typical starch extrudates (Mercier et al., 1998).

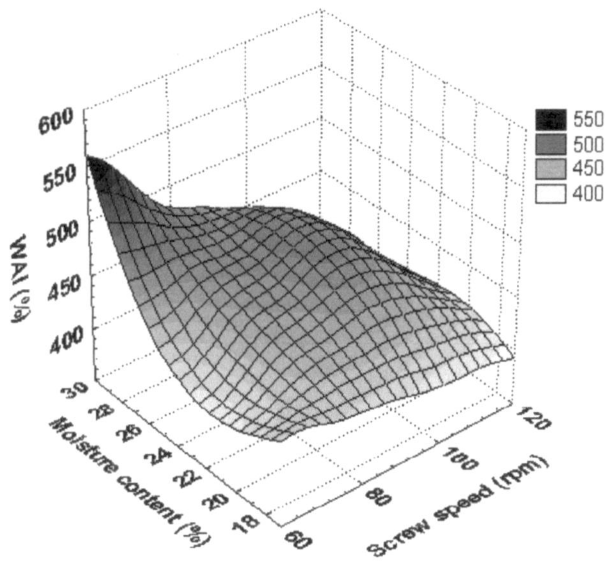

Figure 13. WAI of the potato starch extrusion-cooked at 100 °C

The highest values of water absorption were observed for starch extruded at 100 °C, the lowest for starch processed at 120 °C. It was connected with the progress of the degree of gelatinization. The increase in extrusion temperature to 140 °C caused a re-appreciation of WAI. Most likely due to lower glass transition temperature, the starch was more rapid melting, and liquefaction, which could limit the degree of its degradation. Analysis of variance of the WAI, depending on the moisture content and screw speed, showed statistically significant differences, with significance level of 0.05. Only for extrudates obtained at 140 °C there were no significant statistical differences in the relationship between water absorption and the screw speed.

The research showed that the value of the WSI increased with screw speed increase. The starch moisture content increase caused reduction of extruded starch solubility (Fig. 14). The highest values of solubility (40%) were obtained for the modified starch at 120 °C. Process temperature growth caused an initial increase (120 °C) and than decrease (140 °C) in starch WSI. The changes of the solubility of starch were related with changes in the process of gelatinization and starch degradation due to starch moisture content increase. At low to intermediate moisture content and high temperature, the water contained in starch might behave like a lubricant (Igura et al., 2001). Degradation of starch progressed by increasing extruder screw speed at low moisture content because less lubricant (water) was available. Analysis of variance of the WSI, depending on the moisture content and screw speed, showed statistically significant differences, with significance level of 0.05.

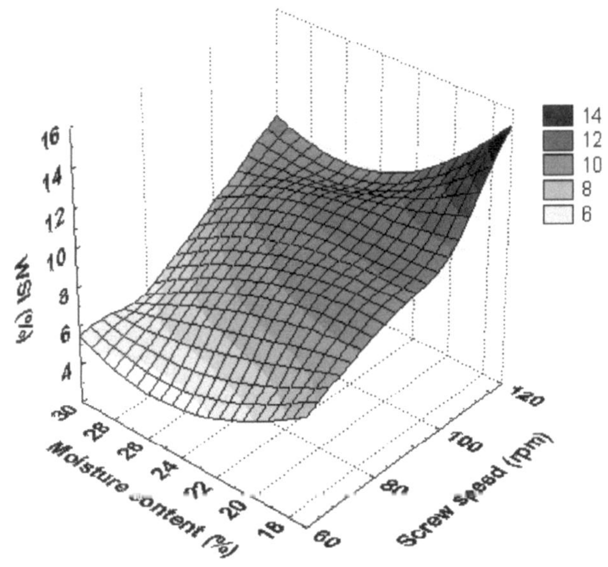

Figure 14. WSI of the potato starch extrusion-cooked at 100 °C

5. Thermoplastic starch

The interest to use starch as a basis for packaging material originates to the 1970's when environmental awareness increased drastically. Since then a steady development of new products can be seen. The possibility to compete in price with traditional materials, like plastics, has always been indispensable for the general acceptance of these new materials.

Starch biodegrades to carbon dioxide and water in a relatively short time compared with most synthetic polymers. Considering some drawbacks of the existing technologies of biodegradable materials manufacture, in the recent years there have been started large-scale researches to increase amount of starch in starch-plastic composites to the highest possible level. The final objective of these investigations is to obtain commercial items for one-time use, produced from pure starch and to exclude synthetic polymers from the formulation. Thermoplastic starch (TPS) seems to be a perfect solution because it can be processed with conventional technologies used in synthetic plastic manufacture (extrusion, injection moulding).

To obtain thermoplastic starch, thermal and mechanical processing should disrupt semi crystalline starch granules. As the melting temperature of pure starch is substantially higher than its decomposition temperature there is a necessity to use plasticizers, for example water. Under the influence of temperature and shear forces, disruption of the natural crystalline structure of starch granules and polysaccharides form a continuous polymer phase is reported. TPS produced from starch plastified only with water becomes very brittle at room temperature. To increase material flexibility and improve processing other plasticizers are also used, e.g. glycerol, propylene glycol, glucose, sorbitol and others. To improve the mechanical properties of TPS based materials also other additives can be applied, like emulsifiers, cellulose, plant fibres, bark, kaolin, pectin and others.

5.1. Materials and methods

5.1.1. Injecting moulding

The blends of a potato starch Superior type, produced by AVEBE Company (NL), a glycerol (98,5% purity) produced by Polfa Odczynniki (PL) and cut flax fibres supplied by Polish producer BELKO Ltd Co., were used as a basic material during processing. The raw materials were mixed in a ribbon blender; a glycerol content varied from 18 to 30 wt % blend mass, and the fibres from 5% to 10% at the selected blends. After mixing the bland samples were packed in airtight plastic bags and stored for 24 hrs to intensify glycerol penetration into starch grains. Immediately before the extrusion the blends were remixed.

Production of thermoplastic starch (TPS) was made in two steps. In first step starch, glycerol and flax fibre blends were extruded in a twin-screw extruder PASQUETTI, an Italian design, characterized by L/D = 5, and the screws diameter - 45mm. The extruder's die was fitted with a bronze matrix having one hole of the diameter ϕ = 3 mm. Thermoplastic starch was made with the screw's speed of 1,5 s^{-1}. The temperature of extrusion in particular barrel parts and a die were between 80 °C and 140 °C; pressure in the die fluctuated between 10 MPa and 18 MPa. The product was cut by high-speed knife, which helped obtain pellets of previously set small size (Oniszczuk, 2006).

In the second step, the TPS pellets were processed on the injection moulding machine ARBURG 220H90-350, L/D = 20,5. The injection speed was maintained at the level of 0,07-0,09 ms^{-1}, injection time - 3s, and the temperature of the processes reached from 100 °C to 180 °C. The samples of injection mouldings (Fig. 15) were used during further examination of mechanical properties of mouldings (Oniszczuk, 2006).

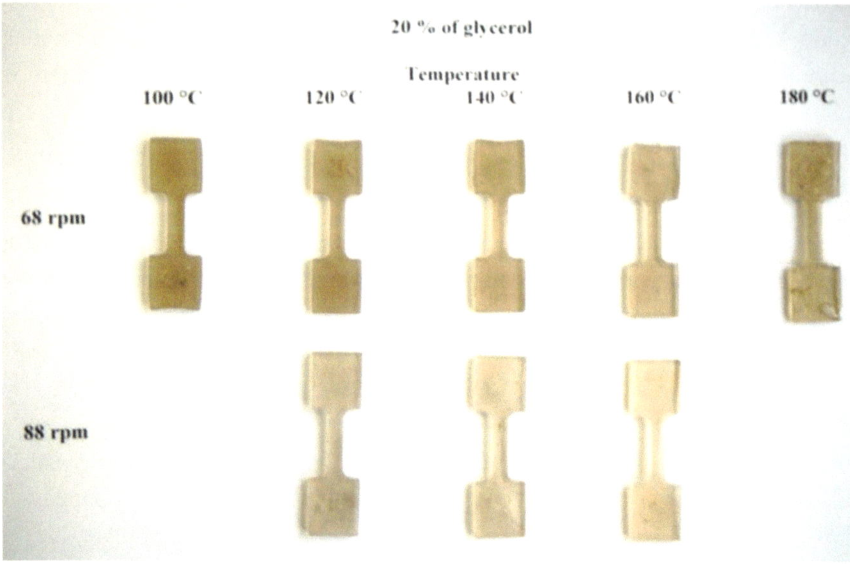

Figure 15. Samples of TPS mouldings

5.1.2. Film blowing

Similar procedure of the TPS granulates production was used to obtain half product for the film blowing. The basic materials were the blends of 2 main components: potato starch and a glycerol. Selected blends were enriched with the addition of the emulsifiers: polyoxyethylene sorbitan monolaurate (Tween 20) and glycerol monostearate at the amount up to 2%.

The process of film extrusion with film blowing method was conducted on a line specially designed for film manufacture in the Department of Food Process Engineering, Lublin University of Life Science (PL), based on a single screw plastic extruder of L/D=35 (see Fig. 16). That line was produced by SAVO Ltd Co., Poland. TPS film was produced using 2 screws of varied geometry (a compression ratio: 2,0 and 3,5), and the screw rotational speed ranged from 50 till 90 rpm. The film was extruded at the barrel and crossdie temperature ranging 70 – 155 °C.

5.1.3. Measurement of physical properties

The measurements of glass transition temperature T_g of thermoplastic starch were done with the use of the Differential Scanning Calorimetry on the Perkin Elmer DSC 7. The samples of thermoplastic starch 7 – 10 mg mass were heated from the temperature of 25 °C to 180 °C at 10 °Cmin^{-1} speed, and next cooled at the same rate down to 25 °C and finally heated up to 180 °C. The thermal transitions were calculated from the second heating cycle. To confirm the obtained results the tests were repeated in a DSC 2920 modulated DSC TA Instruments. The samples were heated from 0 °C up to 150 °C at the rate 1 °Cmin^{-1} and then cooled at the same rate to 0 °C (Mitrus, 2005b).

Figure 16. TPS film blowing

The examination of the mechanical properties of TPS mouldings was performer on a universal texture appliance Zwick type BDO-FBO0, 5TH equipped in the head 0,5kN. The travel speed of the head was 3 mmmin^{-1}. The test focused on the maximum stress and maximum elongation during stretching (Oniszczuk, 2006).

The assessment of the original shrinkage of the biopolymer mouldings was done by means of the micrometer screw initiating the measurement 24 hours after producing the samples.

The tensile testing of film proceeded at the tester Zwick type Z2.5/TN1S using the standard tensile test ISO 527. There was applied the crosshead 1 kN, sample length 100 mm, sample width 17 mm, each sample thickness was measured with micrometer. The crosshead speed for initial stress was 10 mmmin^{-1}, speed up to flexibility limit – 50 mmmin^{-1}, test speed – 200 mmmin^{-1}. The measurements were made on maximum strength, tensile strength, elongation at the maximum strength and tensile.

5.2. Results and discussion

DSC technique has been used often to study the glass transition temperature of thermoplastic starch. The results differ from one another significantly (Shi et al., 2007, van Soest, 1996, Talja et al., 2007). In accordance to Yu & Christi (2001) some key factors, such as sample preparation, type of pan and measurements conditions, have an affect on the results of thermal behavior of starch as measured by DSC. The tests show that both, amylose and amylopectin had a higher T_g`s in absence of glycerol. The estimates demonstrated that the T_g of dry amylose and amylopctin is 227 °C, while Bizot et al. (1997) assessed the dry starch T_g – 332 °C. What's more, to lower the T_g of potato starch closer to the ambient temperature 0,21 g of water should be used for 1g of starch (Bizot et al., 1997, Myllarinen et al., 2002). Myllarinen et al. (2002) confirmed that the T_g of amylose and amylopectin can be equal to the ambient temperature when the water content is 21%, however at the same glycerol level the T_g can be still as high as 93 °C. It can be concluded that glycerol is a less effective plasticizer than water. On the basis of computations they claim that in order to lower a T_g value to the ambient temperature 35% glycerol should be applied.

The glass temperature measurements revealed that with glycerol content growth in material blend, the T_g of the obtained material decreases almost linearly. The highest observed T_g was 132 °C for 15% glycerol, the lowest was 18 °C at a glycerol level of 30%. Figure 17 shows the changes of the glass transition temperature with changing glycerol content. The moisture content of all the mixtures was 15%. This results are similar to this obtained by Graff et al. (2003).

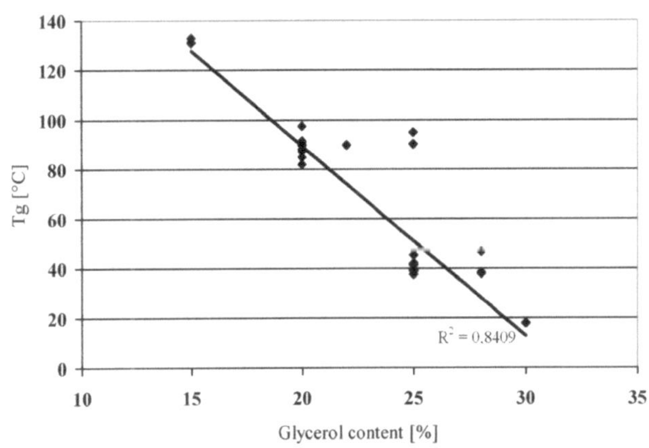

Figure 17. Influence of the glycerol content on the T_g of thermoplastic starch

The mechanical properties of biocomposites depend on a number of factors. These are the quantity and type of fibre added to the material, but also type and amount of plasticizers, finally - the production temperature plays also important role. A predominant influence on the mechanical properties of biocomposites have both: natural fibres and plasticizer. During our investigations it has been observed that the addition of fibres enhances the mechanical strength of mouldings samples. The addition of extra plasticizer causes the decline in its maximum stress.

Figure 18 shows the relationship between the maximal stress and the content of flax fibres in the samples containing 20, 22 and 25% of glycerol and produced at the material injection temperature of 120 °C. On the basis of the performed examination, it can be stated that the addition of flax fibres positively influenced the mechanical properties of biopolymer mouldings. It was noted that together with the percentage growth of the fibres content the mechanical strength of samples was improved. The samples with 20% of glycerol content and 10 % of flax fibres were of the highest strength (26,5 MPa). In the case of mouldings obtained from granulate containing 22% of glycerol, the addition of 5% and 10% of fibres slightly influenced the improvement in strength. It was noted that together with the growth of the glycerol content in mouldings containing flax fibres their mechanical strength dropped. Glycerol acts like diluent and weakens the intermolecular bonds between flax fibres and starch. The lowest mechanical strength was observed with mouldings produced from granulates containing 25% of glycerol. The same tendency was registered in the case of samples obtained at different production temperatures the highest mechanical strength was displayed by mouldings produced at the material injection temperature of 140 °C (maximal stress 27,8 MPa), and the the lowest mechanical strength was noticed for mouldings obtained at the temperature of 180 °C (10,8 MPa).

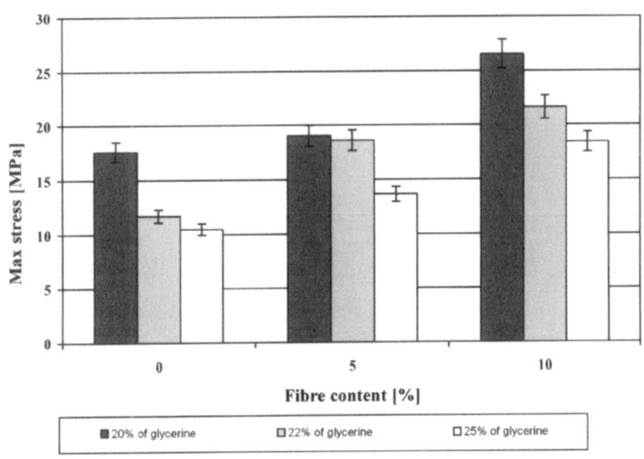

Figure 18. Relationship between the maximal stress and the fibre content in samples (sample injection temperature 120 °C)

Another important parameter is the maximal elongation of the samples during stretching. For TPS without fibres added, it can be stated that for the full spectrum of injection temperatures and for all glycerol concentrations investigated the maximal elongation coincides with increasing injection temperature and with the increase of the glycerol content of the sample. The addition of fibres affected the elongational behaviour of the material considerably.

During the research it was noticed that the maximal elongation values coincided with the growing temperature of material injection and with the increase of the glycerol content in the sample. The largest maximal elongation (58,5%) was recorded at the level of 25% of glycerol content in mouldings and the material injection temperature of 180 °C. The lowest maximal elongation of 10% was recorded with mouldings having 20% of glycerol. To conclude, it corroborates the positive influence of plasticizer on the improvement of moulding flexibility.

Addition of flax fibres influenced the drop of the maximal elongation of biopolymer mouldings. The growth of the percentage share of fibres in the mixture worked upon the decrease of the flexibility of samples during the tension test (Fig. 19).

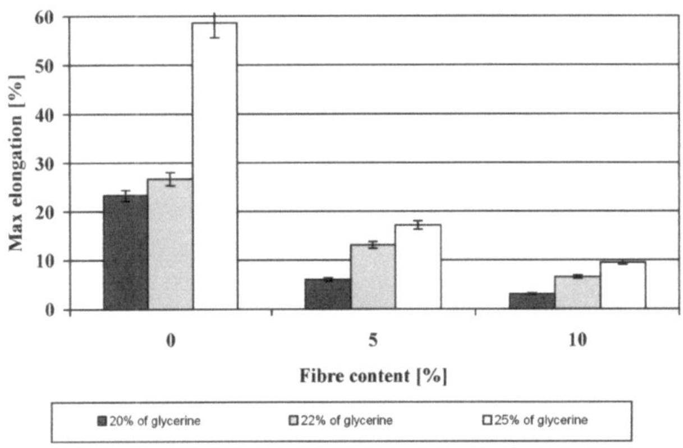

Figure 19. Relationship between the maximal elongation and the flax fibre content in samples (injection temperature 180 °C).

With the application of the production temperature of 180 °C, the largest maximal elongation (17,1%) was observed with the samples containing 25% of glycerol and 5% of flax fibres. The lowest maximal elongation (3%) was observed with mouldings produced from granulate containing 20% of glycerol and 10% of flax fibres. In the case of mouldings produced at the injection temperature of 120 °C, the addition of fibres barely influenced the reduction of the maximum elongation and amounted to 10,5% for mouldings containing 5% of flax fibres and 9,7% for mouldings of double fibre content. It proves the negative impact of fibres on the flexibility of biopolymer mouldings containing such material.

Besides the improvement of durability, the fibres used for the mouldings manufacture stabilize the shape and decrease the original shrinkage of ready products. Oniszczuk & Janssen (2009) describes that an increased linen fibre content is decreased the values for original shrink. The addition of plasticizer (glycerol) had an adverse effect on the original shrinkage.

The examination of original shrinkage of biopolymer mouldings showed that in the whole range of temperatures of sample production (100 °C, 120 °C, 140 °C, 160 °C and 180 °C) and after the time of shrinkage measurement (measured after 24 h) a similar tendency was observed. Together with the increase of the share of flax fibres, the value of the original shrinkage was dropping, regardless of the time when measurement of the shrinkage was made. However, together with the growth of the share of glycerol, the value of the original shrinkage of mouldings slightly increased (Fig. 20).

When examining the shrinkage of mouldings produced with 25% of glycerol share, considerable differences. After 24 h from producing mouldings, the shrinkage level was almost 11,7% in samples without fibres, 1,75% when applying 5% of flax fibres and 0,67% when 10% of fibres were added. For mouldings obtained with 20% of glycerol content, the shrinkage values were lower and least dependent on the addition of fibres. The shrinkage values for these trials did not exceed 1,5%, which indicates a significant stability of the material obtained in specific production conditions. The original shrinkage of mouldings without the addition of fibres was high and amounted to about 6% when using 22% of glycerol and about 12% with 25% of plasticizer. This indicates a stabilizing role of fibres, which maintained the original moulding structure in an almost unchanged shape. It is a very desirable feature when it comes to keeping the shape of ready packaging if produced from such biopolymers.

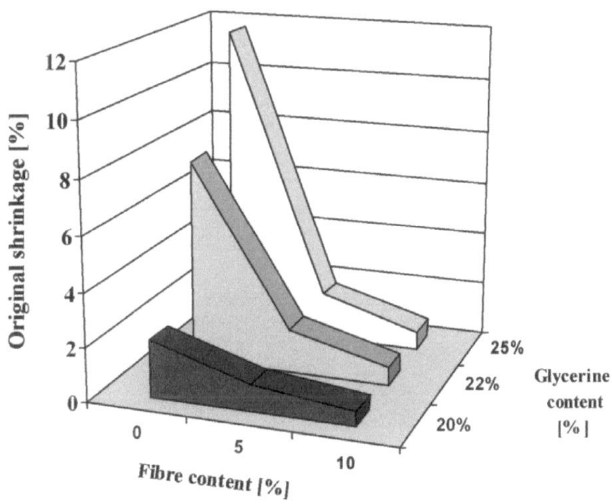

Figure 20. Influence of flax fibres content and glycerol content on the size of original shrinkage of (measured after 24 h); injection temperature 120 °C

The tests of the film extrusion showed that the best results were recorded for the starch mixtures with 20-25% glycerol. At processing temperature below 120 °C the material was not fully processed and some granulate residuals appeared at the film surface. The obtained films were thick, opaque and semi-transparent, that after some time lost flexibility and got brittle due to their drying up. When the higher pressing temperatures were applied as well as a screw with a mixer arm, films of good quality were obtained. They were flexible semi-transparent films readily put to the blow moulding. Unfortunately because of the crosshead with 1mm slit application and a low powered compressor the resulting film had minimum thickness 120μm, yet the production of much more thinner ones is possible.

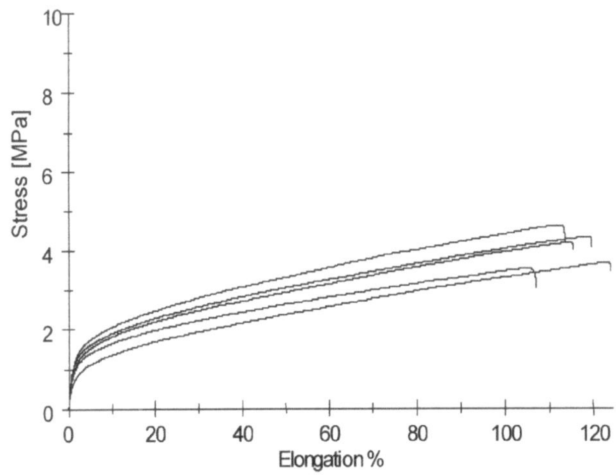

Figure 21. Example of the tests results of the TPS film strength

The addition of the emulsifiers: polyoxyethylene sorbitan monolaurate (Tween 20) and glycerol monostearate at the amount up to 2% has considerably improved film flexibility. Moreover, it was found that mixture damped up with only some water (2-5%) induced an increase of film flexibility and strength. Example of the tests results of the TPS film strength is shown on figure 21.

The analysis of the mechanical properties measurements of TPS films proved that the extrusion processing parameters, emulsifier presence and water content in the material exert a vital impact on film strength and its elongation. The use of the screw of 3,5 compression ratio, equipped with an extra mixing section affected film strength more, while a screw of 2,0 compression ratio influenced the film elongation in a greater measure. According to the expectations, the application of the screw of 3,5c.r. resulted in higher energy-consumption during extrusion processing.

6. Conclusive remarks

Extrusion-cooking technique allows creating the degree of gelatinization of processed starch. It is possible to achieve low or high level of gelatinization depending on the process parameters. This is especially important for food and feed applications.

Expansion index of starchy extrudates largely depends on the parameters of the extrusion process. Its value decreased with moisture content increase while increased with screw speed increase.

The extrusion process of starch increased the water absorption and cold water solubility. These changes are closely related to the course of the process of starch gelatinization; degradation and their extent depends on the extrusion parameters used

Modification of starch by extrusion-cooking technique is characterized by relatively low specific mechanical energy consumption. SME values were within a range from 298.8 to 990 $kJkg^{-1}$. Significant impact on the values of the SME had a screw speed, very little impact had a moisture content of raw material.

When the glycerol content in TPS increases from 15 to 30% the glass transition temperature decreases almost linearly from 132 to 18 °C at moisture contents 15%.

The performed examination of TPS injection moulding showed that it's possible to produce shaped biodegradable packaging materials. The addition of flax fibres to raw material mixture positively influenced the improvement of the mechanical strength of mouldings in the entire range of glycerol addition to mixtures (mouldings produced from the granulate containing 10% of flax fibres exhibited the greatest mechanical strength).

The increased glycerol content in the material and the growth of injection temperature resulted in better elasticity of mouldings. Its drop was caused by the presence of fibres in granulate.

The increase of the percentage share of glycerol in the mixture visibly contributed to the growth of the value of the original shrinkage of mouldings. The lowest values of the original shrinkage of samples were observed at the material injection temperature of 180 °C. The addition of flax fibres to mouldings in the entire range of their production temperatures positively influences the stability of shape and the reduction of shrinkage.

The most advantageous strength properties were recorded for the films with 20 – 25% glycerol. Polyoxyethylene sorbitan monolaurate (Tween 20) and glycerol monostearate in amount up to 2% have substantially improved (even by over 50%) film tensile susceptibility. The analysis of the mechanical properties measurements of TPS films proved that the extrusion processing parameters, emulsifier presence and water content in material exert a vital impact on film strength and elongation. The use of screw of 3,5 compression ratio equipped with extra mixing section affected film strength more, while a screw of 2,0 compression ratio influenced film elongation in a greater measure. According to the expectations, the application of the screw of higher compression ratio increases energy-consumption during extrusion processing.

Author details

L. Moscicki, M. Mitrus, A. Wojtowicz, T. Oniszczuk and A. Rejak

*Address all correspondence to: leszek.moscicki@up.lublin.pl

Department of Food Process Engineering, Lublin University of Life Sciences, Lublin, Poland

References

[1] Anderson RA., Conway HF., Peplinski AK.: Gelatinization of corn grits by roll cooking, extrusion cooking and steaming. Starch 1970;22 130-134.

[2] Barron C., Buleon A., Colonna P., Della Valle G.: Structural modifications of low hydrated pea starch subjected to high thermomechanical processing, Carbohydrate Polymers 2000;43 171-181.

[3] Bizot H., Le Bail P., Leroux B., Davy J., Roger P., Buleon A.: Calorimetric evaluation of the glass transition in hydrated, linear and branched polyanhydroglucose compounds. Carbohydrate Polymers 1997;32 33 – 50.

[4] Cunningham R.: Effect of processing conditions on intrinsic viscosity of extruded cornstarch, Journal of Applied Polymer Science 1996;60 181-186.

[5] Della Valle G.:,Boche Y., Colonna P., Vergnes B. The extrusion behaviour of potato starch, Carbohydrate Polymers 1995;28 255-264.

[6] Della Valle G., Vergnes B., Colonna P., Patria A.: Relations between rheological properties of molten starches and their expansion behaviour in extrusion. Journal of Food Engineering 1997;31 277-296.

[7] Fujio Y., Igura N., Hayakawa I.: Depolymerization of molten-moisturized-starch molecules by shearing-force under high temperature, Starch/Stärke 1995;47 143-145.

[8] Graaf RA., De Karman AP., JanssenLPBM.: Material properties and glass transition temperatures of different thermoplastic starches after extrusion processing. Starch 2003;55 80-86.

[9] Harper JM.: Extrusion of Foods. Boca Raton: CRC Press; 1981.

[10] Igura N., Katoh T., Hayakawa I., Fujio Y.: Degradation profiles of potato starch melts through a capillary tube viscometer, Starch/Stärke2001;53 623-628.

[11] Janssen LPBM., Moscicki L., Mitrus M.: Energy aspects in food extrusion-cooking. International Agrophysics 2002;16 191-195.

[12] Mercier C., Linko P., Harper JM. Extrusion cooking. (2nd ed.). St. Paul: American Association of Cereal Chemists, Inc.; 1998.

[13] Mitrus M.: Changes of specific mechanical energy during extrusion cooking of thermoplastic starch. TEKA Kom. Mot. Energ. Roln. 2005a;5 152-157.

[14] Mitrus M.: Glass transition temperature of thermoplastic starches. International Agrophysics 2005b;19(3) 237-241.

[15] Mitrus M., Moscicki L.: Extrusion-Cooking of TPS. In: Janssen LPBM, Moscicki L. (eds.) Thermoplastic Starch. A Green Material for Various Industries. Weinheim: Wiley-VCH.; 2009. p149-157.

[16] Moscicki L., editor. Extrusion-Cooking Techniques. Weinheim: Wiley-VCH Verlag GmbH; 2011.

[17] Mościcki L., Extrusion, effect on physical and chemical properties. In Gliński J., Horabik J., Lipiec J. (eds.) Encyclopedia of Agrophysics. Dordrecht: Springer; 2011. p284-287.

[18] Mościcki L., Mitrus M., Wójtowicz A. Technika ekstruzji w przemyśle spożywczym. PWRiL: Warszawa; 2009 (in Polish).

[19] Myllärinen P., Partanen R., Seppälä J., Forsella P.: Effect of glycerol on behaviour of amylose and amylopectin films. Carbohydrate Polymers 2002;50 355 – 361.

[20] Nashed G., Rutgers R., Sopade P.: The plasticization effect of glycerol and water on the gelatinisation of wheat starch, Starch/Stärke2003;55 131-137.

[21] Oniszczuk T.: Effect of parameters of injection moulding process on the structural properties of thermoplastic starch packaging materials. PhD Thesis. Agriculture University in Lublinie; 2006 (in Polish).

[22] Oniszczuk T., Janssen LPBM.: Influence of addition of fiber on the Mechanical properties of TPS moldings. In: Janssen LPBM, Moscicki L. (eds.) Thermoplastic Starch. A Green Material for Various Industries. Weinheim: Wiley-VCH.; 2009. p197-208.

[23] Peighambardoust H., van Brenk S., van der Goot A., Hamer R., Boom R.: Dough processing in a Couette-type device with varying eccentricity: effect on glutenin macro polymer properties and dough micro-structure, Journal of Cereal Science 2007;45(1) 34-48.

[24] Peighambardoust H., van der Goot A., Hamer R., Boom R.: A new method to study simple shear processing of wheat gluten-starch mixtures, Cereal Chemistry 2004;81(6) 714-721.

[25] PN-A-79011-11:1998, Food concentrates – Test methods – Determination of degree of starch gelatinization (in Polish).

[26] Rushing T., Hester R.: Intrinsic viscosity dependence on polymer molecular weight and fluid temperature, Journal of Applied Polymer Science2003;89 2831-2835.

[27] Schogren R.L.: Effect of moisture and various plasticizers on the mechanical properties of extruded starch.In: Ching C., Kaplan D., Thomas E. Biodegradable Polymers and Packaging. Lancaster, USA;Technomic Publishing Co.; 1993. p141-150.

[28] Shi R., Zhang Z., Liu Q., Han Y., Zhang L., Chen D. Tian W.: Characterization of citric acid/glycerol co-plasticized thermoplastic starch prepared by melt blending. Carbohydrate Polymers 2007;69 748-755.

[29] Talja RA., Helén H., Roos YH. Jouppila K.: Effect of various polyols and polyol contents on physical and mechanical properties of potato starch-based films. Carbohydrate Polymers 2007;69 288-295.

[30] van den Einde R., Bolsius A., van Soest J., Janssen L., van der Goot A., Boom R.: The effect of thermomechanical treatment on starch breakdown and the consequences for process design, Carbohydrate Polymers 2004;55(1) 57-63.

[31] van den Einde R., van der Goot A., Boom R.: Understanding molecular weight reduction of starch during heating-shearing processes, Journal of Food Science 2003;68(8) 2396-2404.

[32] van den Veen M.: Towards intensification of starch processing, PhD Thesis, Wageningen University, the Netherlands; 2005.

[33] van den Veen M., van Iersel D., van der Goot A., Boom R.:Shear induced inactivation of α-amylase in a plain shear field, Biotechnology Progress 2004;20(4) 1140-1145.

[34] vanSoest JJG.: Starch plastics: structure – property relationships. Phd Thesis. Utrecht University; 1996.

[35] vanSoest J., Benes K., de Wit D., Vliegenthart J.: The influence of starch molecular mass on properties of extruded thermoplastic starch, Polymer 1996a;37(16) 3543-3552.

[36] vanSoest J., Bezemer R., de Wit D., Vliegenthart J.: Influence of glycerol on the melting of potato starch, Industrial Crops and Products1996b;5 1-9.

[37] vanSoest JJG., Bezemer RC., De Wit D., Vliegenthart JFG.: Influence of glycerol on the melting of potato starch. Industrial Crops and Products 1996c;5 1-9.

[38] vanSoest J., Knooren N.: Influence of glycerol and water content on the structure and properties of extruded starch plastic sheets during aging, Journal of Applied Polymer Science 1997;64 1411-1422,.

[39] Wolf B.: Polysaccharide functionality through extrusion processing. Current Opinion in Colloid & Interface Science 2010;15 50-54.

[40] Wójtowicz A.: The melting process in thermoplastic starches. In: Janssen L.,Moscicki L. (eds.) Thermoplastic Starch. A green material for various industries, Weinheim, Germany; Wiley-VCH VerlagGmbH&Co. KGaA; 2009. p105-117.

[41] Wójtowicz A., van der Goot AJ.: The influence of water and glycerol addition on thermal processing rheology and microstructure changes of potato starch, Electronic Journal of Polish Agricultural Univeristies, seria Agricultural Engineering, 2005;8(4) #51.

[42] Yimlaz G.: Thermoplastic starch matrices for encapsulation and controlled release of volatile compounds, PhD Thesis, Utrecht University, the Netherlands; 2003.

[43] You X., Li L., Gao J., Yu J., Zhao Z.: Biodegradable extruded starch blends, Journal of Applied Polymer Science 2003;88 627-635.

[44] Yu L., Christie G.: Measurement of starch thermal transitions using differential scanning calorimetry. Carbohydrate Polymers 2001;46 179-184.

Chapter 14

Determination of Food Quality by Using Spectroscopic Methods

Agnieszka Nawrocka and Joanna Lamorska

Additional information is available at the end of the chapter

http://dx.doi.org/10.5772/52722

1. Introduction

Food is a complex system comprised predominantly of water, fat, proteins and carbohydrates together with numerous minor components. The functional properties of these components, which are governed by their molecular structure and intra- and intermolecular interactions within food system, and the amounts present define the characteristics of food products. Quality of food products refers to the minimum standards for substances to qualify as fit for human consumption or permitted to come in contact with food. Appearance, color, flavor and texture are critical aspects for the sensory quality of food. The food quality includes also chemical, biological and microbial factors, e.g. instability of food products, which limits their shelf life and is connected with irreversible chemical and enzymatic reactions [1]. Recently, public interest in food quality and production has increased, probably related to changes in eating habits, consumer behavior, and the development and increased industrialization of the food supplying chains. The demand for high quality and safety in food production obviously calls for high standards for quality and process control, which in turns requires appropriate analytical tools to investigate food.

Spectroscopic methods have been historically very successful at evaluating the quality of agricultural products, especially food. These methods are highly desirable for analysis of food components because they often require minimal or no sample preparation, provide rapid and on-line analysis, and have the potential to run multiple tests on a single sample. These advantages particularly apply to nuclear magnetic resonance (NMR), infrared (IR), and near-infrared (NIR) spectroscopy. The latter technique is routinely used as a quality assurance tool to determine compositional and functional analysis of food ingredients, process intermediates, and finished products [1]. Additionally, UV–VIS spectroscopy, fluorescence and mid-infrared (MIR) and Raman spectroscopy are used in the food quality monitoring.

The aim of this paper is to demonstrate applicability of four spectroscopic techniques, e.g. UV–VIS spectroscopy, fluorescence, infrared and Raman spectroscopy, as rapid analysis methods to determine the quality of cereals, cereals products and oils. Additionally, physical foundations of the aforementioned methods are described.

2. UV-VIS spectroscopy

Absorption spectroscopy in the UV–VIS region is based on the Lambert-Beer's law, expressed by the following equations (1, 2)

$$I = I_0 10^{-\varepsilon cl} \tag{1}$$

$$\ln\frac{I_0}{I} = \ln\frac{1}{T} = \varepsilon cl = A \tag{2}$$

where: I_0, I – intensity of light coming in and out of the sample, respectively; ε – extinction molar coefficient; c– molar concentration of substance; l– thickness of the sample (cm). The transmission of the light by the sample is shown in figure 1. Absorption of the studied sample depends on the length of the radiation wave, the thickness of the sample and the characteristic extinction coefficient at a given wavelength.

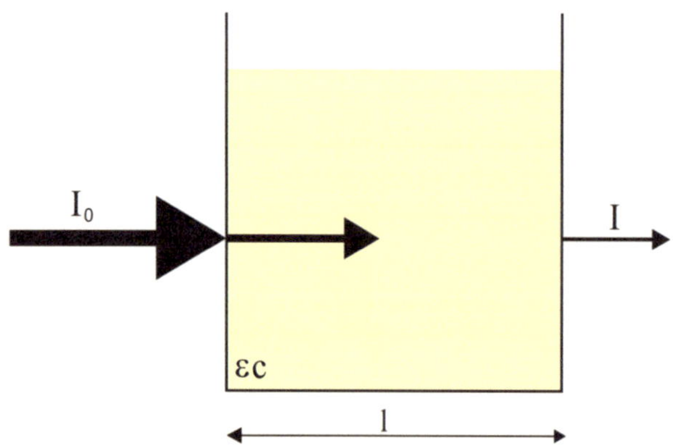

Figure 1. Illustration of Lambert-Beers's law.

The UV-VIS spectroscopy is mainly used to examine the quality of edible oils regarding a number of parameters including the anisidine value. Anisidine value is a measurement of the level of fats oxidation, and is used for the assessment of poorer quality oils. Precisely, it is the measure of aldehyde production during oxidation of fats. The anisidine value (AV) is defined as one hundred-fold value of absorbance of a solution of a fat sample containing aldehydes which have reacted with p-anisidine. The aforementioned aldehydes are dienals or alka-2-enals and both are one of the final products of lipids oxidation. The highest permissible value of AV for edible oils is 8. AV is also an element of Totox (total oxidation value), another factor indicating deterioration level of total fat. The value of Totox is calculated as the sum of two-fold value of AV and peroxide value. According to european standard [2], the measurement of absorbance is performed for three solutions (p-anisidine and sample of fat (A1), acetic acid and sample of fat (A0), and, lastly, blind sample being a mixture of p-anisidine and isooctane (A2)) at the wavelength 350 nm in a 1 cmcuvette. Theanisidine value can be calculated from the formula (3)

$$AV=100QV\left[1.2\left(A_1-A_2\right)-A_0\right]m^{-1} \tag{3}$$

where: Q – content of the sample in the solution based on which the AV is expressed (g cm^{-3}); V – volume in which the fat sample is dissolved (cm^3); m – weight of the fat sample (g).

The anisidine value can be also measured by using Flow Injection Analysis (FIA) combined with UV–VIS spectroscopy. Thanks to the implementation of FIA, the period of time required for analysis can be significantly shortened. Additionally, the number of reagents is also maintained at very reasonable level. Sample of fat dissolved in propanol-2 is injected into continuous flow of p-anisidine with a mixture of solvents: propanol-2 and glacial acetic acid. Spectrophotometer is used as a detector, and the value of absorbance is measured at 350 nm [3].

The process of fat deterioration is also described by the peroxide value (PV). The deterioration takes place during lipids' exposition to some external factors including temperature, daylight and oxygen. It results in production of peroxides and hydroperoxides, which are regarded as products of fatty acids oxidation. The highest value of PV for oil produced through cold press extraction is 10 meq O_2 kg^{-1}, while regarding refined oil it may reach the amount of 5 meq O_2 kg^{-1}. The PV value is measured by employing UV–VIS spectrometer as detector [4, 5]. Method of PV measurement of the frying canola oil was developed by Talpur et al. [6]. The authors used stoichiometric reaction of triphenylphosphine (TPP) with the hydroperoxides to produce triphenylphosphine oxide (TPPO) which shown absorbance maximum at 240 nm. Therefore, the developed method could serve as an alternative to the titration method for the determination of PV in frying oils due to high correlation of peroxide values measured by both methods.

Another parameter of oil quality is general colour which is determined by the saturation of chlorophyll or carotenoid pigments. Unlike oil produced through cold press extraction, refined oil has low saturation intensity colour, as most pigments are removed in the oil refinement. Carotenoid pigments are included to antioxidants. For this reason, oils with high content of carotenoinds are regarded as healthier and of higher quality. The general colour is assayed spectrophotometrically for oil samples diluted in CCl_4, at two wavelengths: 460 nm for carotenoid pigments and 666 nm for chlorophyll pigments for oil samples diluted in

CCl_4 due to polish standard [7]; or 442 nm for carotenoid pigments and 668 nm for chlorophyll pigments when oil samples are dissolved in hexane [7]. Absorbance values obtained for carotenoid and chlorophyll pigments are summed up, multiplied by 1000, and given in the form of an integer or as general colour value:

$$B=\left(A_{460}+A_{666}\right)1000 \tag{4}$$

or

$$B=\left(A_{442}+A_{668}\right)1000 \tag{5}$$

The concentration of carotenoids can also be detected through the use of the UV–VIS method due to the British norm BS 684: Section 2.20:1977 [8]. The absorbance of oil diluted in cyclohexane is measured at the wavelength 445 nm and the proportion of carotenoid expressed by β-carotene content can be calculated using the following equation

$$\beta\text{-}car\left(mg\times kg^{-1}\right)=383E\left(PC\right)^{-1} \tag{6}$$

where: E – the difference in measured absorbance values for oil sample and cyclohexane; P – optical pathlength (cm); C – concentration of the sample (g $100ml^{-1}$).

Chlorophylls also influence general colour of oils, especially pheophytins, which have a pro-oxidative properties. They give some bitter note as well as green colouring to both cold extracted oil and olive oil from green olives. The smaller concentration of chlorophylls in oil sample, the oil has the higher quality. The chlorophyll concentration can be determined with the AOCS Official Method Cc 13d-55 [8, 9]. Following the method, the measurement of chlorophyll absorption is performed at three wavelengths, namely: 630 nm, 670 nm and 710 nm. They refer to the absorbance of the oil sample diluted in carbon tetrachloride at $\lambda = 625.5$, 665.5 and 705.5 nm for oil samples diluted in mixture of ethanol and isooctane, or $\lambda = 630$, 665 and 710 nm for oil samples diluted in mixture of ethanol and heptane. The concentration of total chlorophyll (in carbon tetrachloride) is calculated from the following equation:

$$Chl\left(mg\times kg^{-1}\right)=\left[A_{670}-0.5\left(A_{630}+A_{710}\right)\right]\left(0.901L\right)^{-1} \tag{7}$$

where: A – is the absorbance of the oil at the respective wavelength; L – the cell thickness (cm).

The main chlorophyll pigment, occurring in oils, is pheophytin α. The concentration of this dye can be calculated from the equation (8), as featured in the work of Psomiadou and Tsimidou [10]:

http://dx.doi.org/10.5772/52722

$$Pheop\alpha\left(mg \times kg^{-1}\right) = 345.3\left[A_{670} - 0.5\left(A_{630} + A_{710}\right)\right]L^{-1} \quad (8)$$

where: A – is the absorbance of the oil at the respective wavelength; L - the cell thickness (mm).

3. Fluorescence

Fluorescence is the emission of light subsequent to absorption of ultraviolet or visible light of a fluorescent molecule, called a fluorophore. To typical fluorophores are included quinine, fluorescein, acridine orange, rhodamine B and pyridine 1 [11]. The general principle of the phenomenon is illustrated by Jablonski diagram, as shown in figure 2. The singlet ground, first and second electronic states are depicted by S_0, S_1, and S_2, respectively. The first step of the fluorescence is the excitation of the molecule from the ground state (S_0) to the one of excited states (S_1, S_2) by the absorption of light. This is followed by a vibrational relaxation or internal conversion, where the molecule undergoes a transition from an upper excited state (S_2) to a lower one (S_1), without any radiation. Finally, the fluorescence occurs, typically 10^{-8} s after excitation, when the electron returns to the ground state (S_0). Emitting light has energy equal to the difference between energies of ground and excited states.

Food analysis has exploited the characteristic advantages of fluorescence spectroscopy, i.e. high sensitivity and specificity. However, the technique alone is not usually used in this field. It is mainly linked with liquid chromatography (e.g. HPLC), and the fluorymeter is used as a detector. The combination of these techniques is still advantageous for detecting extremely low concentrations of contaminants such as toxins (mycotoxins), pathogenic microbes (bacterial species: *Salmonella, Escherichia coli*), antibiotics (e.g. penicilin, tetracycline, oxytetracycline), and food additives (e.g. aspartam, salicylates). Other important applications of fluorescence in food area is the analysis of structure changes in proteins, analysis of some carbohydrates and lipids in oils.[12]

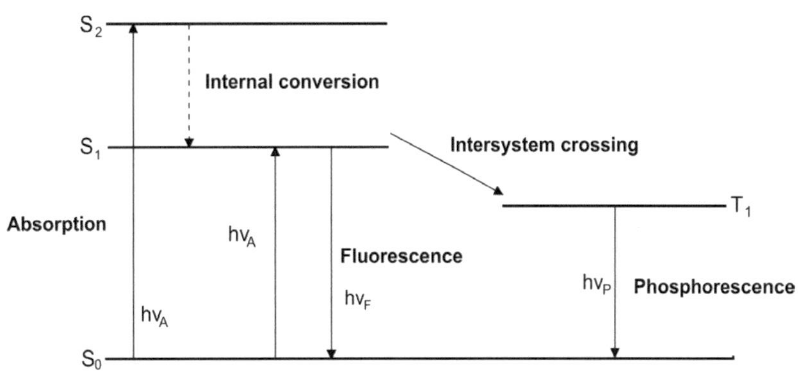

Figure 2. Jablonski diagram.

The high specificity of fluorescence is due to use of two spectra, i.e. excitation and emission spectra, while the high sensitivity of the technique is a result of measuring radiation against absolute darkness. However, the use of only excitation and emission wavelengths could limit the ability of fluorescence spectroscopy to determine the quality of food systems. To comply with this requirement, the variation in the excitation and emission wavelengths allows simultaneous determination of compounds in several food products. This could be realized by using synchronous fluorescence spectroscopy (SFS) [13].

Fluorescence spectroscopy is one of the most valuable instrumental analytical techniques for determining causes of food poisoning by analyzing concentration of toxins, especially mycotoxins. Almost all mycotoxins, apart from aflatoxins (aflatoxins B_1 and B_2 exhibit blue fluorescence, while aflatoxins G_1 and G_2 show yellow-green fluorescence [14]), do not exhibit fluorescence. For this reason, the technique is connected with other analytical techniques and spectrofluorymeter serves as a detector. Corneli and Maragos (1998) used capillary electrophoresis (CE) with a laser induced fluorescence detector to determine ochratoxin A in roasted coffee, corn and sorghum [15]. CE has also been used in corn samples to analyze fumonisin B1, which was fluorescein-labeled due to lack of a UV chromophore [16]. Maragos and Plattner (2002) developed a rapid test for deoxynivalenol (DON) in wheat using the principle of fluorescence polarization (FP) immunoassay. The assay was based on the competition between DON and a novel DON-fluorescein tracer for a DON-specific monoclonal antibody in solution. FP immunoassay utilizes the interaction of a toxin-specific antibody with a toxin-fluorophore conjugate (tracer) to effectively decrease the rate of rotation of the tracer. Binding of the antibody to the tracer increases polarization. In the presence of free toxin less of the antibody is bound to the tracer, reducing polarization [17]. The FP immunoassays have been also developed for the fumonisin mycotoxin detection [18].

Fluorescence has been also used in analysis of cereals and cereals products. Zandomeneghi [19] used fluorescence spectroscopy to differentiate between different cereal flours (e.g. rise, maize). Emission spectra of red and white wheat kernels were recorded by Ram and coworkers (2004) and a clear difference was observed between the two group of samples. This difference has been attributed to the morphological variation in the pericarp and nuclear organization of the two varieties of wheat [20]. The fluorescence spectroscopy has been also used to monitor wheat flour refinement and milling efficiency by recording emission spectra of ferulic acid and riboflavin [21].

One of the most important edible oils is olive oil, which market price depends on its quality. The most expensive is the extra-virgin olive oil (EVOO) owing to its high quality. For economic reasons, it may be adulterated by the addition of cheaper oils such as refined olive oil, residue oil, synthetic olive oil-glycerol products, seed oils and nut oils. For this reason, a rapid method to detect such a practice is important for quality control and labeling purposes [13]. Sayago et al. [22] applied fluorescence spectroscopy for detecting hazelnut oil adulteration in virgin olive oils. Kyriakidis and Skarkalis [23] used excitation wavelength of 360 nm to differentiate between common vegetable oils, including olive oil, olive residual oil, refined olive oil, corn oil, soybean oil, sunflower oil and cotton oil. All the oils studied showed a strong fluorescence band at 430 – 450 nm, except for virgin olive oil, which exhibited a low intensity at both 440 and 455 nm, a medium band around 681 nm and a strong one at 525 nm. The latter two bands have been ascribed to chlorophyll and vitamin E compounds, respectively. All refined oils showed only one intense peak at 445 nm, which is due to fatty acid oxidation products formed as a result of the large percentage of polyunsaturated fatty acids present in these oils. Fluorescence is regarded as the technique that provided the best models for anisidine and iodine values, oligomers and vitamin E content in deteriorated oil after repeated frying cycles [24].

4. Infrared spectroscopy

Infrared (IR) radiation was discovered by F.W Herschel in 1800. This is an electromagnetic radiation extending from 780 nm to 1 mm. The IR range is divided into the following three bands: near-infrared (NIR; 780 nm – 5 μm), mid-infrared (MIR; 5 – 30 μm) and far-infrared (FIR; 30 – 1000 μm).

Infrared (IR) spectroscopy is a technique based on the vibrations of the atoms of a molecule. An infrared spectrum is commonly obtained by passing infrared radiation through a sample and determining what fraction of the incident radiation is absorbed at a particular energy. The energy at which any peak in an absorption spectrum appears corresponds to the frequency of a vibration of a part of a sample molecule. There are different kinds of vibrations observed in the infrared as well as Raman spectra. Vibrations observed in diatomic molecules are shown in fig.3. IR spectroscopy gives information on molecular structure through the frequencies of the normal modes of vibration of the molecule. A normal mode is one in which each atom executes a simple harmonic oscillation about its equilibrium position. All atoms move in phase with the same frequency, while the center of gravity of the molecule does not move. There are 3N-6 normal modes of vibrations (known as fundamentals) of a molecule (3N-5 for linear molecules), where N is the number of atoms. For a molecule with no symmetry, all 3N-6 fundamental modes are active in the IR and may give rise to absorptions. There are also observed overtones in the IR spectra. Overtones has frequencies corresponding approximately to twice, three times etc. that of the fundamental. The frequencies of many overtone bands are in the NIR region [25].

Vibrations of certain functional groups such as –OH, $–NH_2$, $–CH_3$, C=O, C_6H_5–, and so on always give rise to bands in the IR spectrum within well-defined frequency ranges regardless of the molecule containing the functional group. The IR spectrum of any compound that contains a C=O group has a strong band between 1800 and 1650 cm^{-1}. Compounds containing $–NH_2$ groups have two IR bands between 3400 and 3300 cm^{-1}. The spectrum of a compound containing the C_6H_5– group has sharp peaks near 1600 and 1500 cm^{-1} due to stretching modes of the benzene ring. The explanation of these characteristic diatomic group frequencies lies in the approximately constant values of the stretching force constant of a bond in different molecules. Thus, the IR spectrum can be regarded as a 'fingerprint' of the molecule [26].

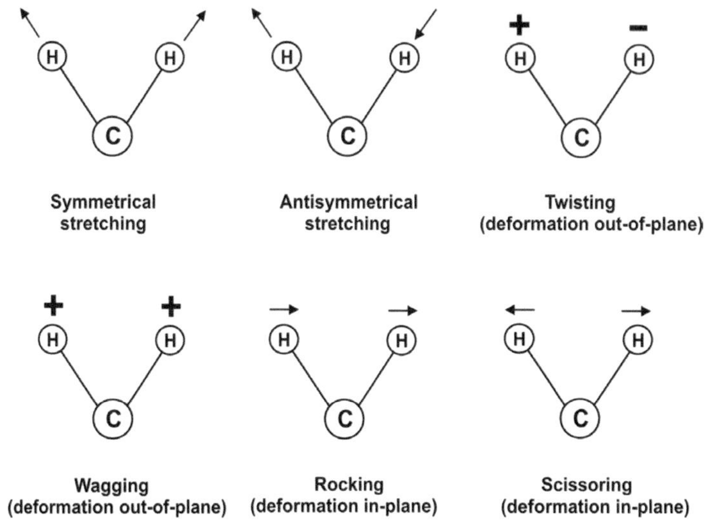

Figure 3. Oscillations observed in diatomic molecules.

4.1. Near-infrared spectroscopy

Near-infrared (NIR) spectroscopy has been primarily employed in the quantitative analysis of foods. The spectroscopy has been applied to measure moisture, fat, protein and carbohydrate content in wide variety of foods. The most significant advantage is its ability to determine simultaneously several components in a food sample within a short time. The precision of NIR analysis for a wide range of applications is comparable to or better than that of the chemical techniques it replaces. On the other hand, the main disadvantage of NIR quantitative analysis is that it requires calibration using samples of known composition. This has seriously limited the use of NIR spectroscopy because of the large amount of time and expense required for the development of calibrations. This disadvantage is compounded by the problem of calibration instability resulting from changes in sample or instrument characteristics over time, which can make frequent recalibration necessary, and the lack of transferability of calibrations owing to optical differences between instruments. Other disadvantages of NIR analysis include the need for high-precision spectroscopic instruments, the complexity of data treatment, and the lack of sensitivity for minor constituents [27].

NIR region of the IR spectrum are due to overtones and combinations of the fundamental vibrations observed in the MIR region. Overtones has frequencies corresponding approximately to twice, three times etc. that of the fundamental, while combination bands arise by interaction of two or more vibrations taking place simultaneously and the frequency of a combination band is the sum of multiples of the relevant fundamental frequencies. Vibrations involving C–H, O–H, N–H and possibly S–H and C=O bonds are responsible for the majority of the observed absorption bands in the NIR region. Table 1 presents principal absorption bands of water, oil, protein and starch, which are observed in the NIR region.

http://dx.doi.org/10.5772/52722

Wavelength [nm]	Assignment
	Water
1454	1st overtone O–H stretching
1932	O–H combinations
	Proteins
1208	2nd overtone C–H stretching
1465	1st overtone H–N and O–H stretching
1734	1st overtone C–H stretching
1932 2058 2180	combinations N–H and O–H stretching
2302 2342	combination C–H stretching
	Oil
1210	2nd overtone C–H stretching
1406	1st overtone H–N and O–H stretching
1718 1760	1st overtone C–H stretching
2114	combinations N–H and O–H stretching
2308 2346	combination C–H stretching
	Starch
1204	2nd overtone C–H stretching
1464	1st overtone H–N and O–H stretching
1932 2100	combinations N–H and O–H stretching
2290 2324	combination C–H stretching

Table 1. Principal absorption bands of water, oil, protein and starch observed in the NIR region.

The application of NIR analysis in wheat and wheat products has included flour yield, damaged starch, water absorption, dough development time, extensibility and loaf volume measurements. The use of NIR technology to determine the protein and moisture contents of both wheat and flour is now routine practice in flour mills worldwide. It is used for testing each delivery of wheat in order to make decisions about acceptance, price and binning; for determination of conditioning time from measurement of hardness; and for analyzing flour to check that it complies with specifications before shipment to the customer [28].

NIR was used in Australia to predict optimum fertilizer requirements of cereal crops by analysis of total nitrogen and carbohydrate (fructan) in plant tissue samples [29]. Wheat hardness had been measured in both meal and whole grain by using NIR [30]. NIR spectroscopy has also been shown to be useful in the study of changes in starches during processing and storage [31]. This utility is primarily due to the sensitivity of the O–H stretching mode overtone absorptions of starch and of the water bound to starch to changes in hydrogen bonding that accompany changes in starch structure. The spectroscopy combined with chemometrics has been applied to discriminate wheat varieties [32]. It was developed a discriminant equation, which gave 94% correctly identified varieties.

Recently, NIR spectroscopy is often connected with hyperspectral imaging system. Canadian wheat classes has been determined by using near-infrared hyperspectral imaging (NIR-HSI) system [33]. Seventy-five relative reflectance intensities were extracted from the scanned images and used for the differentiation of wheat classes using a statistical classifier and an artificial neural network (ANN) classifier. Classification accuracies were 100%. This imaging system has also been used to detect spectral differences between healthy and damaged products, which are connected with different chemical composition of these products. Singh et al. [34, 35] studied insect-damaged kernels. It was observed that insect-damaged kernels had less starch compared to healthy ones, due to consumption of starch by insects during their development.

The study of food authenticity involves establishing whether a sample is genuine in terms of its description, including geographical origin. The application of NIR for authenticity testing of coffee, fruit pulps, milk powders, orange juice, pig carcasses, rice, sausages, sugars, vegetable oils, wheat grain and wheat flour have been reviewed by Downey [36]. This spectroscopy is required to classify within a series of possible classes, to identify a particular kind of adulteration or to quantify adulteration.

4.2. Mid-infrared spectroscopy

Mid-infrared (MIR) spectroscopy can both provide information on structure-functionality relationships and serve as a quantitative analysis tool. For this reason, it is regarded as a highly valuable technique for both food-related research and quality control purposes in the food industry. The Fourier-transform infrared spectroscopy (FTIR) is used the most often.

The mid-infrared spectrum (4000 – 400 cm^{-1}) includes four regions: the X–H stretching region (4000 – 2500 cm^{-1}), the triple-bond region (2500 – 2000 cm^{-1}), the double-bond region (2000 – 1500 cm^{-1}) and the fingerprint region (1500 – 600 cm^{-1}). The fundamental vibrations of the X–H stretching region are generally due to O–H, C–H and N–H stretching. Vibrations of C≡C and C≡N bonds are mainly observed in the triple-bond stretching region. Whereas absorption bands corresponding to C=C, C=O and C=N occurs in the double-bond region. The "fingerprint" bands are connected mostly with bending and skeletal vibrations The important mid-infrared bands associated with major food components (water, proteins, fats and carbohydrates) are summarized in table 2 [25].

The MIR spectroscopy, similar to NIR spectroscopy, is applying to analysis of moisture, protein, carbohydrate and fat content in food products. Owing to the intense absorption of MIR radiation by water, FTIR was used to determination of moisture of food emulsions e.g. butter [37] and mayonnaise [38]. FTIR technique is also well established as a powerful tool for the study of protein secondary structure, based primarily on examination of the amide I region (1600 – 1700 cm^{-1}) [39]. Proteins are widely used as ingredients in the food processing industry because of their useful properties such as emulsification, gelation and thickening. These properties are highly related to the secondary structure of protein, which can change during processing and storage of food products [27]. Examination of amide I region are made by mathematical process called deconvolution. Deconvoluted amide I band of wheat gluten are presented in figure 4 [40].

Wavenumber [cm^{-1}]	Assignment
	Water
3200 – 3600	O–H stretching
1650	H–OH stretching
	Proteins
1600 – 1690	Amide I (C=O stretching)
1480 – 1575	Amide II (C–N stretching and N–H bending)
1230 – 1300	Amide III (C–N stretching and N–H bending)
	Fats
2800 – 3000	C–H stretching
1725 – 1745	C=O stretchnig
970	C=C–H bending
	Carbohydrates
2800 – 3000	C–H stretching
800 – 1400	Skeletal stretching and bending

Table 2. The major bands of food components localized in MIR region.

MIR spectroscopy play a crucial role in research on the chemistry of fats and oils. An important result of early investigations of the IR spectra of fats and oils was the identification of an absorption band at 996 cm^{-1} which is characteristic for isolated trans double bonds. Although the double bonds in naturally occurring fats and oils predominantly have the cis configuration, extensive cis-trans isomerization occurs during industrial catalytic hydrogenation processes, which are widely employed to convert oils to fats and to increase the oxidative stability of polyunsaturated oils. Van de Voort et al. [41] developed a method for the simultaneous determination of trans unsaturation, cis unsaturation and total unsaturation, which is traditionally expressed as the iodine value (IV), as well as saponification number (SN) (a measure of weight-average molecular weight).

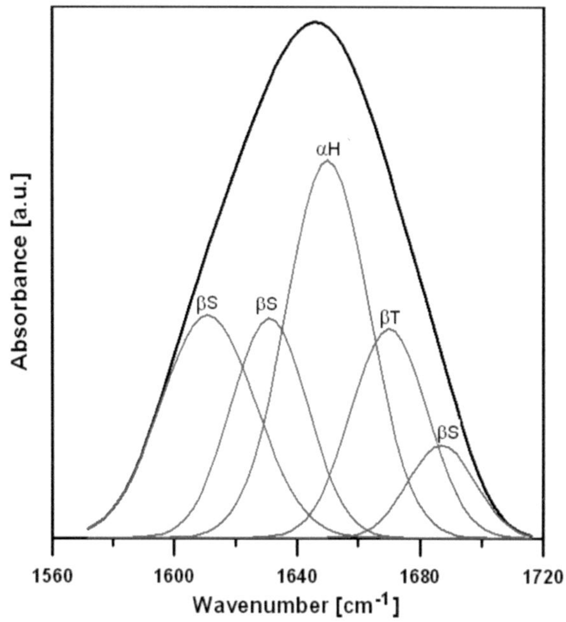

Figure 4. Deconvoluted amide I band of wheat gluten (αH - α-helix, βS - β-sheet, βT - β turns).

Another area of application of FTIR spectroscopy that has been investigated is its use in the assessment of oil quality and stability. The reaction between unsaturated lipids and atmospheric oxygen under ambient conditions, termed lipid autooxidation, is a leading cause of deterioration of fats and oils, as well as of any lipid-containing food, as it gives rise to the off-flavors and unpalatable odors associated with oxidative rancidity. FT-IR spectroscopy proved to be the most direct and accurate method of monitoring gross changes in the frying oil over time [24, 42]. A quantitative FTIR method was used for monitoring the oxidative state of frying oils, based on the determination of anisidine value (AV), a measure of aldehydes that are major secondary oxidation products in polyunsaturated oils, has also been reported [43]. FTIR methods have also been developed to serve as alternatives to the peroxide value (PV) test, which is widely employed by the fats and oils industry to assess the oxidative status and stability of refined oils. This method entails measurement of the hydroperoxide OO–H stretching absorption, which is observed at 3444 cm^{-1} in the spectra of neat oils [44]. FTIR spectroscopy was also used to understand in more detail the mechanisms of thermally induced oxidative processes (thermal oil degradation) in extra-virgin olive oils [45].

FTIR methods for the determination of other minor components present in oils, including free fatty acids in refined [46] and crude oils [47], β-carotene in palm oil [48], and phospholipids in vegetable oils [49], have also been reported.

Likewise fluorescence, MIR spectroscopy is used to study adulteration of olive oil by other edible oils. The edible oils widely employed in virgin olive oil adulteration can be lower quality olive oil (refined or pomace olive oil) or other vegetable or seed oils such as corn, peanut, cottonseed, sunflower, soybean and poppy seed oils [50]. Gurdeniz and Ozen [51] wanted to demonstrate that MIR spectroscopy connected with chemometrics is a rapid method to detect and quantify adulteration of extra-virgin olive oil (EVOO) with vegetable oils (rapeseed, cottonseed, and corn-sunflower binary mixture). The adulteration of EVOO by different concentrations of palm oil was studied by using FTIR technique [52]. Not only EVOO is adulterated by a low-quality edible oils but also virgin coconut oil (VCO), which possesses several biological activities such as antiviral and antimicrobial. Rohman and Che Man used FTIR spectroscopy combined with chemometrics to determine the level of adulteration of VCO with corn and sunflower oils [53].

4.3. Far-infrared spectroscopy

The region below 400 cm^{-1} down to 10 cm^{-1} is defined as the far-infrared. The region below 200 cm^{-1} is not readily accessible and there are not many useful spectra-structure correlations in this region. However, compounds containing halogen atoms, organometallic compounds and inorganic compounds absorb in the far-infrared and torsional vibrations and hydrogen bond stretching modes are found in this region [26].

5. Raman spectroscopy

Raman effect arises from the interaction of incident photons with electrons of the matter under investigation (inelastic scattering of the incident light). During this interaction the photon can lose (Stokes' process; $h(\nu-\nu_R)$) or gain (anti-Stokes' process; $h(\nu+\nu_R)$) energy equal to the vibrational energy of the atoms (see fig.5). Consequently the vibrational energy of the atoms increases or decreases. Such communication is possible for the motions of atoms, which modulate the polarizability of the molecule. Intense Raman bands will be observed from non-polar groups, particularly aromatic rings, the vibrations of which produce considerable modulation of polarizability. The resulting Raman spectrum, presented in wave numbers (cm^{-1}) as the difference between the excited and emitted photon energy, is the vibrational spectrum of the molecule. The effect is very weak, because the energy exchange probability is low [54].

Raman spectroscopy (RS) is a vibrational spectroscopy technique, based on the Raman effect. It is an irreplaceable tool for the study of biological events at the molecular level and for identification of molecules. There are several reasons for this. First of all, RS measures the vibrations of atoms. This implies that the positions of the bands, widths and intensities are sensitive to the molecular structure. Vibrations of some molecular groups are very characteristic and therefore can be used for the identification of certain groups or even whole molecules. Second, water causes weak Raman scattering, and consequently molecules can be studied in their natural environment without solvent interference. The third reason for the growing interest in RS is related to the considerable intensity enhancement (up to 10^6) observable in resonance Raman spectroscopy (RRS) and surface-enhanced Raman spectroscopy (SERS). Fourthly, because the Raman effect is instantaneous (timescale in order of 10^{-14} s), and due to the recent advances in laser technology, the time-resolved action of molecules down to picosecond resolution can be followed, retaining the value of structural information [54].

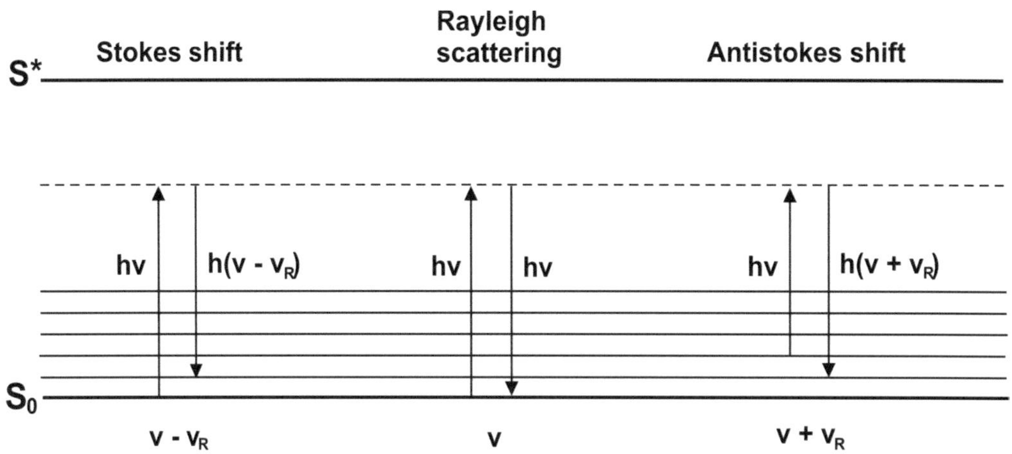

Figure 5. Scheme of scattering of the light.

The limitations of the technique must also be considered. First of all, the probability of Raman scattering is very low, so the effect is weak and high concentrations of samples are required. Secondly, in Raman experiments excitation within the electronic absorption band often causes photodegradation of the molecule. To avoid such problems, low laser powers, moving samples and independent inspection of the sample integrity are often employed. Finally, many molecules, or impurities in the sample, exhibit intense fluorescence, obscuring the Raman spectra [54].

Raman spectroscopy, which is regarded as a complementary technique to IR spectroscopy, is similarly applied to study water, carbohydrate, protein and fat structure in food samples. Raman bands of major food components are shown in table 3. As it was mentioned above, weak Raman scattering of water is the advantage, but it also makes difficult to observe changes in water structure due to weak Raman signals. However, changes in water structure have been observed as decrease of intensity of the O–H stretching band at 3250 cm^{-1} relative to the C–H band at 2938 – 2942 cm^{-1}. It was a result of interaction between water molecules and food proteins [55]. Structure of food proteins may also be analyzed by using RS. The –CO–NH– amide or peptide bond has several distinct and conformationally sensitive vibrational modes, with the amide I and III bands being the most commonly used for secondary structure characterization. [56]. Changes in food carbohydrate structure inducing by processing or storage can be monitored by this technique. Using RS technique interactions of carbohydrates with other food components, particularly with water have also been studied [27].

Wavenumber [cm-1]	Assignment
	Water
3200 – 3600	O–H stretching
	Proteins
510 525 545	S–S stretching
630–670 700–745	C–S stretching
1235 – 1245	Amide III (C–N stretching and N–H bending)
1600 – 1700	Amide I (C=O stretching and N–H bending)
2550 – 2580	S–H stretching
2800 – 3000	C–H stretchnig
	Fats
1441	CH_2 bending
1457	$CH_3 - CH_2$ bending
1656	C=C stretching
2855 – 2960	C – H stretching
	Carbohydrates
836	C – C stretching
1064	C – O stretching
2912 2944	C – H stretching
3451	O – H stretching

Table 3. Raman bands of major food component.

Raman spectroscopy, likewise IR spectroscopy, has been used to quantify and characterized the lipid components of food systems, including quantitative analysis of the degree of unsaturation and the content of cis and trans isomers, identification or detection of adulteration of various oils, characterization of polymorphism and chain packing, and monitoring of interactions with other food components or changes induced by processing or storage, such as autooxidation or isomerization. Rapid quantitative analysis of unsaturation and cis and trans isomers content has been reported by using dispersive laser Raman spectroscopy and FT-Raman spectroscopy [57, 58, 59]. Low-resolution Raman spectroscopy was used to monitor oxidation status of olive oil. Primary and secondary oxidation parameters (e.g. peroxide value) were obtained in a rapid, non-destructive and direct way [60]. FT-Raman spectroscopy combined with multivariate analysis procedures was applied to determine the level of adulteration of virgin olive oil by some vegetable oils (soybean, corn and raw olive residue oils) [61].

Nowadays, Raman spectroscopy is often combined with microscopy (Raman microspectroscopy). The combination results in an analytical method that allows spatially resolved investigation of the chemical composition of heterogeneous food and food ingredients. Both qualitative and quantitative information can be obtained using microspectroscopy. A number of organic compounds and functional groups can be identified by their unique pattern of absorption, and the intensity of the absorption may be used for the calculation of the relative concentration in the sampled entity. Furthermore, samples of microscopic size can be analyzed directly, in air, at ambient temperature and pressure, wet or dry, and in many cases without destroying the sample [62]. Confocal Raman microspectroscopy has been applied to obtain information about microstructure and chemical composition of wheat grain [63]. The work was focused on the protein content and composition of the starchy endosperm and on the aleurone cell walls in arabinoxylan and ferulic acid derivatives. Confocal Raman microscopy was also used to follow the evolution of protein content and structure during grain development of various wheat varieties selected on the basis of hardness level and aptitude to separation of peripheral layers during milling. Raman microspectroscopy is not only a powerful technique to identify cereal components, but it also gives information about secondary structure and configuration of these proteins. For instance, the technique permits to determine the conformation of a non-specific wheat phospholipid transfer protein, and to study the role of disulphide bridges in the stabilization of the α-helical structure. In addition, Raman microspectroscopy was used to determine the secondary structure and conformation of puroindolines, lipid-binding protein of wheat [64].

6. Conclusions

Spectroscopic methods seem to be very successful at evaluating the food quality. They are used to qualitative as well as quantitative analysis of food products. Furthermore, they provide information on structure-functionality relationships (e.g. secondary structure of proteins). These techniques can be used independently or combined with other analytical techniques such as chromatography and served as detector. Some of them (IR and Raman spectroscopy) are regarded as rapid, no sample preparation techniques which additionally can be used on-line.

The above described methods are mainly used to determine technological parameters (anisidine value, peroxide value and general color) and authenticity (kind and level of adulteration) of edible oils. They can also be used to differentiate cereal classes (wheat classes), determine quality of flours, and level of the insects and fungal damage in food products. But first of all, the spectroscopic methods are applied to determine content of the major food components (water, proteins, lipids and carbohydrates). Although some researches claimed that spectroscopy is unsuitable tool for studying food products, nowadays it is used commonly and with a great success.

Author details

Agnieszka Nawrocka[1] and Joanna Lamorska[2]

1 Bohdan Dobrzansky Institute of Agrophysics, Polish Academy of Sciences, Lublin, Poland

2 The Institute of Agricultural Sciences, The State School of Higher Education in Chełm, Chełm, Poland

References

[1] McGorrin, R.J. Food analysis techniques: Introduction. Encyclopedia of Analytical Chemistry 2006.

[2] European standard PE-EN ISO 6885:2008 – Oils and animal and vegetable fats – Determination of anisidine value.

[3] Labrinea E.P., Thomaidis N.S., Georgiou C.A. Direct olive oil anisidine value determination by flow injection. AnalyticaChimicaActa 2001; 448, 201-206.

[4] Gray J.I. Measurement of Lipid Oxidation: A Review. Journal of the American Oil Chemists' Society1978;55, 539-546.

[5] Lezerovich A. Determination of Peroxide Value by Conventional Difference and Difference-Derivative Spectrophotometry. Journal of the American Oil Chemists' Society 1985; 62, 1495-1500.

[6] Talpur M.Y., Sherazi S.T.H., Mahesar S.A., Bhutto A.A. A simplified UV method for determination of peroxide value in thermally oxidized canola oil. Talanta 2010; 80, 1823-1826.

[7] Polish standard PN-A-86934:1995 – Oils and animal and vegetable fats – Spectrophotometric determination of general color (in Polish).

[8] Mińkowski K., Grześkiewicz S., Jerzewska M., Ropelewska M. Characteristic of chemical composition of vegetable oil about high contents of linoleic acids ŻYWNOŚĆ. Nauka. Technologia. Jakość 2010; 73, 146-157 (in Polish).

[9] Dirman H., Dibeklioğlu H. Characterization of Turkish Virgin Olive Oils Produced from Early Harvest Olives. American Oil Chemists' Society (AOCS) 2009; 86, 663-674.

[10] Psomiadou E., Tsimidou M. Pigments in Greek virgin olive oils: occurrence and levels. Journal of the Science of Food and Agriculture 2001; 81, 640-647.

[11] Lakowicz, J.R. Principles of Fluorescence Spectroscopy. New York; Springer Science + Business Media, LLC; 2006.

[12] Nakai, S.; Horimoto, Y. Fluorescence spectroscopy in food analysis. Encyclopedia of Analytical Chemistry 2006.

[13] Karoui, R.; Blecker, C. Fluorescence spectroscopy measurement for quality assessment of food systems – A review. Food and Bioprocess Technology 2011; 4, 364-386.

[14] Li, P.; Zhang, Q.; Zhang, W. Immunoassays for aflatoxins. Trends in Analytical Chemistry 2009; 28, 1115-1126.

[15] Corneli, S.; Maragos, C.M. Capillary electrophoresis with laser-induced fluorescence methods for the mycotoxinochratoxin A. Journal of Agricultural and Food Chemistry 1998; 46, 3162-3165.

[16] Maragos, C.M. Detection of mycotoxinfumonisin B1 by a combination of immunofluorescence and capillary electrophoresis. Food and Agricultural Immunology 1997; 9, 147-157.

[17] Maragos, C.M.; Plattner, R.D. Rapid fluorescence polarization immunoassay for the mycotoxindeoxynivalenol in wheat. Journal of Agricultural and Food Chemistry 2002; 50, 1827-1832.

[18] Maragos, C.M.; Jolley, M.E.; Plattner, R.D.; Nasir, M.S. Fluorescence polarization as a means for determination of fumonisins in maize. Journal of Agricultural and Food Chemistry 2001; 49, 596-602.

[19] Zandomeneghi, M. Fluorescence of cereal flours. Journal of Agricultural and Food Chemistry 1999; 47, 878-882.

[20] Ram, M.S.; Seitz, L.M.; Dowell, F.E. Natural fluorescence of red and white wheat kernels. Cereal Chemistry 2004; 81, 244-248.

[21] Symons, S.J.; Dexter, J.E. Computer analysis of fluorescence for the measurement of flour refinement as determined by flour ash content, flour grade color, and tristimulus color measurements. Cereal Chemistry 1991; 68, 454-460.

[22] Sayago, A.; Morales, M.T., Aparicio, R. Detection of hazelnut oil in virgin olive oil by a spectrofluorometric method. European Food Research and Technology 2004; 218, 480-483.

[23] Kyriakidis, N.B.; Skarkalis, P. Fluorescence spectra measurement of olive oil and other vegetable oils. Journal of the American Oil Chemists' Society 2000; 83, 1435-1439.

[24] Engelsen, S.B. Explorative spectrometric evaluation of frying oil deterioration. Journal of the American Oil Chemists' Society 1997; 74, 1495-1508.

[25] Stuart, B.H. Infrared spectroscopy: Fundamentals and applications. John Wiley & Sons, Ltd., 2004.

[26] Shurvell, H.F. Spectra-structure correlations in the mid- and far-infrared. Handbook of Vibrational Spectroscopy 2006.

[27] Li-Chan, E.C.Y.; Ismail, A.A.; Sedman, J.; van de Voort, F.R. Vibrational Spectroscopy of Food and Food Products. Handbook of Vibrational Spectroscopy 2006.

[28] Osborne, B. G. Near-Infrared Spectroscopy in Food Analysis. Encyclopedia of Analytical Chemistry 2006.

[29] McGrath, V.B.; Blakeney, A.B.; Batten, G.D. Fructan to nitrogen ratio as an indicator of nutrient status in wheat crops. New Phytologist 1997; 136, 145-152.

[30] Norris, K.H.; Hruschka, W.R.; Bean, M.M., Slaughter, D.C. A definition of wheat hardness using near infrared reflectance spectroscopy. Cereal Foods World 1989; 34, 696-705.

[31] Osborne, B.G. Near infrared spectroscopic studies of starch and water in some processed cereal foods. Journal of Near Infrared Spectroscopy 1996; 4, 195-200.

[32] Miralbes, C. Discrimination of European wheat varieties using near infrared reflectance spectroscopy. Food Chemistry 2008; 106, 386-389.

[33] Mahesh, S.; Manickavasagan, A.; Jayas, D.S.; Paliwal, J.; White, N.D.G. Feasibility of near-infrared hyperspectral imaging to differentiate wheat classes. Biosystems Engineering 2008; 101, 50-57.

[34] Singh, C.B;,Jayas, D.C.; Paliwal, J.; White, N.D.G. Detection of insect-damaged wheat kernels using near-infrared hyperspectral imaging. Journal of Stored Products Research 2009; 45, 151-158.

[35] Singh, C.B., Jayas, D.C., Paliwal, J., White, N.D.G. Identification of insect-damaged wheat kernels using short-wave near-infrared hyperspectral and digital colour imaging. Computers and Electronics in Agriculture 2010; 73, 118-125.

[36] Downey, G. Authentication of food and food ingredients by near infrared spectroscopy. Journal of Near Infrared Spectroscopy 1996; 4, 47-61.

[37] van de Voort, F.R.; Sedman, J.; Emo, G., Ismail, A.A. A rapid FTIR quality control method for fat and moisture determination in butter. Food Research International 1992; 25, 193-198.

[38] van de Voort, F.R.; Sedman, J.; Ismail, A.A. A rapid FTIR quality-control method for determining fat and moisture in high-fat products. Food Chemistry 1993; 48, 213-221.

[39] Mejri, M., Roge, B., BenSouissi, A., Michels, F., Mathhlouti, M. Effects of some additives on wheat gluten solubility: A structural approach. Food Chemistry 2005; 92, 7-15.

[40] Nawrocka, A.; Cieśla, J. Influence of silver nanoparticles stabilized by sodium citrate on gluten structure in wheat grain. 2012 (in press).

[41] van de Voort, F.R.; Ismail, A.A.; Sedman, J. A rapid automated method for the determination of cis and trans content of fats and oils by Fourier transform infrared spectroscopy. Journal of the American Oil Chemists' Society 1995; 72, 873-880.

[42] Moros, J.; Roth, M.; Garrigues, S.; de la Guardia, M. Preliminary studies about thermal degradation of edible oils through attenuated total reflectance mid-infrared spectroscopy. Food Chemistry 2009; 114, 1529-1536.

[43] Dubois, J.; van de Voort, F.R.; Sedman, J.; Ismail, A.A.; Ramaswamy, H.R. Quantitative Fourier transform infrared analysis for anisidine value and aldehydes in thermally stressed oils. Journal of the American Oil Chemists' Society 1996; 73, 787-794.

[44] van de Voort, F.R.; ISmail, A.A.; Sedman, J.; Dubois, J.; Nicodemo, T. The determination of peroxide value by Fourier transform infrared (FTIR) spectroscopy. Journal of the American Oil Chemists' Society 1994; 71, 921-926.

[45] Navarra, G.; Cannas, M.; D'Amico, M.; Giacomazza, D.; Militello, V.; Vacarro, L.; Leone, M. Thermal oxidative process in extra-virgin olive oils studied by FT-IR, rheology and time-resolved time luminescence. Food Chemistry 2011; 126, 1226-1231.

[46] Bertran, E.; Blanco, M.; Coello, J.; Iturriaga, H.; Maspoch, S.; Montoliu, I. Determination of olive oil free fatty acid by Fourier transform infrared spectroscopy. Journal of the American Oil Chemists' Society 1999; 76, 611-616.

[47] Che Man, Y.B.; Moh, M.H.; van de Voort, F.R. Determination of free fatty acids in crude palm oil and refined-bleached-deodorized palm olein using Fourier transform infrared spectroscopy. Journal of the American Oil Chemists' Society 1999; 76, 485-490.

[48] Moh, M.H.; Che Man, Y.B.; Badlishah, B.S.; Jinap, S.; Saad, M.S.; Abdullah, W.J.W. Quantitative analysis of palm carotene using Fourier transform infraredand near infrared spectroscopy. Journal of the American Oil Chemists' Society 1999; 76, 249-254.

[49] Nzai, J.M.; Proctor, A. Determination of phospholipids in vegetable oil by Fourier transform infrared spectroscopy. Journal of the American Oil Chemists' Society 1998; 75, 1281-1289.

[50] Harwood, J.; Aparicio, R. Handbook of olive oil. Analysis and properties. Gaithersburg: Aspen 2000.

[51] Gurdeniz, G.; Ozen, B. Detection of adulteration of extra-virgin olive oil by chemometric analysis of mid-infrared spectral data. Food Chemistry 2009; 116, 519-525.

[52] Rohman, A.; Che Man, Y.B. Fourier transform infrared (FTIR) spectroscopy for analysis of extra virgin olive oil adulterated with palm oil. Food Research International 2010; 43, 886-892.

[53] Rohman, A.; Che Man, Y.B. The use of Fourier transform mid infrared (FT-MIR) spectroscopy for detection and quantification of adulteration in virgin coconut oil. Food Chemistry 2011; 129, 583-588.

[54] Niaura, G. Raman spectroscopy in analysis of biomolecules. Encyclopedia of Analytical Chemistry 2008.

[55] Li-Chan, E.; Nakai, S. Raman spectroscopic study of thermally and/or dithiothreitol induced gelation of lysozyme. Journal of Agricultural and Food Chemistry 1991; 39, 1238-1245.

[56] Belton, P.S. New methods for monitoring changes in protein. Food Reviews International 1993; 9, 551-557.

[57] Sadeghi-Jorabchi, H.; Hendra, P.J.; Wilson, R.H.; Belton, P.S. Determination of total unsaturation in oils and margarines by Fourier transform Raman spectroscopy. Journal of the American Oil Chemists' Society 1990; 67, 483-486.

[58] Sadeghi-Jorabchi, H.; Wilson, R.H.; Belton, P.S.; Edwards-Webb, J.D.; Coxon, D.T. Quantitative analysis of oils and fats by Fourier-transform Raman spectroscopy. SpectrochimicaActaA 1991; 47, 1449-1458.

[59] Yang, H.; Irudayaraj, J.; Paradkar, M.M. Discriminant analysis of edible oils and fats by FTIR, FT-NIR and FT-Raman spectroscopy. Food Chemistry 2005; 93, 25-32.

[60] Guzman, E.; Baeten, V.; Fernandez Pierna, J.A.; Garcia-Mesa, J.A. Application of low-resolution Raman spectroscopy for the analysis of oxidized olive oil. Food Control 2011; 22, 2036-2040.

[61] Baeten, V.; Meurens, M.; Morales, M.T.; Aparicio, R. Detection of virgin olive oil adulteration by Fourier transform Raman Spectroscopy. Journal of Agricultural and Food Chemistry 1996; 44, 2225-2230.

[62] Thygesen, L.G.; Lokke, M.M.; Micklander, E.; Engelsen, S.B. Vibrational microspectroscopy of food. Raman vs. FT-IR. Trends in Food Science & Technology 2003; 14, 50-57.

[63] Piot, O.; Autran, J.-C.; Manfait, M. Spatial distribution of protein and phenolic constituents in wheat grain as probed by confocal Raman microspectroscopy. Journal of Cereal Science 2000; 32, 57-71.

[64] Le Bihan, T.; Blochet, J.E.; Desormeaux, A.; Marion, D.; Pezolet, M. Determination of the secondary structure and conformation of puroindolines by infrared and Raman spectroscopy. Biochemistry 1996; 35, 12712-12722.

Chapter 15

Plant Fibres for Textile and Technical Applications

M. Sfiligoj Smole, S. Hribernik,
K. Stana Kleinschek and T. Kreže

Additional information is available at the end of the chapter

http://dx.doi.org/10.5772/52372

1. Introduction

Recently natural and made-man polymer fibres are used for preparation of functionalised textiles to achieve smart and intelligent properties. There are numerous application possibilities of these modified materials. Main pathways for functionalizaton of fibres are: inclusion of functional additives (inorganic particles, polymers, organic compounds); chemical grafting of additives on the surface of fibres and coating of fibres with layers of functional coatings. A new approach to produce new materials is by nanotechnology, which offers a wide variety of possibilities for development of materials with improved properties. Composites of cellulose fibres with nano-particles combine numerous advantageous properties of cellulose with functionality of inorganic particles, hence yielding new, intelligent materials. For preparing cellulose composite materials profound knowledge about fibres properties is needed. Besides, new fibre qualities are demanded to guaranty the modification efficiency. Therefore non-standard methods are involved to determine physical properties of fibres.

In addition to, manufacture, use and removal of traditional textile materials are now considered more critically because of increasing environmental consciousness and the demands of legislative authorities. Natural cellulose fibres have successfully proven their qualities when also taking into account an ecological view of fibre materials. Different cellulose fibres can be used for textile and technical applications, e.g. bast or stem fibres, which form fibrous bundles in the inner bark (phloem or bast) of stems of dicotyledenous plants, leaf fibres which run lengthwise through the leaves of monocotyledenous plants and fibres of seeds and fruits. Flax, hemp, jute, ramie, sisal and coir are mainly used for technical purposes. Recently, the interest for renewable resources for fibres particularly of plant origin is increasing. Therefore several non-traditional plants are being studied with the aim to isolate fibres from plant leaves or stems.

A review of different conventional and non-conventional fibres is presented. For extraction of fibres different isolation procedures are possible, e.g. using bacteria and fungi, chemical and mechanical methods. The procedure used influences fibres surface morphology. By fibre isolation procedures mainly technical fibres are obtained, which means that cellulose fibres are multicellular structures with individual cells bound into fibre bundles.

2. Plant fibres

Many useful fibres have been obtained from various parts of plants including leaves, stems (bast fibres), fruits and seeds. Geometrical dimensions of these fibres, especially the fibre length depends mainly on fibre location within the plant. Fibres from fruits and seeds are few centimetres long, whereas fibres from stems and leaves are much longer (longer than one meter) [Blackburn 2005].

With an exception of seeds' and fruits' fibres, plant fibres are sclerenchyma elongated cells which occur in different parts of plants, mainly in the stems and leaves. These are elongated cells with tapering ends and very thick, usually heavily lignified cell walls. Sclerenchyma gives mechanical strength and rigidity to the plant, since it is usually a supporting tissue in plants. Fibres are also associated with the xylem and phloem tissue of monocotyledonous and dicotyledonous plant stems and leaves.

All plant cells have a primary wall. During cell growth and after it has stopped, the cytoplasm in sclerenchyma cells dries while the cell wall becomes thickened by addition of a thick and rigid secondary cell wall which is formed inwards of the primary cell wall and constructed of cellulose fibrils. The secondary cell wall is formed by successive deposition of cellulose layers, which are divided in three sub-layers (S1, S2 and S3), of which the middle layer is the most important for fibres mechanical properties. It consists of helically arranged microfibrils. The diameter of microfibrils is between 10-30nm [John 2008]. An important parameter of the structure of the secondary wall is the angle that the cellulose microfibrils are making with the main fibre direction. Actually each of three fibres sub-layers has a different microfibrillar orientation [Krässig 1992, John 2008, Cuissinat 2008] which is specific for the fibre type. Due to the formation of a thick secondary wall, the lumen becomes smaller.

The cell wall in a fibre is not a homogeneous layer. The walls of plant cells (the primary and the secondary cell wall) can be considered as a composite consisting of cellulose fibrils embedded within a matrix of lignin and hemicellulosic polysaccharides [Krässig 1992].

Vegetable fibres are generally composed of three structural polymers (the polysaccharides cellulose, and hemicelluloses and the aromatic polymer lignin) as well as by some minor non-structural components (i.e. proteins, extractives, minerals) [Marques 2010]. Cellulose forms a crystalline structure with regions of high order i.e. crystalline regions and regions of low order i.e. amorphous regions. Middle lamellas composed of pectic polysaccharides are connecting individual cells in bundles [Caffall 2009].

Retting which is the process of separating fibres from non-fibre tissues in plants, involves bacteria and fungi treatments and mechanical and chemical processes for fibres extraction. Despite good quality of fibres, dew retting is usually replaced by other more economic methods because the process is very time consuming and weather dependent. Instead of atmospheric retting chemical methods and enzyme retting with pectinases, hemicellulases and cellulases is used, however fibre properties depend on extraction conditions significantly.

2.1. Morphology of lignocellulosic fibres

Sclerenchyma cells possess fibre like form and are arranged longitudinally. The cells are long and narrowed at the cell ends and surrounded and protected by a cell wall which is a complex macromolecular structure. During cells growth the wall is thickened and further strengthened by addition of a secondary wall. Usually fibre cells are occurring in strands or bundles which are called technical fibres [Caffall 2009]. The cells are polygonal in transverse section and connected between themselves by sclerenchyma middle lamellas. The lumen or cavity inside mature, dead fibre cells is usually very small when viewed in cross section [Lewin 1998, Cook, 1993].

2.2. Fibre structure

The cellulose, hemicellulose and lignin content in plant fibres vary depending on the plant species, origin, quality and conditioning [Blackburn 2005].

Chemically unmodified cellulose is generally recognised to occur in four polymorphic forms. There is some evidence for the existence of others [Krässig1992, Lewin 1998]. The monoclinic spatial model for the unit cell of native cellulose is cellulose I crystal modification. The unit cell houses the cellobiose segments of two cellulose molecules, one being part of the 002 corner plane and the second being part of the 002 centre plane [Lewin 1998, Hu 1996]. The monoclinic unit cell has dimensions of 0.835 nm for the a – axis, 1.03 nm for the b-axis or fibre period, 0.79 nm for the c-axis, and 84^0 for the ß angle according to Meyer, Mark and Misch [Krässig 1992]. For natural cellulose a typical x-ray diffraction diagram is observed, that is, three equatorial diffraction peaks at the angles of about 14°, 16° and the strongest diffraction peak at an angle of 22° [Yueping 2010].

However, the crystalline dimorphism of cellulose and the existence of two families of native cellulose were confirmed lately. The crystalline phases I_α and I_β can occur in variable proportions according to the source of the cellulose. Phase I_β is a monoclinic unit cell having space group $P2_1$ and dimensions a = 0.801nm, b = 0.817nm, c = 1.036 nm, ß = 97.3º and very close to the cell proposed by Meyer, Mark and Misch. Phase I_α corresponds to a triclinic unit cell with space group P_1 and dimensions a = 0.674nm, b = 0.593nm c = 1.036nm, α= 117º, γ = 113º and ß = 97.3º [O'Sullivan 1997]. The celluloses produced by primitive organisms (bacteria, algae etc.) are classified by the I_α phase whereas the cellulose of higher plants (woody tissues, cotton, ramie etc.) consists mainly of the I_β phase. I_α and I_β were found to have the same conformation of the heavy atom skeleton, but to differ in their hydrogen bonding patterns [O'Sullivan 1997].

Regenerated cellulose II is obtained when native cellulose is treated with strongly alkaline solutions or precipitated from solutions, such as when producing man-made cellulose fibres. The cellulose III crystal structure is formed after treating the cellulose with liquid ammonia and cellulose IV lattice structure is obtained by treating regenerated cellulose fibres in a hot bath under stretch.

Furthermore, cellulose molecules are, during the course of biosynthesis, arranged in morphological units elementary fibrils. The fibrillar structure model is accepted for cellulose native and man made fibres however, there are some differences in the structural arrangement between different types of these fibres [Krässig 1992]. Elementary fibrils are strings of elementary crystallites which are associated in a more or less random fashion into aggregations. Isolated segments of the fibrils fringing from aggregations are forming a fibrillar network. By transition of cellulose molecules from crystallite to crystallite the longitudinal connections are achieved and coherence of the fibrils by hydrogen bonds at close contact points or by diverging molecules [Krässig 1992].

Microfibrillar orientation is different for different types of cellulose native fibres. It is a very important influence factor for fibres mechanical properties. Microfibrillar angle MFA of bamboo is 2^0-10^0, of coir 41^0-45^0, of flax 10^0, of jute 8^0, of ramie 7.5^0, of sisal fibres 20^0 [Blackburn 2005] and of cotton 20-30^0 [Morton 1993]. Besides microfibrillar orientation, fibres strength and stiffness depend on fibres constitution, cellulose content, crystallinity and degree of polymerisation. In addition to, fibres maturity and part of the plant from which fibres are obtained plays an important role.

Due to the imperfect axial orientation of the fibrillar aggregates, interfibrillar and intrafibrilar voids and less ordered interlinking regions between the crystallites inside the elementary fibrils the pore system of cellulose fibres is formed.

3. Conventional plant fibres

Textile fibres are broadly classified as natural fibres and man-made fibres, as shown in Figure 1. Natural fibres refer to fibres that occur within nature, and are found in vegetables respectively plants (cellulose fibres), animals (protein fibres) and minerals (asbestos). Man-made fibres are those that are not present in nature, although they may be composed of naturally-occurring materials. They are classified into three main groups: those made by transformation of natural polymers (regenerated fibres), those made from synthetic polymers (synthetic fibres), and those made from inorganic materials (fibres made of metal, ceramics, and carbon or glass) [BISFA.2006].

Nature in its abundance offers us a lot of materials that can be called fibrous. Plant fibres are obtained from various parts of plants, such as the seeds (cotton, kapok, milkweed), stems (flax, jute, hemp, ramie, kenaf, nettle, bamboo), and leaves (sisal, manila, abaca), fruit (coir) and other grass fibres. Fibres from these plants can be considered to be totally renewable and biodegradable. Plant fibres, which have a long history in human civilisation, have

http://dx.doi.org/10.5772/52372

gained economic importance and are now cultivated on a large scale globally [Blackburn 2005, Mather 2011, Hearle 1963, Mwaikambo 2006].

Fibres that are produced on the seeds of various plants have been called seed hair or seed fibres. The most important fibre of this class is cotton. Other fibres of this group (kapok, floss from milkweed, dandelion, and thistle fibres) are not generally spun into yarns, but are utilized mainly as staffing in pillows and mattresses, and for life belts [Hearle1963].

NATURAL FIBRES	MAN-MADE FIBRES
VEGETABLE ORIGIN - **CELLULOSE FIBRES**	**NATURAL POLYMER BASED**
▪ **SEED FIBRES:** COTTON, KAPOK ▪ **BAST FIBRES:** JUTE, FLAX, HEMP, RAMIE, KENAF ... ▪ **LEAF FIBRES:** ABACA, SISAL, HENEQUEN ...	▪ **REGENERATED CELLULOSE**: ▪ VISCOSE, MODAL, LYOCELL, CUPRO ▪ **REGENERATED PROTEIN**: CASEIN, ARACHIN ZEIN ▪ **CELLULOSE ESTERS**: ACETATES ▪ **RUBBER:** ELASTODIENE ▪ **ALGINATE**
ANIMAL ORIGIN – **PROTEIN FIBRES**	**SYNTHETIC POLYMER BASED**
▪ **WOOL** ▪ **HAIR FIBRES:** ANGORA, MOHAIR, ALPACA ... ▪ **SILK**	ACRYLIC, ARAMID, CHLOROFIBRE, FLUOROFIBRE, MODACRYLIC, POLYAMIDE, POLYESTER, POLYETHYLENE, POLYIMIDE, POLYPROPYLENE, VINYLAL, POLYLACTIDE ...
INORGANIC:	**INORGANIC:**
ASBESTOS	CARBON, CERAMIC, GLASS, METAL

Figure 1. Classification of textile fibres

3.1. Seed fibres

3.1.1. Cotton

Due to fibres properties and low cost, cotton represents the most used textile fibre in the world. Fibres are obtained from seeds of the plant species *Gossypium*, which belongs to the *Malvaceae* family. Cotton fibres consist of unicellular seed hairs of the bolls of the cotton plant. Cotton fruit bursts when mature, revealing a tuft of fibres with the length from 25 to 60 mm and diameters varying between 12 and 45 μm. Cotton fibres have a pronounced three-wall structure. The cuticle layer consists of wax and pectin materials. This outer wax layer protects the primary wall, which is composed of cellulose crystalline fibrils. The secondary wall of the fibres consist of three distinct layers, which include closely packed parallel fibrils with spiral winding of 25 – 30° and represent the majority of cellulose within the fibres. Lumen is surrounded by the tertiary wall. The cross section of fibres is bean-shaped; however by swelling it is almost round when moisture absorption takes place (Figure 2).

Cotton fibres consist of 80-90% cellulose, 6-8% water, 0.5-1% waxes and fats, 0-1.5% proteins, 4-6% hemicelluloses and pectins and 1-1,8% ash [Lewin 1998, Hu 1996].

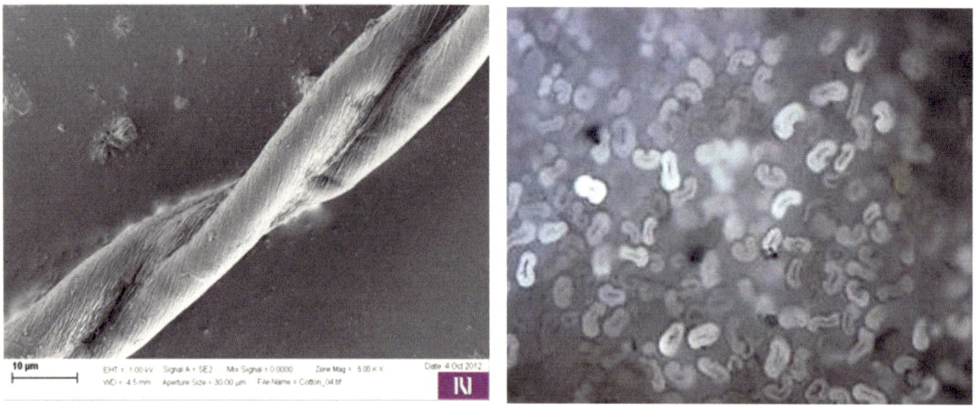

(a) (b)

Figure 2. a) Longitudinal view (5000x magnification) and b) cross-section of cotton fibre

Cotton is hydrophilic and the fibres swell considerably in water. Fibres are stable in water and its wet tenacity is up to 20% higher then its dry tenacity (25-40cN/tex). The toughness and initial modulus of cotton are lower compared to hemp fibres, whereas its elongation at break (5-10%) and its elastic recovery are higher. The fibres are resistant to alkali but degraded by acids. The microbial resistance of cotton is low, it burns readily and quickly, can be boiled and sterilized, and does not cause skin irritation or other allergies [Lewin 1998, Cook 1993].

3.1.2. Kapok

Kapok *(Ceiba pentandra)* is a highly lignified organic seed fibre, containing 35-50% of cellulose, 22–45% of hemicelluloses, 15–22% of lignin and 2–3% of waxes. It also contains smaller quantities of starch, about 2.1% of proteins, and inorganic substances, notably iron (1.3–2.5%). Kapok contains 70–80% of air and provides excellent thermal and acoustic insulation. The absolute density of a kapok cell wall is 1.474 g/cm^3, whilst the density of fibres by considering about 74% of lumen is only 0.384 g/cm^3 [Cook 2006]. Kapok is a smooth, unicellular, cylindrically shaped, twist less fibre. Its cell wall is thin and covered with a thick layer of wax. A wide lumen is filled with air and does not collapse like cotton. By the microscope observation kapok fibres are transparent with characteristic air bubbles in the lumen. The cross section of fibres (Figure 3) is oval to round. The kapok cell wall structure differs from other natural cellulosic fibres. A primary cell wall, which is directly related to the superficial properties of fibres, consists of short microfibrils, which are oriented rectangular to the surface of fibres. In the secondary cell wall microfibrils run almost parallel to the fibre axis. [Hearle 1963, Rijavec 2008, Fengel 1986, Khalili 2000, Fengel 1986/2]. Considering the content of alpha cellulose, kapok is more like wood than flax and other plant fibres. The average degree of polymerisation is 6600 [Fengel 1986]. Kapok fibres are 10–35 mm long, with a diameter of 20–43 μm. The cell wall thickness is about 1–3 μm. The tensile strength is 0.84 cN/dtex (93.3 MPa), Young's module 4 GPa, and breaking elongation 1.2% [Mwaikamno 2001].

Due to its wide lumen, kapok has an exceptional capability of liquids retention. Its excellent thermal and acoustic insulating properties, high buoyancy, and good oil and other non-polar liquids absorbency distinguish kapok from other cellulosic fibres. Kapok is mainly used in the form of stuffing and nonwovens; it is rarely used in yarns, mostly due to low cohesivity of its fibres and their resilience, brittleness, and low strength. New potentials of kapok are in the field of technical textiles, yachts and boats furnishing, insulating materials in refrigeration systems, acoustic insulation, industrial wastewaters filtration, removal of spilled oil from water surfaces, and reinforcement components in polymer composites [Rijavec 2008].

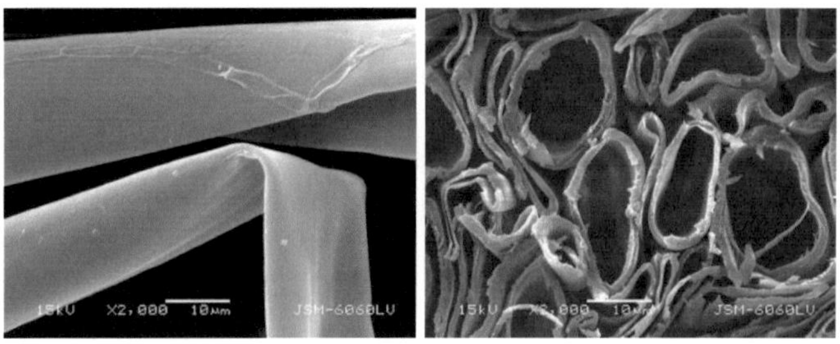

(a) (b)

Figure 3. SEM image of longitudinal view (a) and cross section (b) of kapok (2000x magnification) [Rijavec 2008].

3.2. Bast fibres

Bast fibres i.e. flax, jute, hemp, ramie, kenaf, and abaca are soft woody fibres, which are obtained from stems or stalks of dicotyledonous plants. The fibres occur in bundles or aggregates [Hearle 1963]. The bundles consist of 10 to 25 elementary fibres, with the length of 2 to 5 mm and a diameter of 10 to 50 μm. The bundles are connected by lateral ramification, which forms a three dimensional network. The elementary fibrils and bundles are cemented by lignin and pectin intercellular substances, which must be removed during the processing of fibres extraction [Mohanty 2005]. Bast fibres have a long utilization tradition. They have been used for more than 8000 years. Currently bast fibres are raw materials not only used for the textile industry but also for modern environmentally friendly composites used in different areas of applications like building materials, particle boards, insulation boards, food, cosmetics, medicine and source for other biopolymers etc.

3.2.1. Flax

Flax fibres are obtained from the stems of the plant *Linum usitatissimum*. Fibres are running at the surface of the plant stem, which is about 1 m height and 2 – 3 mm thick in the diameter [Blackburn 2005]. Like cotton, flax fibre is a cellulose fibre, however its structure is more crystalline, making it stronger, and stiffer to handle, and more easily wrinkled. Flax fibre properties are controlled by the molecular fine structure, which is affected by the plant growing conditions and the retting procedure that is applied.

The process of retting tends to separate the bundles of flax fibres into individual fibres, although many fibres remaining together in bundles [Hearle 1963]. Flax fibres are not as pure as cotton in terms cellulose content; indeed they contain only about 60 - 70% of cellulose. In addition they contain other substances such as hemicelluloses 17% and lignin 2-3%, as well as waxes 2%, pectins 10% and natural colouring matters [Mather 2011, Mohanty 2005]. Flax fibres have a soft handle and have fairly lustrous appearance. The length of fibres varies between 6 – 65 mm, but on average they are about 20 mm long. Their diameter is about 20 μm. Flax fibres are not as twisted as cotton fibres, but both have a lumen in the centre. Several dislocations that are areas of the cell wall in natural fibres where the direction of the microfibrils (the microfibril angle) differs from the microfibril angle of the surrounding cell wall, are observed on longitudinal images of fibres (Figure 4). These deformations are due to extraction procedures [Thygesen 2006]. The shape of fibres varies from polygonal to oval and irregular. Fibres cross-section form depends on variety, plant growth conditions and maturity. Flax fibres are amongst the strongest in the group of naturally occurring fibres (55 cN/tex and about 20% stronger in wet state), but they do not stretch much. Flax fibres elongation at break is only 1.8% and their moisture regain is 12% [Lewin 1998, Cook 1993].

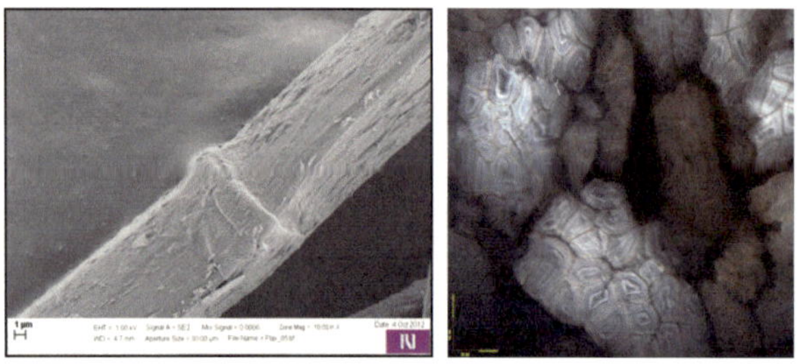

(a) (b)

Figure 4. a) Longitudinal view (10000x magnification) and b) cross-section (30x magnification) of flax fibre

3.2.2. *Jute*

Jute is a natural fibre obtained as an extract from the bark of the white jute plant *Corchorus capsularis* and to a lesser extent from tossa jute (*Corchorus olitorius*) [Mohanty 2005]. Jute is a long, soft and shiny fibre that can be spun into coarse, strong threads and is one of the cheapest natural fibres. It is also the most versatile, eco-friendly, natural, durable and antistatic fibre available. The plants are retted by the same method used for flax. The resulted jute strand, which are up to 3 m long, are composed of many very short fibres, elementary fibres (length between 0.5-6.0 mm, diameter 26-30 μm) held together by lignocelluloses. The fibres contain between 61-71% cellulose, large amount of hemicelluloses (14-20%) and lignin

(12-13%) and pectin (0.2%) [Mather 2011]. The cross-sections of bundles of jute fibres show a range in the size and number of fibres per bundle, in the thickness of the wall and in the shape and diameter of lumens. The fibre is generally smooth, with some dislocations. The individual fibres are mainly polygonal, with rounded corners and oval to round lumens (Figure 5) [Hearle 1963]. Jute has a moderate strength (30-45 cN/tex), however it is not as strong as flax or hemp. For fibres low extension at break (1-2%) is characteristic. Moisture regain of jute fibres is 12.6%, but it can absorb up to 23% of water under conditions of high humidity. Jute has high insulating and anti-static properties and low thermal conductivity [Cook 1993, Mwaikamno 2009].

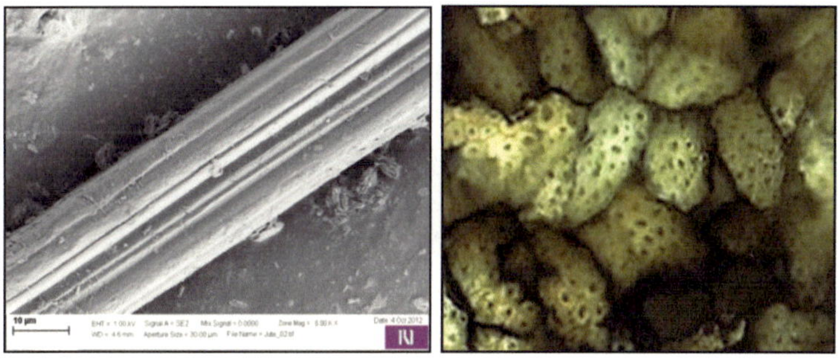

(a) (b)

Figure 5. a) Longitudinal view (5000× magnification) and b) cross-section (180× magnification) of jute fibre

3.2.3. Hemp

Hemp is the bast fibre obtained from stems of *Cannabis sativa* L plants. It grows easily to a height of 4 m without agrochemicals and captures large quantities of carbon. The most important components of fibres are cellulose (77%), pectin (1.4%) and waxes (1.4%). Pectin is found in the middle lamellae and glues the elementary fibres to form bundles. The lignin (1.7%) is an incrusting component of the fibre. It is incrusting cellulose and contributes to the hardness and strength of fibres. It is located in the middle lamellae and fibre primary cell wall. Other components of hemp fibres are tannin, resins, fats, proteins etc. The content of these components is much higher in hemp than in cotton.

Therefore the processing of those fibres requires different technology [Blackburn 2005]. The diameter of the cell varies considerably from 16 to 50 μm, with broad flat lumen. The length of the individual or elementary fibres is ranging from 2 to 90 mm (average length is 15 mm). Elementary fibres are thick walled and the cross-section of fibres is polygonal with rounded edges (Figure 6). In longitudinal view, the fibre is roughly cylindrical, with surface irregularities and lengthwise deformations caused by dislocations. The ends of fibres are slightly tapered and blunt [Hearle 1963]. Hemp fibres are coarser when compared to flax and rather difficult to bleach. The fibres have an excellent moisture resistance and rot only very slowly in water. Hemp fibres have high tenacity (53-62 cN/tex); about 20% higher than flax, but low elongation at break (only 1.5%) [Mohanty 2005].

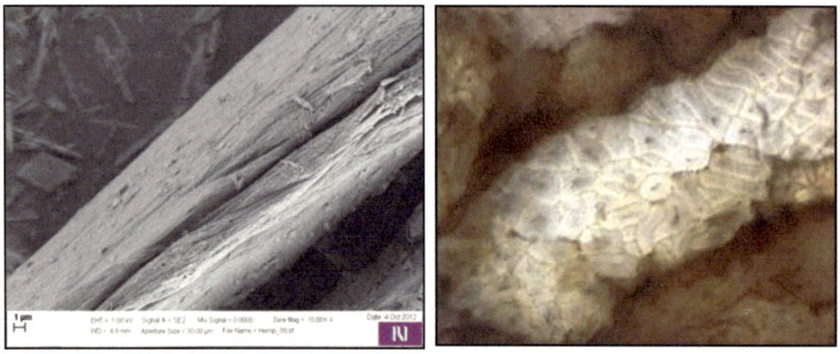

(a) (b)

Figure 6. a) Longitudinal view (10000x magnification) and b) cross-section (200x magnification) of hemp fibre

In recent years because of the interest for alternative renewable resources, hemp gained again relevance. Beside the traditional textile application of hemp numerous new directions emerge: building and isolation materials, composite materials, special cellulose materials (papers), technical textile, geotextiles and agricultural textile, oil based products, items for agriculture and horticulture etc. [Blackburn 2005].

3.2.4. Ramie

Ramie is a herbaceous perennial plant in the nettle family *Urticaceae*, native to eastern Asia. Ramie fibres are extracted from the stem of the plant *Boehmeria nivea* of the nettle family. Individual fibre cells in stems are bound together in fibre bundles by waxes, hemicelluloses, lignin and pectins that are difficult to remove (Figure 7). Therefore the efficiency of the retting process usually used for e.g. hemp fibres extraction is not sufficient to extract ramie fibres from stems. But, a combined microbial and chemical treatment is very effective and economical. Chemical composition of ramie fibres is: cellulose (91-93%), hemicelluloses (2.5%), pectin (0.63%) and lignin (0.65%). Ramie fibres exhibit excellent mechanical properties, i.e. the best in the group of bast fibres (45-88 cN/tex) and, as most of the natural cellulose fibres the strength increases by 25% when fibres are wet. The ultimate fibre length is between 120-150mm and fibre diameter is 40-60 μm. Fibres are durable and they have good resistance to bacteria, mildew and insect attack. The main disadvantage of ramie is its low elasticity (elongation at break is 3-7%), which means that it is stiff and brittle [Mather 2011]. Fibres are oval to cylindrical in shape and their colour is white and high lustrous. Fibres surface is rough and characterized by small ridges, striations, and deep fissures. Ramie fibre can be easily identified by its coarse, thick cell wall, lack of twist, and surface characteristics [Hearle 1963].

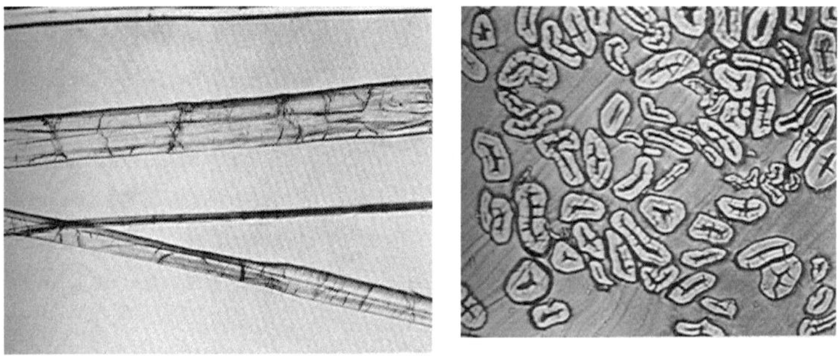

(a) (b)

Figure 7. a) Longitudinal view and b) cross-section (100x magnification) of ramie fibre

3.2.5. Kenaf

Kenaf fibres are obtained from *Hibiscus cannabinus*. Kenaf contains two fibre types: long fibre bundles situated in the cortical layer and short fibres located in the ligneous zone. Elementary fibres are short; their fibre length ranges from 3 to 7 mm, with average diameter of 21 μm. The cross-sections are polygonal with rounded edges and the lumens are predominantly large and oval to round in shape [Hearle 1963]. The lumen varies greatly in thickness along the cell length and it is several times interrupted. Kenaf fibres contain about 45-57% of cellulose, 21.5% hemicelluloses, 8-13% lignin and 3-5% pectin. Kenaf fibres are coarse, brittle and difficult to process. Their breaking strength is similar to that of low-grade jute and is weakened only slightly when wet. There are many potential specific utilization possibilities for kenaf whole stalk and outer bast fibres, including paper products, textiles, composites, building materials, absorbents, etc. [Mohanty 2005].

3.3. Leaf fibres

Leaf fibres are often referred to as hard fibres, and have limited commercial value, mainly because they are generally stiffer and coarser texture than the bast fibres. The fibres are usually obtained from the leaves by mechanically scraping away the non fibrous material. Above all the leaves fibres are used for production of cordage and ropes. The most important fibres of this group are sisal, henequen and abaca.

3.3.1. Sisal

The sisal fibre is a "hard" fibre extracted from fresh leaves of sisal plant *Agave sisalana*. It is usually obtained by a decortication process, in which the leaf is crushed between rollers and then mechanically scraped. The length of the sisal fibre varies between 0.6 and 1.5 m and its diameters range from 100 to 300 μm [Mohanty 2005]. Cellulose content in sisal fibres is about 70%. The fibre is composed of numerous elongated fibre cells that are narrowed towards both ends. Fibre cells are linked together by middle lamellae, which consist of hemicelluloses, lignin and pectin. A sisal fibre in cross-section is built up of about 100 fibre cells. The cross section of sisal fibres is neither circular nor fairly uniform in dimension. The lumen varies in size but is usually well defined. The longitudinal shape is approximately cylindrical. Longitudinal view and cross-section of sisal fibres is demonstrated on Fig.8. Physically, each fibre cell is made up of four main parts, namely the primary wall, the thick secondary wall, the tertiary wall and the lumen. The fibrils are, in turn, built up of microfibrils with a thickness of about 20 μm. The microfibrils are composed of cellulose molecular chains with a thickness of 0.7 μm and a length of a few μm [Joseph 1999]. Sisal fibre is fairly coarse and inflexible. The tensile properties of sisal fibres are not uniform along its length. The fibres extracted from the root or lower parts of the leaf have a lower tensile strength and modulus. The fibres become stronger and stiffer at midspan, and the fibres extracted from the tip have moderate properties. The lower grade fibre is processed by the paper industry because of its high content of cellulose and hemicelluloses. The medium grade fibre is used in the cordage industry for making ropes, baler and binders twine. The higher-grade fibre after treatment is converted into yarns and used by the carpet industry.

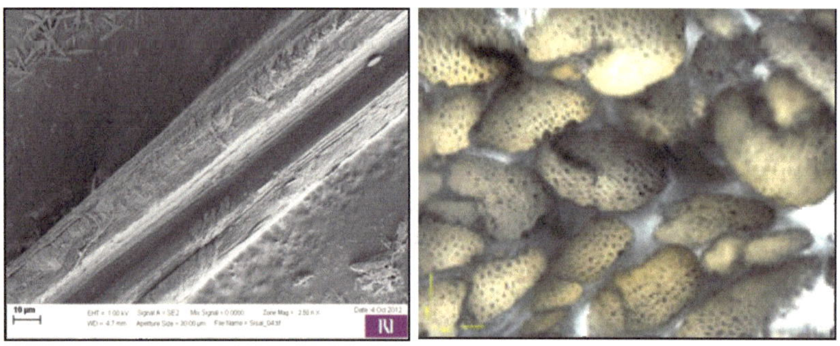

(a) (b)

Figure 8. a) Longitudinal view (2500x magnification) and b) cross-section of sisal fibre

3.3.2. Abaca

Abaca or Manila hemp is extracted from the leaf sheath around the trunk of the abaca plant (*Musa textilis*). The commercial fibres are utilized in the form of strands, and the strands in turn are composed of bundles of individual fibres. Individual fibres, when removed from the strands by boiling in an alkali solution, are smooth and fairly uniform in diameter. The

http://dx.doi.org/10.5772/52372

lumens are large in relation to wall thickness. Cross-marking is rare, and fibre tips pointed and often flat and ribbon –like. The technical fibres are 2 to 4 m long. The single fibres are relatively smooth and straight and have narrow pointed ends. Individual fibre diameters range from 14 to 50 μm and the lengths from 2.5 to 13 mm [Hearle 1963]. Chemically, abaca comprises 76.6% cellulose, 14.6% hemicelluloses, 8.4% lignin, 0.3% pectin and 0.1% wax and fat. Abaca is considered as one of the strongest of all natural fibres, being three times stronger than cotton and twice that of sisal, and is far more resistant to saltwater decomposition than most of the vegetable fibres. Abaca is a lustrous fibre and yellowish white in colour. Abaca fibres are used manly to manufacture ropes and handicraft goods [Blackburn 2005].

3.3.3. Henequen

Henequen (*Agave fourcroydes*) plant of the family agave is a close relative to the sisal plant. The henequen plant is native to Mexico, where it has been a source of textile fibre since pre-Columbian times. Many factors can influence the properties of the fibre including weather conditions, age of the plant, type of soil, extraction method, etc. Henequen fibre is composed of approximately 77% cellulose, 4-8% hemicelluloses, 13% lignin and 2-6% pectin and waxes by weight [Blackburn 2005, Aguilarvega 1995]. Fibres have variable diameter, being larger at the butt end and the smaller at the tip end of the fibre. Also, the diameter is connected with the fibre's origin, fibres cultivated at different locations have different diameters. The fibre cross-section changes from a beanlike shape at the butt end to rounded form at the tip end of the fibres. Like sisal, henequen fibres are smooth, straight, yellow, and easily degraded in salt water. Compared to other leaf fibres, henequen has low elongation at break and low modulus, but relatively high tenacity which makes them suitable as reinforcement for polymers [Blackburn 2005].

4. Non-conventional plant fibres

Lignocellulosic agricultural by-products are a promising and beneficial source for cellulose fibres. Due to the chemical and physical properties, composition and sustainability agro-based biofibres represent a potential for use in textile and paper industry for fibres, chemicals, enzymes and other industrial products. Annually renewable resources, e.g. corn, wheat, rice, sorghum, barley, sugarcane, pineapple, banana and coconut, etc. by-products are utilized as agro-based biofibres [Reddy 2005].

Also in non-conventional fibre plants elongated sclerenchyma cells are organized in a similar manner than traditional fibre cells like flax, hemp etc. These cells provide strength and support and are located next to the outer bark in the bast or phloem and serve to strengthen the stems. The fibres are in strands running the length of the stem.

To extract the fibre strands from other plant tissues the natural gum binding them must be removed by retting. The most common way is a biological treatment by an enzymatic or bacterial action on the pectinous matter of the stem.

Several techniques are used for extraction of conventional bast fibres: (i) Dew retting by the action of dew, sun, and fungi on the plants spread out on the ground, (ii) Water retting is conducted in rivers or pools through bacterial action and takes 2–4 weeks, (iii) For chemical retting solutions of different chemicals are used, e.g. sodium hydroxide, sodium carbonate, soaps, or mineral acids. The process takes only a few hours, (iiii) controlled biological or biochemical retting by addition of enzymes. The differences between the procedures are not only in expenses and process duration but the most important the quality and uniformity of retted fibres.

Ultimate fibres extracted from agricultural by-products are round, polygonal or elliptical in cross section and have a lumen in the centre. Their geometrical properties are conditioned by fibres origin and are different.

Reddy and Yang have collected structural characteristics and biofibres properties (Table1) [Reddy 2005]. Fibres obtained from pineapple leaves are the longest in this group and because of high crystallinity and high content of cellulose (70-82%) they express good mechanical properties (Young's modulus 400–627 MPa) [John 2008].

Fibre	Length (mm)	Width (μm)	Crystallinity (%)
cornhusk	0.5-1.5	10-20	48-50
pineapple leaf fibre	3-9	20-80	44-60
coir	0.3-1.0	100-450	27-33
bagasse	0.8-2.8	10-34	-
banana	0.9-4.0	80-250	45
wheat straw	0.4-3.2	8-34	55-65
rice straw	0.4-3.4	4-16	40
sorghum stalks	0.8-1.2	30-80	-
barley straw	0.7-3.1	7-24	-

Table 1. Properties of some non-conventional plant fibres [Reddy 2005]

4.1. Fibres from corn stover

As a kind of abundant and renewable agricultural residue, corn (Zea mays L.) stover, that refers a combination of corn stalk (stem) and leaf, could be a low-cost and sustainable source for energy and chemicals in future. For a long time (since 1929) fibres obtained from corn waste materials have been studied and utilized for pulp and papermaking [Li 2012].

Cornstalks as a potential for fibres extraction were studied by Reddy and Yang [Reddy 2005/2]. They have found, that natural cellulose fibres obtained from cornstalks have the structure and properties required for textile and other industrial applications.

The fibres obtained from cornstalks are composed of single cells bound together in cell bundles. Stronger fibres extraction conditions remove most of the binding substances resulting in single cells that are too small to be used for high value fibrous applications. Elementary fibres with the length of 0.7 -1.5mm and cell diameter of 15 – 35 μm which is comparable to rice and wheat straw fibres were extracted and analysed. Fibres contain about 80% cellulose, 8% lignin and 8% moisture. The rest are minerals and pectin. The most important parameters for fibres properties, i.e. crystallinity and microfibrillar angle MFA condition fibres properties. The typical cellulose I structure is observed with the crystallinity of 52% and MFA of about $11^{0.}$ MFA is lower than that of cotton which has MFA in the range of 20–30^{0} depending on the maturity and cotton species. Due to high fibrils orientation tensile properties of fibres are good, i.e. they have high strength but low elongation. Elementary fibres form bundles with mechanical properties similar to that of kenaf and with moisture regain of about 7.9%, which is similar to that of cotton but lower than flax (12%) and kenaf (17%), respectively, are suitable for blending and processing with other common textile fibres to produce various products [Reddy 2005/2].

Although fibre properties of corn stover have been studied for decades, the first systematic investigation of cell morphology and fibre quality of different corn stover fractions was performed by Li et al. [Li 2012]. Individual fibres were connected in bundles by middle lamella with the highest lignin concentration. Obvious differences in cell morphologies and chemical compositions between four different plant fractions, i.e. stalk rind and stalk pith, and leaf blade and leaf sheath were observed. Fibres were shorter and finer in stalk pith and parenchyma and vessel content was the highest in this part of the plant. Therefore it was not suitable for papermaking, while morphological characteristics of fibres from corn stalk rind were appropriate as papermaking materials. There were also differences in lignification and hemicellulose content. Sclerenchyma cells in stalk rind were more lignified than those in other tissues. The lowest hemicellulose content was observed in stalk rind [Li 2012].

4.2. Wheat straw fibres

The microstructure, thermal and mechanical properties of wheat straw fibres have been examined and compared to flax straw fibres with an idea of using these natural fibres as reinforcing additives for thermoplastics [Hornsby 1997]. Of crucial importance in this regard is the manner by which their inherent mechanical properties alter on exposure to elevated temperatures, which are encountered during melt processing of the polymer. Under all test conditions flax straw was significantly stronger and stiffer than wheat straw. The tensile strength and elastic modulus decreased with increasing temperature up to 200^{0}C. This effect was minor for wheat straw than flax straw. The differences are due to fibres structural form. The form of wheat straw is much more cellular than flax. Due to different lignin content the thermal stability of flax fibres was significantly higher than it was for wheat straw [Hornsby 1997].

4.3. Fibres from hop stems

Hop (*Humulus lupulus* L.) belongs to the family *Cannabaceae* and genus cannabis that includes hemp. The use, production or properties of natural cellulose fibres from hop stems was studied by Reddy and Yang [Reddy 2009].

The single sclerenchyma cells in hop stem fibres are small. Their length is 2.0 ± 1.0 mm and width 16.5 ± 5.5μm. Fibres extracted from hop stems contain 84% of cellulose, 6% of lignin in 2% of ash. From the diffraction patterns of cellulose in hop stem fibres cellulose crystalline structure was determined. The crystallinity index is 44 ± 5% (65–70% for cotton and 81–89% for hemp cellulose) and microfibrillar angle of cellulose fibrils 8 ± 0.7^0. The diffraction pattern is very similar to the diffraction pattern of hemp. Cellulose crystallites in hop and hemp fibres are regularly distributed and are also parallel to the fibre axis and to each other.

Mechanical properties of hop stem fibres are close to that of hemp fibres. Shorter single cells and low crystallinity of cellulose in the fibres should be the major reasons for the lower breaking tenacity of hop fibres compared to hemp. Sorption properties of hop fibres are comparable to cotton properties and slightly lower than that of hemp [Reddy 2009].

4.4. Banana, sugarcane bagasse and sponge gourd fibers

Fibres from *Musaceae maturate rachis:* The fibrous structure of agro-industrial residues of two different types of *Musaceae* maturate waste rachises (banana -Musa AAA, cv 'Valery' and Musa AAB, cv 'Dominico Harton') has been studied by Gañán et al. Using an up-bottom approach, rachises, fibre bundles and conducting tissues, elementary or ultimate fibres, microfibrils bundles and cellulose microfibrils have been isolated [Gañán 2008]. For fibre bundles extraction biological retting followed by chemical treatment was used.

An important amount of vascular bundles that were formed by conducting tissues and fibre bundles was observed on rachis cross sections. Researchers are suggesting two groups of fibrous structures: the first at the microscopic level formed by conducting tissues, fibre bundles and their elementary fibres, and the second at nanoscopic or ultrastructural level where cellulose microfibrils are grouped in microfibril bundles. In addition to, on fibre surface calcium oxalates crystal structures were observed. Their occurrence on residue surfaces is related to the maturate state of samples.

The diameter of elementary fibres was 10-20μm and diameter of macrofibrils with helicoidal arrangement inside the secondary cell wall was less than 1μm. In addition to microfibril bundles with the diameter 40 – 60nm and cellulose microfibrils with the diameter 5-10nm were identified [Gañán 2008].

Three types of fibres, namely banana fibres (*Musa sapientum*) obtained from the pseudo stem of the plant, sugarcane (*Saccharum officinarum*) bagasse fibres and Brazil sponge gourd (*Luffa cylindrica*) fibre were studied by Guimarăes and co-workers [Guimarăes 2009].

The chemical structure of extracted fibres was determined. The cellulose content is the highest in Sponge gourd (66.59%±0.61%), Bagasse follows (54.87%±0.53%) and the lowest cellulose content was determined for Banana fibres (50.15%±1.09%). Cellulose crystallinity degree was between 39% and 50% for the analysed fibres. The most crystalline structure was observed in Sponge gourd fibres (50%), cellulose in Bagase was 48% crystalline and in banana fibres only 39%.

A high content of lignin was observed for all types of fibres (17.44%±0.19% Banana, 23.33% ±0.02% Bagasse and 15.46%±0.02% Sponge gourd). Sorption properties of these fibres (banana and bagasse: 8.57±0.19 and 9.21±0.01, respectively) are very similar as cotton fibres, however moisture content in Sponge gourd fibres at standard climate conditions is significantly lower (4.79±0.02) [Guimarães 2009].

4.5. Bamboo fibres

Bamboo is an abundant resource and it has always been used in agriculture, handicraft, paper-making, furniture and architecture. Recently attempts have been made to produce textile fibre from bamboo. Since a single bamboo fibre is 2 mm in length, it is used in textile production in the form of a fibre bundle. Bamboo is a very-fast growing grass. Environmental friendly fibres extracted from bamboo, which is renewable, fast growing, degradable, and does not occupy cultivated land are economically efficient and especially useful to grow in hilly areas.

After degumming through a chemical treatment, the cellulose content in the bamboo fibre reached more than 70%. Comparing chemical structure of hemp, jute and bamboo, lignin and hemicellulose contents in bamboo are far higher than that of the flax fibres, and almost as much as that of the jute fibres (Hemicellulose content: bamboo 12.49%, jute 13.53%, flax 11.62; lignin content: bamboo: 10.15%, jute 13.30%, flax 2.78%). Lignin in bamboo fibre bundles is reason for yellow colour of fibres and coarse fibre fineness [Yueping 2010].

Cross section of single bamboo fibre is round with a small round lumen. The bamboo single fibre width is 6–12 μm and the length 2–3 mm and it is smaller than that of flax (12–20 μm, 17–20 mm, respectively) [Yueping 2010].

By the x-ray analysis of bamboo fibres a similar x-ray diffraction pattern is obtained as it is of jute fibres. Two diffraction peaks are observed at the angles of 15–16° and 22° for the bamboo fibre and jute fibre. It is known that the crystalline structure of natural cellulose from various plants belongs to cellulose I with typical diffraction maxima at scattering angles 14°, 16° and 22°, respectively. Due to a high content of amorphous hemicellulose and lignin in the bamboo fibre and jute fibre an overlapping of the two peaks at angles of 14° and 16° on the diffraction pattern is observed [Yueping 2010].

The fine structure and mechanical properties of fibres within a maturing vascular bundle of moso bamboo Phyllostachys pubescens was studied by Wang with co-workers [Wang 2012].

Almost axially oriented cellulose fibrils were found in the fibre cell walls. This fibrilar arrangement maximizes the longitudinal elastic modulus of the fibres and their lignification increases the transverse rigidity [Wang 2012].

Because of high and different content of non-cellulose substances in various plant fibres the fibres' crystallinity is different. When comparing crystallinities of some plant fibres, the crystallinity of ramie is the highest, follows flax and cotton and the lowest crystallinity is observed for bamboo fibres and jute fibres. These structural differences are reflected on fibre properties, i.e. density, moisture regain, tenacity, dyeing and thermal properties, etc. [Yueping 2010].

4.6. Quinoa fibres

Quinoa originates from Andes in South America and it belongs to the family Chenopodiaceae (Chenopodium quinoa Willd). It is a grain-like crop grown primarily for its edible seeds and it has become highly appreciated for its nutritional value. It has been recognized as a complete food due to its protein quality. It has remarkable nutritional properties; not only from its protein content (15%) but also from its great amino acid balance. It is an important source of minerals and vitamins, and has also been found to contain compounds like polyphenols, phytosterols, and flavonoids with possible nutraceutical benefits [Abugoch 2009]. The plant is not problematic and it can be cultivated everywhere. Quinoa has a high nutritional value and has recently been used as a novel functional food because of all these properties; it is a promising alternative cultivar.

The elementary fibres can be isolated from Quinoa stems. It is possible to use different processes for fibre isolation. Sfiligoj et al. reported about fibres which were obtained from untreated stems by mechanical isolation. Besides, stems were subjected to chemical treatment in alkaline medium (1%NaOH; different treatment times and temperatures were used; sample A – 1day treatment, room temperature; sample B – 11days treatment, room temperature; sample C – 1 hour T = 100^{0}C). In addition to they were water treated, respectively. Thereby the pectin structures connecting fibres with other plant tissues were loosed and the mechanical separation of the elementary fibres or fibre bundles was performed [Sfiligoj-Smole 2011].

Figure 9. Quinoa – ripe plant and the stem

Morphological characteristics of fibres were microscopically observed. Light microscopy tests were performed on whole stems and on ultimate fibres and fibre bundles. Different structures were observed on cross- sections and on longitudinal views of stems. Quinoa plant and its stem are shown on Figure 9. In addition to, stem's cross-section is demonstrated on Figure 10. Quinoa technical fibres, i.e. bundles of elementary cells were isolated from untreated and differently treated stems.

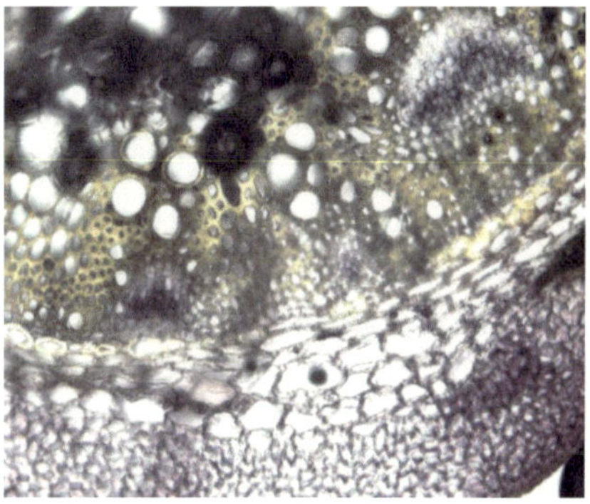

Figure 10. Cross – section of the quinoa stem

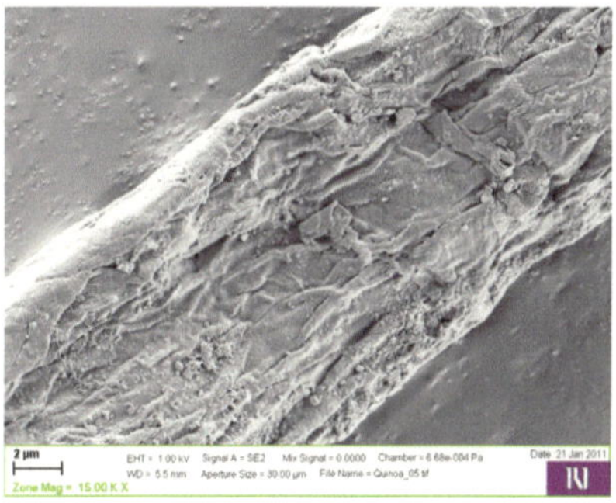

Figure 11. SEM image of surface morphology of isolated fibres from quinoa (fibres obtained by decortication from untreated stems)

The fibre bundles were mainly inhomogeneous and sclerenchyma cells were often accompanied by tracheary elements. Fibres surface morphology was strongly dependent on isolation process (Fig. 11, 12, 13 and 14). Fibres obtained by decortication, i.e. by only mechanical isolation show totally different surface morphology when compared to the fibres obtained from water and alkaline treated stems. In addition to, thermal conditions of the treatment influenced the surface morphology (cf. Fig. 13 and Fig.14).

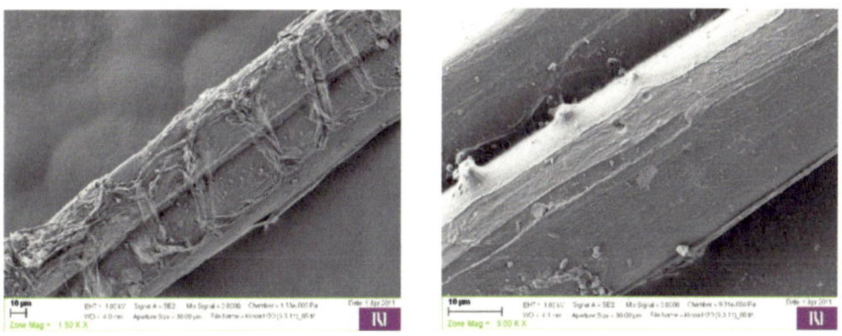

Figure 12. SEM images of surface morphology of differently isolated fibres from quinoa (fibres from water treated stems)

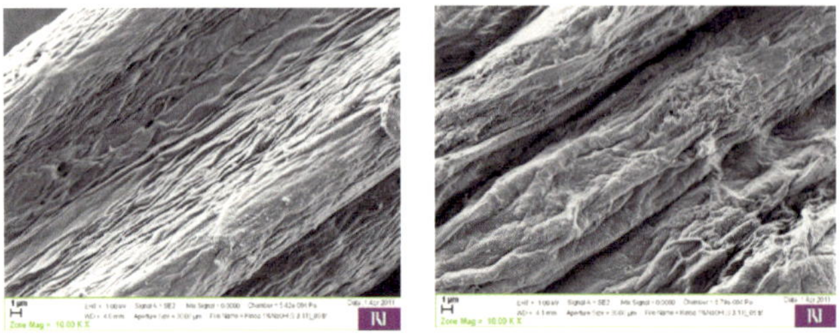

Figure 13. SEM images of surface morphology of differently isolated fibres from quinoa (fibres from NaOH treated stems)

Fibre dimensions were measured on microscopy images. Fibres' diameters are dependent on the procedure of fibres isolation. When untreated stems were processed fibre bundles were formed from a bigger number of cells and a mean diameter of 164μm was determined for these fibres. Pre-treatment of stems facilitates sclerenchyma cells separation from other plant tissues, and fibres' diameter for fibres isolated from pre-treated stems was 42.61μm. The variation of fibres' diameter is very high (variation coefficient is 43.76%).

In addition to, geometrical and mechanical properties of isolated fibres and fibre bundles were determined. The measurements were performed on Lenzing apparatus Vibrodyn and Vibroskop according to standard test methods. Ten parallel samples were measured. Fineness of fibre bundles was between 24.66 and 96.84 dtex depending on the isolation method used for fibres extraction. The fineness variation is related to different number of cells in the bundle and quality of fibre extraction process which is connected with the presence of different non-cellulose compounds on fibres.

It was important to obtain a representative sample for testing due to the inherent variability of most biological materials and extensive mechanical damage due to the isolation process. As mechanical and geometrical properties vary considerably according to temperature and humidity, all samples for testing were conditioned and prepared in the ISO standard atmosphere for textile testing of 65 ±2% relative humidity and 20 ±2ºC according to ISO 5079 was used [ISO 5079 (1995)].

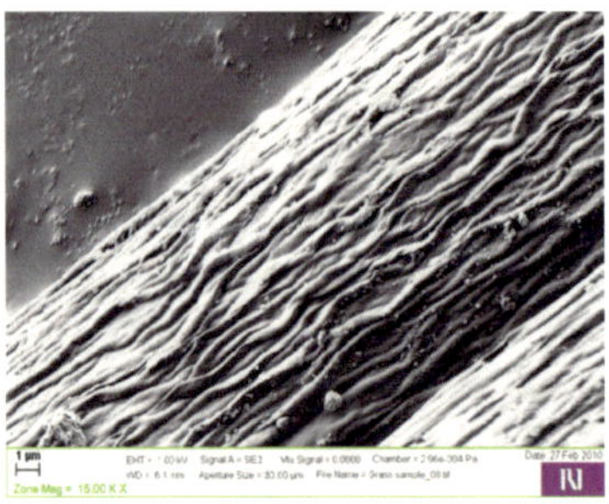

Figure 14. SEM images of surface morphology of differently isolated fibres from quinoa (fibres from stems, treated in NaOH at T = 100ºC)

Fibres' elongation is low and breaking strength high. The elongations vary between 1.41 % to 2.11 % and tenacities from 13.78 to 32.19cN/tex. The obtained values are comparable with the mechanical properties of some textile bast fibres, e.g. jute, hemp or coir. Ultimate fibre cell in hemp is 13 – 26 mm long; its diameter is between 5 and 32 μm, tenacity 29 – 47 cN/tex and elongation 1.8% [Sfiligoj-Smole in press].

In addition to, the powder X-ray diffraction spectra of quinoa fibres, which were obtained by the fibres extraction by water treatment and mechanical isolation, exhibit a diffraction pattern typical of cellulose I, with a diffraction peak of the 2θ angle at about 22^0, which can be assigned to the 002 reflection. However, the two diffraction maxima of 101, 10-1 reflections at diffraction angles 14 and 16^0, respectively, typical for native cellulose are not pronounced. The diffraction pattern is very similar to the pattern obtained by x-ray scattering of bamboo and jute fibres [Yueping 2010]. It is assumed that accompanying substances were not removed sufficiently and therefore the remaining amorphous hemicellulose and lignin are origin of overlapping of these two peaks [Sfiligoj-Smole in press].

4.7. Grass fibres

Grass because of its huge available amounts represents a great potential. It is an annual plant with bundles of elementary fibre cells bound by pectin middle lamellae. Parenchyma cells separate fibre bundles from each other.

The most important representatives in the group of grasses are: Perennial Ryegrass (*Lolium perenne*), Italian ryegrass (*Lolium multiflorum*), Hybrid ryegrasses (*Lolium perenne x multiflorum*), tetraploid varieties of perennial and Italian ryegrass, Timothy (*Phleum pratense*), Cocksfoot (*Dactylis glomerata*), Fescues (Meadow fescue - *Festuca pratensis*; tall fescue – *F.arundinasea*; red fescue – *F.rubra*), Bromes (*Bromus willdenowii*) [Holmes 1989, Petersen 1981]. Legumes are presented by: White clover (*Trifolium repens*), Red clover (*Trifolium pratense*), Lucerne (*Medicago sativa*) [Holmes 1989].

The elementary grass fibres were studied. They were isolated from different grass and legumes sorts, i.e. Ryegrass (*Lolium hybridum* Gumpenstein), Trefoil (*Trifolium pratense*) and Lucerne (*Medicago sativa*) [Sfiligoj-Smole 2005, Sfiligoj-Smole 2004]. The fibre-samples were obtained in a bio-refinery, after the liquid phase containing proteins and lactic acid was eliminated from the ensiled and green grasses, respectively. For the isolation of elementary grass fibres different processes were used. Cross section of a Trefoil stem is presented on Fig.15.

On the microscopy images of grasses cross sections stem area, lumen area, fibre area and fibre content was determined. A high content of fibres was detected in stems regardless the fibres origin. The highest fibre content was determined in Ryegrasses (39.5%), Lucerne followed (34.5%) and the lowest content of fibres was observed in the cross-section of Trefoil (20.2%) [Sfiligoj-Smole 2005, Sfiligoj-Smole 2004].

4.8. Alfa or esparto fibres

Esparto fibres, esparto grass or Alfa are cellulose based fibres extracted from esparto *Stipa tenacissima* leaves. It is a fast growing perennial plant from poaceae family that grows in

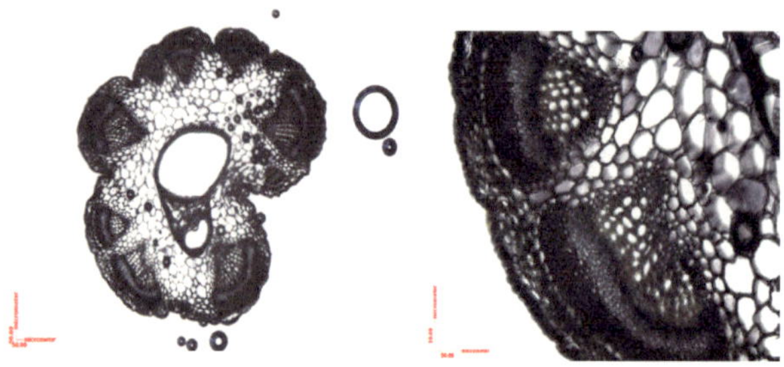

Figure 15. Cross section of a Trefoil stem [Sfiligoj-Smole 2005].

North Africa and southeast Spain. Leaves which reach up to 1m are rich with fibres [Belkhir 2012]. Fibres are extracted for cellulose pulp and paper manufacturing and therefore fibres and pulp were extensively studied. Pulp properties, chemical composition and cell wall architecture was researched. It was found that fibres morphological variability (length and width) is related to growth conditions, i.e. growth location, season and leaf level. Average length of fibres is 1-2 mm and fibres width varies from 14-17μm [Belkhir 2012].

4.9. Sea grass – Zostera marina

Researchers report about different new cellulose sources, however mainly from terrestrial plant origin. But fibres from marine sources offer addition options when appropriate species are identified. Sea grasses belong to angiosperm and are found in most of the oceans. Among sixty different species *Zostera marina* called eel-grass is the most widespread. [Davies 2007].

P.Davis et al. reported about Baltic species of *Zostera marina* which was collected on the German Baltic coast. The diameter of the plant stem was about 2-5 mm and it was 3-8 times branched. The plant was up to 1.2 m long.

In *Zostera marina* a very interesting plant structure was observed. Fibres were reinforcing a matrix and thereby forming a composite structure. Fibres were organized in bundles. Individual fibres with the diameter around 5μm and approximately circular cross section were mechanically extracted from sea grass *Zostera marina*. Fibres are composed of 57% cellulose, 38% of non-cellulosic polysaccharides (10%pectins and 28% hemicellulose) and 5% of residual matter [Davies 2007]. Single fibre stiffness was determined. It was 28 GPa [Davies 2007].

Due to sea-grass fibres mechanical properties and its low density fibres present an attractive reinforcement for composite materials, especially when bio-degradability is required.

5. Applications of non-conventional cellulose fibres

Depending on their physical properties and cellulose content lingocellulose fibres can be used for various applications. The typical fibre morphology with a lumen in the centre, reduces the bulk density, thereby acoustic and thermal insulation properties of biofibres are increased and therefore these fibres are preferable for lightweight composites for noise and thermal automobile insulators.

In addition to insulation, these materials are used in Civil Engineering as building materials. From industrial hemp Cannabis Sativa L useful cellulose fibres to manufacture fibre cement products for roofing are obtained. The disadvantages of some cellulose fibres are: lower modulus of elasticity, high moisture absorption, decomposition in alkaline environments, they are susceptible to biological attack, variable mechanical and physical properties. Hemp fibres with a higher durability than traditional cellulose fibres are more suited for this kind of application, and therefore a lot of research was performed about the use of hemp fibres as reinforcement for building materials based on cement. In addition to, hemp core fibres from agricultural waste industrial hemp straw with the length between 5-10 mm were studied by Jarabo et al. [Jarabo 2012].

An important aspect of natural fibres is associated with their hierarchically built anatomies developed and optimized in a long term evolution process. A variety of non-wood plants offer multiple possibilities in dimensions, composition and morphology of fibrous structures that can be useful for pulp and paper making industries [Gañán 2008]. Therefore based on high cellulose content they are replacing wood pulp in paper and fibres production. Grass stems and leafs fibres could be utilized for this purpose [Saijonkari – Pahkala 2001].

Natural fibres are currently attracting a lot of attention for reinforcement. Fibre reinforced composites consists of fibre as reinforcement and a polymer as a matrix. Their special advantage is their low cost, low density, good mechanical properties, biodegradability, etc. The advantage of natural fibre composites includes lack of health hazards and non-abrasive nature [Sreenivasan 2012]. Natural fibres provide stiffness and strength to the composite and are easily recyclable. Hemp fibres represent a good potential for this utilization. The use of hemp fibres as reinforcement in composite materials has increased in recent years as a response to the increasing demand for developing biodegradable, sustainable and recyclable materials [Shahzad 2012]. Hemp fibres are used for reinforced thermoplastics (composites hemp fibres - polypropylene PP, polyethylene PE, polystyrene PS, hemp fibres - maleated polypropylene MAPP, kenaf-hemp nonwoven impregnated with acrylic matrix., etc.), thermosets (polyester, epoxy resin, vinylester, phenolics) [Shahzad 2012] and biodegradable polymers (thermoplastic starch, polyhydroalkanoates (PHA), polyactides (PLA), lignin based epoxy, soy based resin, etc [Shahzad 2012].

Also other natural cellulose fibres have been used for composite preparation. Polymers including high density polyethylene (HDPE), low density polyethylene (LDPE) polypropylene (PP) polyether ether ketone (PEEK), have been reported as matrices [Li 2007].

A major disadvantage of cellulose fibres is their highly polar nature which makes them incompatible with non-polar polymers. These fibres therefore are inherently incompatible with hydrophobic thermoplastics, such as polyolefins [John 2008]. This characteristic results in compounding difficulties leading to non-uniform dispersion of fibres within the matrix which influences composite properties. To achieve strong adhesion at the interfaces which is needed for an effective transfer of stress and load distribution through out the interface, sometimes surface modification is needed. Surface modifications include (i) physical treatments, such as solvent extraction; (ii) physico-chemical treatments, like the use of corona and plasma discharges or laser, and UV bombardment; and (iii) chemical modifications, both by direct condensation of the coupling agents onto the cellulose surface and by its grafting by free-radical or ionic polymerizations [John 2008].Therefore different coupling agents which introduce chemical bonds between the matrix and fibre are involved (e.g. silane, isocyanate and titanate based products, alkaline treatment, acetylation, benzoylation, acrylation, maleated coupling agents, permanganate, etc) [51]. or methods of physical fibre treatments (e.g. surface fibrillation, plasma treatment) are used [George 2001]. An additional possibility is to impregnate cellulose fibres in monomer solution, follows the in-situ catalyst, heat or UV polymerisation [George 2001].

Different natural fibres species have been used for preparation of composites. Some examples are: aspen fibre, abaca fibre, bagasse fibres, bamboo fibre (BF), banana fibre, etc.

Unidirectional isora fibre reinforced polyester composites were prepared by compression moulding. Isora is a natural bast fibre separated from the *Helicteres isora* plant by a retting process. Untreated and alkali treated fibres were used for composite preparation and influence of fibre content on composite properties was studied. It was observed that the pretreatment process conditions the fibre content for achieving optimum composites mechanical properties [Joshy 2007].

Green composites were prepared from pineapple leaf fibres and soy-based resin. The addition of polyester amide grafted glycidyl methacrylate (PEA-g-GMA) as compatibilizer increased the mechanical properties of composites. For preparing composites from pineapple leaf fibres in natural rubber fibres were pre-treated in NaOH solutions and benzoyl peroxide (BPO) of different concentrations. It was found that all surface modifications enhanced adhesion and tensile properties [Joshy 2007].

Elephant grass (*Pennisetum purpureum*) is available abundantly in nature and is renewable. It is a tall grass growing in dense clumps along lake and riverbeds up to 3 m height. The diameter of the stem is 25 mm and leaves are 0.6 to 0.9 m long and about 25 mm wide. It represents a potential and economic source compared to other natural fibers, however it is still underutilized, therefore K. Murali Mohan Rao with co-workers suggest fibres from the grass for reinforcement of polyester composites [55]. The density of the elephant grass fiberres is very low compared to other lignocellulose fibres.

This property is a good base for designing lightweight material from these fibres. The diameter of fibers is between 70 lm to 400 μm. Fibres mechanical properties are: tensile strengh is 185 MPA, tensile modulus is 7.40 GPa and elongation at break 2.50% [55]. The positive impact of elephant grass fibres on tensile strength of fiber reinforced

composites was determined and it was found that composite mechanical properies increase with percentage volume of fibers. Whereas the fibre extraction is simple, fibres are cheap and of appropriate properties elephant grass is also suitable for composites used for lightweight structures preparation [55].

Cellulose nanofibres and crystals have gained a large interest, not only in the academic research society but also in industries, during the last few years [Oksman 2012].

It is well known that isolation of nanocrystals from cellulose is possible by strong acid hydrolysis. Under controlled conditions, acid hydrolysis allows removal of the amorphous regions of cellulose fibres whilst keeping the crystalline domains intact in the form of crystalline nanoparticles [Sheltami 2012].

The diamensons of nanofibres are usually around 20–30 nm in diameter with the length of few μm. Nanocrystals are much smaller. Their length is about 200nm and diameter about 3–5 nm [Oksman 2012]. Cellulosic nanomaterials are obtained form different resources, i.e. wood, bioresidues and annual plants, e.g. wood fibres, sisal, pineapple leaves, coconut husk fibres and bananas, mengkuang leaves (*Pandanus tectorius*) [Sheltami 2012], mulberry bark [Li 2009]. The use of isolated cellulose nanocrystals as reinforcements in the field of nanocomposites has attracted considerable attention since it was first reported in 1995 [Sheltami 2012]. Natural fibre reinforced composites can be applied in the plastic, automobile and packaging industries [Li 2007].

6. Conclusions

Lignocellulosic natural fibres have a very long tradition for textile materials manufacturing. Especially are these fibres important for technical textiles production. The series of plants yielding conventional textile fibres, e.g. flax, hemp, etc. has been recently extended by several abundant plant species traditionally not-connected with fibres extraction. Of huge interest are especially agricultural wastes from cultures which are primary grown for food industry, and their plant wastes additionally containing fibres. Different fibres have been studied by several authors; their properties were determined and compared to the properties of conventional fibres. Regardless of the origin fibre cells are elongated sclerenchyma cells of different geometrical characteristics, associated in fibre bundles with adequate mechanical properties. Several plant species were suggested for utilization from different geographic areas.

Natural fibres from conventional and unconventional source are considered as potential replacement for man-made fibres in composite materials for their reinforcement. Natural fibres from annual plants have advantages of being low cost and low density and therefore they are light. They are a renewable material. In addition to, an important advantage of these materials is their biodegradability and low toxicity. It was confirmed by many researchers that properties of natural fibres of different origin improve composites properties, e.g. the mechanical properties of natural fibres - polymer composites are superior to those of the unreinforced materials.

Author details

M. Sfiligoj Smole[*], S. Hribernik, K. Stana Kleinschek and T. Kreže

*Address all correspondence to: majda.sfiligoj@uni-mb.si

University of Maribor, Faculty of Mechanical Engineering, Department for Textile Materials and Design, Maribor, Slovenia

References

[1] Abugoch James L.E. Quinoa (Chenopodium quinoa Willd.): Composition, Chemistry, Nutritional, and Functional Properties; Advances in Food and Nutrition Research 2009; 58 1–31.

[2] Aguilarvega M., Cruzramos C.A. Properties of cellulosic henequen fibres. Journal Of Applied Polymer 1995; 56 (10) 1245-1252

[3] Belkhir S., Koubaa A., Khadhri A. Ksontini, M,., Smiti, S. Variations in the morphological characteristics of Stipa tenacissima fiber: The case of Tunisia. Industrial Crops and Products 2012; 37 (1) 200-206 .

[4] BISFA. Terminology of man-made fibres, 2006 Edition. Brussels: BISFA; 2006.

[5] Blackburn R.S., editor. Biodegradable and sustainable fibres. Cambridge: Woodhead Publishing Series in Textiles: 47, The Textile Institute; 2005.

[6] Caffall K.H., Mohnen D. The Structure, function, and biosynthesis of plant cell wall pectic polysaccharides.; Carbohydrate Research 2009; 344 1879-1900. Cellulose 2008; 15 (1) 67-74.

[7] Cook, J.G. Handbook of Textile Fibres, Natural Fibres: Merrow Publishing; 1993.

[8] Cuissinat C., Navard P. Swelling and dissolution of cellulose, Part III: plant fibres in aqueous systems.

[9] Davies P., Morvan C., Sire O., Baley C. Structure and Properties of fibres from Seagrass (Zostera marina). J.Mater.Sci 2007; 42 4850-4857.

[10] Fengel D., Wenzkowski M. Studies on kapok. 1. Electronmicroscopic observations. Holzforschung 1986; 40 137–141.

[11] Fengel D., Przyklenk M. Studies on kapok. 2. Chemical Investigation. Holzforschung 1986; 40 325–330.

[12] Gañán P., Zuluaga R., Cruz J., Velez J. M., Retegi A., Mondragon I. Elucidation of the fibrous structure of Musaceae maturate rachis. Cellulose 2008; 15 131–139.

[13] George J., Sreekala M. S., Thomas S. A Review on Interface Modification and Characterization of Natural Fiber Reinforced Plastic Composites. Polymer Engineering and Science 2001; 41 (9) 1471 – 1485.

[14] Guimarăes J.L., Frollini E., da Silva C.G., Wypych F., Satyanarayana K.G. Characterization of banana, sugarcane bagasse and sponge gourd fibers of Brazil. Industrial Crops and Products 2009; 30 407–415.

[15] Hearle J.W.S., Peters R.H. Fibre structure. London: The Textile Institute Butterworths; 1963.

[16] Holmes W. Grass, its Production and Utilization. Oxford, London, Edinburgh: The British Grassland Society, Blackwell Scientific Publications; 1989.

[17] Hornsby PR., Hinrichsen E., Tarverdi K. Preparation and properties of polypropylene composites reinforced with wheat and flax straw fibres .1. Fibre characterization. Journal of Materials Science 1997; 32 (2) 443-449.

[18] Hu X.P., Hsieh Y.L. Crystalline Structure of Developing Cotton Fibers. Journal of Polymer Science: Part B Polymer Physics 1996; 34 1451-1459.

[19] ISO 5079: (1995) Textile fibres -- Determination of breaking force and elongation at break of individual fibres

[20] Jarabo R., Fuente E., Monte M.C. Use of cellulose fibers from hemp core in fiber-cement production. Effect on flocculation, retention, drainage and product properties. Industrial Crops and Products 2012; 39 89-96.

[21] John M. J.; Anandjiwala R. D. Recent developments in chemical modification and characterization of natural fiber-reinforced composites Polymer Composites 2008; 29 (2) 187-207.

[22] John M. J.; Anandjiwala R.D. Recent developments in chemical modification and characterization of natural fiber-reinforced composites. Polymer Composites. 2008; 29 (2) 187-207.

[23] John M.J., Thomas S. Biofibres and biocomposites. Carbohydrate Polymes 2008; 71 343-364.

[24] Joseph P.V., Joseph K., Thomas S. Effect of processing variables on the mechanical properties of sisal-fiber-reinforced polypropylene composites. Composites Science and Technology 1999; 59 (11) 1625-1640.

[25] Joshy, M.K., Lovely M., Rani J. Studies on interfacial adhesion in unidirectional isora fibre reinforced polyester composites. Composite Interfaces 2007; 14 (7-9) 631-646.

[26] K. Murali Mohan Rao, A. V. Ratna Prasad, M. N. V. Ranga Babu, K. Mohan Rao and A. V. S. S. K. S. Gupta Tensile properties of elephant grass fiber reinforced polyester composites. Journal of Materials Science 2007; 42 (9) 3266-3272.

[27] Khalili S., Daniel G., Nilsson T. Use of soft rot fungi for studies on the microstructure of kapok (Ceiba pentandra) fibre cell walls. Holzforschung 2000; 54 229–233.

[28] Krässig H.H. Cellulose, Structure, Accessibility and Reactivity. Switzerland, USA: Gordon and Breach Science Publishers; 1992.

[29] Lewin, M.; Pearce, E.M. Handbook of Fiber Chemistry. New York: Marcel Dekker; 1998.

[30] Li R., Fei J., Cai Y., Li Y., Feng J., Yao J. Cellulose whiskers extracted from mulberry: A novel biomass production. Carbohydrate Polymers 2009; 76 94–99.

[31] Li X., Tabil G.L:, Panigrahi S. Chemical Treatments of Natural Fiber for Use in Natural Fiber-Reinforced Composites: A Review Journal of Polymers and the Environment 2007; 15 (1) 25-33.

[32] Li Z.Y. ; Zhai H.M. Zhang Y., Yu L . Cell morphology and chemical characteristics of corn stover fractions. Industrial. Crops and Products 2012; 37 (1) 130-136.

[33] Marques G., Rencoret J., Gutiérrez A., Río J C. del. Evaluation of the Chemical Composition of Different Non-Woody Plant Fibers Used for Pulp and Paper Manufacturing. The Open Agriculture Journal 2010; 4 93-101.

[34] Mather R.R., Wardman R.H. The Chemistry of Textile fibres. Cambridge: RSC Publishing; 2011.

[35] Mohanty A.K., Manjusri M., Drzal L.T. Natural fibres, Biopolymers and Biocomposites. Boca Raton: CRC Press, Taylor & Francis Group; 2005.

[36] Morton W.E., Hearle J.W.S.Physical Properties of Textile Fibres. Manchester: The Textile Institute; 1993.

[37] Mwaikambo L.Y. Review of the history, properties and application of plant fibres. African Journal of Science and Technology, Science and Engineering Series 2006; 7 120–133.

[38] Mwaikamno, L.Y., Ansell M.P. The determination of porosity and cellulose content of plant fibers by density methods. Journal of Material Science Letters; 2001; 20 2095–2096.

[39] Mwaikamno, L.Y. Tensile Properties of Alkalized Jute Fibres. Bioresources 2009; 4(2) 566-588.

[40] Oksman K. Nanocelluloses and their use in composite materials, eXpress Polymer Letters 2012; 6 (9) 687.

[41] O'Sullivan A.C. Cellulose: The Structure Slowly Unravels. Cellulose 1997; 4 (3) 173-207.

[42] Petersen Asmus. Die Gräser als Kulturpflanzen und Unkräuter auf Wiese, Weide und Acker. Berlin: Akademie Verlag; 1981.

[43] Reddy N., Yang Y. Biofibers from agricultural byproducts for industrial applications. Trends in Biotechnology 2005; 23 (1) 22-27.

[44] Reddy N., Yang Y. Structure and properties of high quality natural cellulose fibers from cornstalks. Polymer 2005; 46 5494–5500.

[45] Reddy N., Yang Y. Properties of natural cellulose fibers from hop stems. Carbohydrate Polymers 2009; 77 898–902.

[46] Rijavec T: Kapok in technical textiles. Tekstilec 2008; 51 (10–12) (In Slovene) 319–331.

[47] Saijonkari – Pahkala K. Non-wood plants as raw material for pulp and paper. Dissertation. Faculty of Agriculture and Forestry, University of Helsinki; 2001.

[48] Sfiligoj-Smole M., Kreže T., Hribernik S., Pointner B., Stana-Kleinschek K., Ribitsch V. Characterization of natural cellulose fibres from quinoa. In: Valant M., Gardonio S., Fabbretti E., Pirnat U.(ed) Slovenian-Italian Conference on Materials and Technologies for Sustainable Growth: Proceedings book., 4-6 May 2011, Ajdovščina, Slovenia. Nova Gorica University; 2011.

[49] Sfiligoj-Smole M., Kreže T. Hribernik S., Kurečič M., Stana-Kleinschek K. Properties of Quinoa fibres; in press

[50] Sfiligoj-Smole M., Kreže T., Strnad S., Stana-Kleinschek K., Hribernik S. Characterisation of grass fibres; J. Mater. Sci. 2005; 40 (20) 5349-5353.

[51] Sfiligoj-Smole M., Stana-Kleinschek K., Kreže T., Strnad S., Mandl M., Wachter B. Physical properties of grass fibres. Chem. biochem. eng. q. 2004; 18 (1) 47-53.

[52] Shahzad A. Hemp fiber and its composites - a review, Journal Of Composite Materials 2012; 46 (8) 973-986.

[53] Sheltami R.M., Abdullah I., Ahmad I., Dufresne A., Kargarzadeh H. Extraction of cellulose nanocrystals from mengkuang leaves (Pandanus tectorius). Carbohydrate Polymers 2012; 88 (2) 772–779.

[54] Sreenivasan V.S., Ravindran D., Manikandan V., Narayanasamy R. Influence of fibre treatment on mechanical properties of short Sansevieria cylindrica / polyester composites; Materials and design 2012; 37 111-121.

[55] Thygesen L.G., Bilde-Sørensen J.B., Hoffmeyer P. Visualisation of dislocations in hempfibres: A comparison between scanning electron microscopy (SEM) and polarized light microscopy (PLM). Industrial Crops and Products 2006; 24 (2) 181–185

[56] Wang X., Ren H., Zhang B., Fei B., Burgert I. Cell wall structure and formation of maturing fibres of moso bamboo (Phyllostachys pubescens) increase buckling resistance. J. R. Soc. Interface 2012; 9 (70) 988-996.

[57] Yueping W., Ge W., Cheng H., Tian, G.L. Liu, Z , Xiao, Q.F. , Zhou, X.Q., [4]; Han, X.J. Gao, X.S. Structures of Bamboo Fiber for Textiles. Textile Research Journal 2010; 80 (4) 334-343.